Hominid Culture in Primate Perspective

Hominid Culture in Primate Perspective

DUANE QUIATT
and
JUNICHIRO ITANI
Editors

UNIVERSITY PRESS OF COLORADO

Published by the University Press of Colorado
P.O. Box 849
Niwot, Colorado 80544

The University Press of Colorado is a cooperative publishing enterprise supported, in part, by Adams State College, Colorado State University, Fort Lewis College, Mesa State College, Metropolitan State College of Denver, University of Colorado, University of Northern Colorado, University of Southern Colorado, and Western State College of Colorado.

Library of Congress Cataloging-in-Publication Data

Hominid culture in primate perspective / Duane Quiatt and Junichiro
    Itani, editors.
        p. cm.
    Includes bibliographical references and index.
    ISBN 0-87081-313-7 (alk. paper)
    1. Primates—Behavior. 2. Human behavior. 3. Behavior evolution.
4. Social behavior in animals. I. Quiatt, Duane D. II. Itani,
Jun 'ichirō, 1926– .
QL737.P9H765 1994
599.88'0451—dc20   93-47400
                                                          CIP

10   9   8   7   6   5   4   3   2   1

We dedicate this book to Gordon W. Hewes, whose contributions to comparative studies in primate and human behavior, and strong advocacy of such studies as aids to understanding hominid evolution, helped reform anthropological understanding of the nature of culture.

Hewes's theory of the gestural origins of language, remarkable both for intellectual plenitude and philosophic rigor, revived the study of language origins as a respectable pursuit. A masterful synthesis of information gleaned from a wide range of scientific disciplines, it retains its influence and its power to generate fresh insights into the evolution of language and cognition.

Contents

Contents

Contributors

Duane Quiatt has studied the behavior of nonhuman primates in nature, in provisioned settings, and in captivity. He is interested in the socioecology of human as well as nonhuman primate groups and populations; although he continues to pursue primatological research, for the past several years he has been engaged in a study of energy exchange in a prehistoric culture, that of the Anasazi of Mesa Verde. Quiatt teaches primatology and paleoanthropology at the University of Colorado, Denver.

Junichiro Itani was born in 1926 and earned his doctorate in zoology from Kyoto University. On the faculty of Kyoto University until his retirement in 1990, Professor Itani was director of the Laboratory of Human Evolution Studies (1981–1986) and the center for African Area Studies (1986–1989). He is currently a professor on the Faculty of Humanistic Sciences, Kobe-Gakuin University. In 1984 Itani was awarded the Thomas Huxley Memorial Medal by the Royal Anthropological Institute of Great Britain and Ireland.

Stephen E. Harper is a research associate in Project Chantek. He has a master's in English and research interests in semiotics, language origins, and critical theory. He is currently working on a critical theory of space and, with H. Lyn White Miles is coediting "Aliens, Apes, and Artificial Intelligence: Who is a Person in the Postmodern World?" a volume based on a symposium presented by the Southern Humanities Council.

Gordon W. Hewes was educated at the University of California, Berkeley, receiving his Ph.D. there in 1947. He has taught at the University of North Dakota, the University of Southern California, and the University of Colorado and was a Fulbright lecturer in Japan and Peru. His studies in cultural anthropology include aboriginal fishing, postural habits, comparative civilizations, and language origins.

Tim Ingold is professor of social anthropology at the University of Manchester. He has carried out fieldwork among the Skolt Saami (Lapps) of northeastern Finland and in a Finnish farming community and has written extensively on the comparative anthropology of hunter-gatherer and pastoral societies, on the relations between human beings and other animals, and on evolutionary theory. He is currently working on aspects of the anthropology of technology, including its relation to language, and on issues of environmental perception.

Jo Liska earned her Ph.D. at the University of Colorado at Boulder. She is especially interested in the evolutionary foundations of communication and is currently a research professor in the Department of Anthropology at the University of Colorado, Denver.

H. Lyn White Miles is UC Foundation Professor of Anthropology at the University of Tennessee, Chattanooga, and director of Project Chantek, a comparative developmental study of an orangutan learning to use sign language, with special emphasis on understanding the origins of language. Miles studied primate behavior at Yale University and the University of Oklahoma and received her Ph.D. in anthropology from the University of Connecticut.

Richard G. Milo received his M.A. in anthropology at the University of Colorado, Denver. He is presently a doctoral candidate in paleoanthropology and archeology at the University of Chicago. His interests include zooarcheology, modern human origins, and the origin and evolution of modern human behavior; he is now conducting fieldwork in South Africa, investigating changing patterns in human exploitation of animals across the middle to later Stone Age boundary. He has collaborated with Duane Quiatt on several papers that consider the evolution of human language and cognition, as well as the socioecology of the prehistoric Anasazi of the American Southwest.

Robert W. Mitchell earned both a B.S. in general science and a B.A. in experimental psychology at Purdue University, where he assisted in research into the linguistic abilities of a parrot. At the University of Hawaii he examined the linguistic abilities of dolphins, and he studied dog-human play for his Ph.D. in psychology at Clark University. With Nick Thompson he coedited the first book on deception in nonhuman animals. He teaches psychology at Eastern Kentucky University and is coediting two books, one on self-recognition in humans and nonhumans, the other on anthropomorphism.

Mary Ellen Morbeck studies primate anatomy, behavior, ecology, and evolution. A professor of anthropology and anatomy at the University of Arizona, she investigates how an individual's teeth and bones can be used to determine information on phylogeny, sequence and timing of growth, locomotion, and unique life experiences, including nutrition, disease, and injury. Morbeck is currently studying the skeletal biology and life history of chimpanzees observed by Jane Goodall and others in Gombe National Park, Tanzania.

Vernon Reynolds studied rhesus monkeys for his Ph.D. at London University and subsequently studied chimpanzees in the Budongo Forest, Uganda. He has lectured in anthropology at Bristol University and since 1972 has been lecturing at Oxford University. He also directs the Budongo Forest Project, which studies the primates of Budongo in relation to the logging history of the forest.

E. Sue Savage-Rumbaugh is a professor of biology at Georgia State University and director of the Language Acquisition Project at Georgia State's Language Research Center. She received her Ph.D. in philosophy from the University of Oklahoma. Her work has focused upon comparative behavior studies of human and nonhuman primates, with special emphasis on cognition and language.

Jo A. Myers Thompson studied primate behavior at the University of Colorado, Denver, where she earned her M.A. in anthropology with a multidisciplinary specialization in animal behavior. She is currently working toward a Ph.D. at Oxford University, continuing her study of *Pan paniscus* at a new forest research site in Zaire. Her interests include the study of ecology, demography, and behavior, and assessing the role of time and geographic distance in producing differences across fragmented populations and isolated groups.

Preface: Culture, Nature, and the Nature of Culture

CULTURE AS THE OUTCOME OF BIOLOGICAL EVOLUTION

The intent of this preface is not to explain, in a general way, how biological evolution gave rise to culture, nor to reveal the specific pathway by which *Homo sapiens* ended up with a sapient culture (cf. Itani 1985). Authors of the chapters that follow in most cases offer varying details of some portion of that pathway and in all cases present fresh and commanding views of the country through which it winds. Our job in introducing those chapters is to ask what is the view of culture we are likely to take when we speak of it as the outcome of biological evolution.

Biologists whose teachers insisted early on that they study human evolution within the broader context of general mammalian evolution, and whose interest in human social behavior led them subsequently to investigate the complex social behaviors of other primates, looking for commonalities and interesting differences, are inclined to take a comparative view of culture. Indeed, a comparative approach is the very essence of evolutionary biology, and it would be strange indeed if an evolutionary biologist were to abandon that approach when it came to the examination of culture in evolution.

There is, however, a stumbling block in the way of studying human culture from an evolutionary or cross-specific comparative standpoint. It is, to give it a name, the familiar attribution of uniqueness to human behavior, the stubborn persistence with which we insist on recognizing qualitative differences between ourselves and other animals in practically every avenue of behavior. We are what we are, goes the claim, in one or another of its conventional variants, because of our ability to use tools, to manufacture tools, to invent symbols, to make words, to use words grammatically, to dissemble, to seek revenge, to apply justice — and so on, as if there could never be enough similarities between ourselves and even our closest primate relatives to make up for the great and overwhelming *difference* that separates us.

But that defining difference, the difference that supposedly makes us unique, has proved increasingly difficult to isolate.

TOWARD A COMPARATIVE DEFINITION OF CULTURE

The concept of culture — for anthropologists who wish to apply it to evolutionary comparisons or, the other side of the coin, who wish to employ evolutionary comparisons in examining the concept — ought not be tied to consideration of just our own species or a few ancestral hominid species. We cannot take a restrictive view of culture if we hope to understand its evolution. As psychologists do not restrict their study of *intelligence* to that of human beings, so a proper anthropological concept of *culture* should be applicable to other primates, to the more social mammals, and, in principle at least, to all forms of animal life (cf. Imanishi 1952, also Itani and Nishimura 1975).

We argue for an informational view of culture, and we want to note at the outset that such a view is dependent on an information-based theory of knowledge. An information-based theory of knowledge should also be applicable if not to all forms of life, including cabbages and carrots, certainly to all forms of intelligent life. But this brings us up against a second stumbling block, for a current and highly influential theory of the communicative behavior that is integral to social exchange of knowledge (Dawkins and Krebs 1978, Krebs and Dawkins 1984) explicitly denies informational content to signals transmitted by nonhuman animals.

Our strategy with this stumbling block will be to leave it in place. It is like the first in that it postulates a basic difference between human and nonhuman animals, but it does so by the way; the difference is not the reason that the stumbling block is there: it wasn't set up to hinder an understanding of human evolution but as a definitional demarcation designed to ensure that, as a rule, nonhuman animal communication be interpreted strictly and consistently as manipulative behavior. For many purposes this is a very useful interpretation to make. Much social behavior, human or nonhuman, clearly is manipulative, and if one's interests lie in the contexts in which specific kinds of manipulation are practiced and in the efficiency with which they are conducted, the focus of analysis ought to be on effects produced by signals, not on their presumed informational content.

The issue, it is important to note, is not whether a signal can have content that is somehow separate from the perceptual effects produced by its physical structure. Dawkins and Krebs are less concerned with information *content* than with information *sharing*, with the question whether a signal can have the same meaning for its sender as for its receiver. This is in some respects a philosophical issue (see Quine 1960), but as a practical problem it appears to be not much different for human users of language than for animal communicators in general. Reduced to a problem of translation, "It is only *less* of a problem for our own species, in which ways of processing socially accessible information have been refined and conventionalized by adaptation both genetic and cultural — primarily but not solely in connection with culture" (Quiatt & Reynolds 1992).

CULTURE AND INFORMATION

This leads to a narrow but powerful definition of culture as the outcome of such processing: *culture = socially processed information* (Quiatt and Reynolds 1992). The definition applies, as should be clear, to a distinct subset of all the environmental (not genetically coded) information that animals of a given species have access to and utilize in their day-to-day lives, and it directs attention to the behaviors that enable social processing. Implicit in this definition is the idea that any species whose members make use of socially processed information has a culture that is describable in terms of that particular set of information. This in no way devalues human culture; it simply provides us with a definitional foundation for systematically comparing human culture with chimpanzee, rhesus monkey, or for that matter wild dog or naked mole rat culture, without resorting to stonewall contrasts ("human culture as compared to animal behavior") or sleight-of-hand transformations ("the behavioral roots of human culture"). There is a great deal, obviously, that is special in human culture, unique to it. What is not special *about* human culture is its uniqueness. As we have been much reminded lately (e.g., Foley

1987), every species is unique in its morphology, in its behavior, and, as asserted by definition here, in its culture.

APPLICATIONS

It appears, then, to return for a moment to Dawkins and Krebs, that the answer to the question whether or not signals carry information that can be shared must depend on underlying problems. Dawkins and Krebs were concerned with how animals use signals to manipulate others. We are not concerned in this book with animals per se but with human culture, its evolution, and its relation to the cultural behavior of other animals, especially those most closely related to us. However we may want to define culture (many anthropologists prefer not to define a concept so central to our existence, as if to do so would be to rob it of its magic), it is hard to see how discussion of the cultural practices, artifacts, and expectations of a given people could proceed without referring at some point to shared information. So it seems we may and perhaps must take an informational view of human language and human culture; and if we want to look at human language and human culture in evolutionary perspective — that is, in comparison with the inferred behavior of hominids now extinct and with the observable behavior of our closest living relatives, the great apes — then we will want to ask whether for these purposes it may not be better to apply to our comparisons an information-based theory of knowledge than to rule it out.

The choice we face is one of analytic method, not ontology. If our concern were a bottom-up comparison of animal behavior, perhaps *excluding* that of human beings, we might well want to see how far a manipulative view of communication and social interaction would take us. It won't take us very far working from the top down. Human culture, whatever else it may be, is an informational system, and some of the most interesting questions to ask about it have to do with how the information gets loaded and moved around, presumably by ordinary processes of individual and social cognition.

Consequently, unless and until we discover that there is irrefutably a cognitive Rubicon between our own and other species (say, human language as conceived by genetic "essentialists") — at which point of course we would hasten to replace that first stumbling block so lightly removed from our path — we advocate moving freely back and forth around the second stumbling block, examining processes of cognition and of behavior in general to see how this and that process may contribute to storage, retrieval, and exchange of information.

Duane Quiatt and *Junichiro Itani*

PART I
Hominid Origins

Introduction

Paleoanthropologists and human paleontologists, nurtured in an academic tradition to which a "hunter" theory of human origins has been central for some thirty years, have been reluctant to abandon models linking the manufacture and use of stone tools with male cooperation in hunting mammals of medium to large size. Arguably, toolmaking and cooperative hunting, primarily though not exclusively by male members of the group, along with food transport and reciprocal sharing (not only of butchered meat but of plant foods and animals that come in small packages) at an activity center or "home base" *are* the key elements of a dual reciprocal economic system made possible by our forebears' conversion from quadrupedal to hindlimb bipedal locomotion. More problematic are suggestions that the key behaviors of this subsistence system could have been established very early in hominid evolution (Washburn and Lancaster 1968; Isaac 1978a, 1978b; Quiatt and Kelso 1985), quickening in feedback relation the locomotor conversion to habitual bipedalism (Kennedy 1978) and perhaps shaping subsequent neural evolution. While cooperative hunting by males may well have influenced the development of human culture in important ways at some *later* stage, there is very little evidence for a Pliocene or lower to mid-Pleistocene linkage among male cooperation, hunting, and stone toolmaking. We need to reexamine the links among those behaviors that we find in traditional models of hominid origins, and primatologists, particularly students of chimpanzee and bonobo behavior, have recently begun encouraging us to do just that (see, e.g., Boesch and Boesch 1989, Stanford and Allen 1991, Quiatt and Huffman 1993).

Some may object that the "hunter" theory has already been if not discredited at least forced from center stage (in the arena of paleoanthropological speculation) by alternative theories of origin. The "gatherer" theory (Zihlman 1978, Zihlman and Tanner 1978,

Tanner 1981) holds that collecting plant foods has priority over hunting and that the earliest tools were likely to have been digging sticks, containers, carrying slings, and other implements used primarily by women. The "scavenger" theory in its several varieties (including "scavenger/hunter" and "hunter/scavenger") holds that stone tools were first used by hominids of carefully unspecified gender not for killing and butchering game but for secondary processing of animal remains — for extracting the last shreds of meat and for breaking bones to get at the marrow (see, e.g., Binford 1981, Brain 1981, Shipman 1983, Potts 1984, Blumenschine 1987). Both are clearly intended as replacement theories, but we would argue that they are essentially corrective, that they leave the notion of man as hunter or man as scavenger/hunter intact and, apart from positing a marked increase in his willingness to cooperate with woman the gatherer, unchanged. In particular, they appear to accept without question the close relation between *stone* tools and an animal protein supplement to the diet, and they leave unexamined the seemingly obvious connection between bipedalism and tool manufacture, transport, and use.

E. Sue Savage-Rumbaugh does not take this connection for granted. She approaches it indirectly, via the mother-offspring relationship, noting the implications for social cognition and communication of a shift in the ventrally held infant's position when its primate mother makes the hominid conversion to hindleg bipedalism. She conducts a congruent analysis of the difference, from the mother's perspective and in terms of demands placed on memory and cognition, between carrying an infant and transporting an inanimate object (a tool, for instance), which will not cry out if it is left behind. The analyses are illuminating in themselves, and they lead to a surprising conclusion that the adaptive conversion, which Savage-Rumbaugh places in the late Miocene, may have been made not by an ape grounded behaviorally by a previous quadrupedal "choice" but by an ape more arboreal, more catholic in its substrate preferences, and, especially, more frequently inclined toward suspensory forelimb locomotion than current speculation would have it.

African apes and orangutans share an ancestor that was adapted to arboreal locomotion, though without the highly specialized suspensory adaptations characteristic of modern gibbons. Orangutans stuck to the arboreal way of life and became specialized for quadrumanual clambering. Meanwhile, in the African lineage, gorillas underwent bipedal adaptation to a more terrestrial forest life — this in competition with a smaller arboreal competitor, the chimpanzee/bonobo/hominid forerunner. Subsequently, protohominids underwent their own bipedal adaption to ecotonal habitats in a woodland (or savanna-mosaic?) biome, as did bonobos to a lesser degree, whereas chimpanzees remained essentially arboreal though not quadrumanual like orangutans. In other words, gorillas diverged earlier, hominids later, following adaptation to a more general use of arboreal/terrestrial substrates.

Savage-Rumbaugh's scenario is provocative, going against the current grain of speculation founded on assumptions that the immediate ancestor common to both hominids and the African apes was a Miocene ape as much at home on the ground as are gorillas and chimpanzees, and perhaps more — that is, a rather generalized quadruped capable of moving about with ease on both arboreal and terrestrial substrates.

Note that the term *Hominidae* is applied with increasing frequency by paleoanthropologists to a taxonomic grouping that includes contemporary African apes along with extinct species of australopithecines and that more "advanced" subfamily of hominids, the Homininae, of which our own species is the latest and probably the last representative. Formal redefinition of the family Hominidae would extend its time depth by as many as 10 million years (Figure I.1). (See Klein 1989, pp. 42–43, for discussion of why the old usage is retained.) What Savage-Rumbaugh urges us to consider is whether species ancestral to the Hominidae, so defined, may not have been closer in their locomotor adaptations to an arboreal gibbonlike ancestor than we are accustomed to believe.

As is evident in Figure I.1, the problem for either view is that fossil evidence currently available for that critical period, the late Miocene, does not allow us to say with certainty either who our immediate "prehominine" ancestors were or how they got around. Whether or not *Oreopithecus* is an appropriate "reference model" (Savage-Rumbaugh does not go so far as to propose that *Oreopithecus is* that ancestor) is debatable, but the debate concerning hominid origins is better served by a scenario designed to be provocative and to direct attention to critical assumptions concerning ancestral behavior than it is by scenarios that perpetuate assumptions that too often are, as Savage-Rumbaugh notes, "misconstrued as facts."

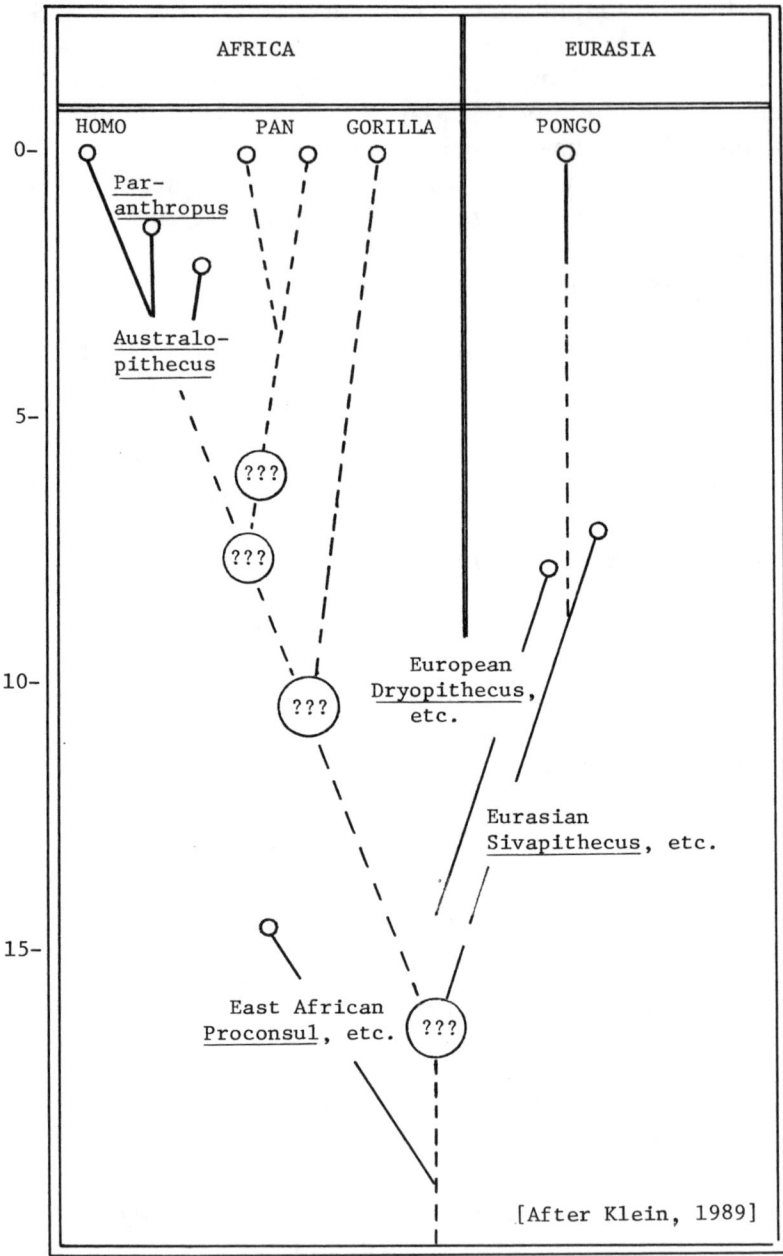

Figure I.1   Taxonomic relationships within the Hominoidea.

1

Hominid Evolution: Looking to Modern Apes for Clues

E. SUE SAVAGE-RUMBAUGH

ABSTRACT

The ability to plan ahead has been a significant driving force in hominid evolution. An individual who plans ahead must monitor the current environment while also contemplating a future environment. I argue that the ability to simultaneously process information from a large number of different channels permits one to respond to the present and simultaneously contemplate the future. This skill possibly arose as mothers needed to monitor infants that did not cling, as well as themselves. Such "dual monitoring" would place a heavy cognitive burden on any primate whose infant did not cling. Clinging behavior ceased as a consequence of becoming a confirmed biped, as an infant was not able to hold onto a bipedal creature who did not support the infant with her thighs. I suggest that a bipedal suspensory ape was the common ancestor of ape and hominids and that as this ape had to travel greater and greater distances on the ground, the problem of infant transport became the limiting factor. Apes solved this problem by becoming knuckle walkers who supported their infants with their thighs as they walked, and so their infants continued to cling. Hominids could not provide thigh support, however, and consequently began to carry their infants with their arms. This revolutionary manner of infant transport within the order Primates permitted the relaxation of the clinging reflex and allowed the mother to put the infant down when she was not traveling. Once the infant was not in contact with the mother's body, she was required to monitor its well-being, and her own, simultaneously. This requirement, essential for the infant's survival, led to the development of joint attention and

7

joint regard, as well as the ability to carry and monitor items other than infants — most importantly, tools.

I discuss the implications of tool carriage for tool manufacture and the development of planning skills with regard to separate evolutionary paths taken by human and ape.[1]

INTRODUCTION

Who are we, where did we come from, and where are we going? These questions lie at the basis of both science and religion. Currently, science provides only partial answers, and those who want complete ones must turn elsewhere. Of these questions, science as a discipline is equipped to attack just one: where did we come from? For it is we humans, not a discipline, who define who we are, and it is humans in all their guises, not just that of science, who will determine where we go. But science has a major role to play by helping us understand where we came from and the forces that shaped our bodies and our minds.

Humans are in the odd position of sharing an evolutionary heritage with all living creatures on the planet yet alone understanding this fact. All other creatures lack the concept of a world prior to their own existence and that of one beyond their own existence. Humans are also alone in the awareness of their own individual, inevitable demise. Perhaps this is partly why they have developed such a strong desire to alter their contemporary world, so that their future world will be a better place.

KNOWING THAT THERE IS A FUTURE TIME

Humans spend time making tools for which they have no immediate need because they know that they will find them to be of value in the future. At first these tools were crude stone implements, now they have become computers and nuclear-tipped rockets. Yet the principle remains the same: behavior is not shaped by the exigencies of that moment alone but rather by humans' vision of imagined future moments.

How did a nervous system that was able to act in the present while imagining the future evolve? Certainly any nervous system that could plan for the future would have an evolutionary advantage over one that could not. In a sense all nervous systems plan for the future. The brain of an orb-weaving spider comes equipped with the ability to make a web each day that it will use for catching insects that night. Yet more complex brains, especially those of higher mammals, are faced with the task of learning a vast array of behaviors that are dependent upon the experiences encountered within the lifetime of a given individual. The behavior of such organisms varies so dramatically from day to day that planning for the future must be accomplished upon the basis of information learned within a given life span. Planning skills at that level cannot be passed from generation to generation in the genome but rather must be spun by each brain anew for each occasion.

It is not difficult to envision means by which selection would operate to produce increasingly efficient planners who could plan anew for each occasion once the process was bootstrapped. When planners pitted their planning skills against one another, those who could plan further in advance would clearly have an advantage. Individuals would then need to plan their future actions not only by anticipating the consequences of their own actions but also by assuming that other individuals were similarly making plans and that these plans (imagined but not observed) could also affect the future. For example, a young hominid might imagine making a tool and throwing it to knock down an animal. However, another hominid might observe him making the tool and imagine stealing the tool before it could be used. If the toolmaker were to imagine the intent of the thief as well as his own plans, he could hide the tool and prevent this from happening. At this point, his behavior would no longer be driven by the exigencies of the real world nor even by an understanding of future events but rather behaviors of imagined plans of others. Such visions of the imagined plans of others would not be based in real fact but in imagined fact. Each hominid's view of the world would be constructed from an imagined view of others, and societies would be shaped as much by these imagined worlds as by the real events around them.

Why is it that the adaptation of making a stone tool when it is not needed, for an imagined future time, has arisen only once in the evolutionary history of our planet? Other equally advantageous skills such as flight and adaptation to water have arisen independently among different genera on more than one occasion. The making of stone tools requires a longer time-based extrapolation from the behavior of the present to potential future needs than do other learned behaviors. Indeed, long-term extrapolations (such as migrations, reproduction, etc.) are generally handled by genes, not neurons. What conditions gave rise to long-term planning skills in early hominids?

What follows is an attempt to reconstruct some of the salient events that may have generated planning skills. In contrast to approaches based primarily upon knowledge of the anatomy of fossilized bones, this perspective relies extensively upon what is known about living primates. For dates of divergence, the techniques of molecular biology serve as the guideposts, for they, like the behavioral perspective that follows, are based upon extant organisms.

HAND AXES: TOOLS FOR THE FUTURE

If we wish to understand why we, as a species, behave the way we do today, we need to know how it was that imagined future contingencies came to control our behavior in the present time. The appearance in the fossil record of stone tools like the Acheulean hand ax tells us that the hominids who constructed these tools must have been able to plan ahead. These elegant, symmetrical tools take time to make and are found in areas where suitable stone is absent, indicating that their makers manufactured and transported them in advance of their use. Modern toolmakers require forty-five minutes to construct an Acheulean hand ax, even when the needed stone is nearby, which would not have been the case every time a hominid needed a hand ax (Toth, personal communication).

By contrast, modern apes require only a few minutes to fashion a termiting tool, utilizing whatever vegetation is at hand. Occasionally chimpanzees form termiting tools en route to the mounds, though sometimes they may carry them as they travel from one mound to another. But they discard the tools once the impetus to fish for termites is either fulfilled or abandoned (Goodall 1968; McGrew, Tutin, and Baldwin 1978). Such observations suggest that chimpanzees do not think about making termiting tools until they have already decided to engage in the activity of termiting and are in the preparatory stages of so doing. Similarly, they may pick up rock hammers and carry them to a nut-cracking site if they are already on their way to a known nut-cracking location (Boesch and Boesch 1984). However, chimpanzees have not been observed to collect good rock hammers and pile them up around an appropriate site in advance of having ripe nuts available for hammering. As with termiting, it appears that it is the availability of nuts and the decision to consume them that serves to initiate rock transport.

It seems that in modern apes the recognition that a tool is needed follows from a decision to obtain the food item that requires that tool. Early hominids, with their Acheulean hand axes, knew that they would need these tools in the future and prepared them in advance, perhaps days before they actually used such a tool. It is as though the need for the tool is coupled with the need for a food item in the mind of the ape, but in the mind of the Acheulean hand ax maker these needs have become uncoupled. A general recognition of the need for the tools, independent of plans for their immediate usage, has arisen.

PLANNING SKILLS IN APES

How did the uncoupling of present and future needs arise? In order to answer that question a more complete understanding of the planning limitations evidenced by contemporary apes is needed. For such information we must turn to the data provided by captive studies, particularly those that have focused upon the cognitive and linguistic capacities of apes. As these studies have moved beyond the "gee whiz" stage followed by the "is it really language?" stage, they have begun to paint a clearer picture of the competencies of our closest living relatives, a picture that is changing our perspective of early hominids.

We now know that bonobos (*Pan paniscus*) can come to understand language at the level of a 2-to-2½-year-old child and that they are able to do so with essentially the same type of language exposure that a child encounters. Although they learn more slowly than a child, they nonetheless become able to interpret novel sentences and even to decode syntactical devices that rely upon word order and recursion (Savage-Rumbaugh and Rubert 1992). Bonobos can understand and appropriately carry out the instructions of five- to seven-word sentences such as "Go scare Matata with the monster mask" or "Pour the milk in the jelly" on the first occasion they hear them.

Yet bonobo language production lags far behind bonobo language comprehension. Productive communications are limited to one- or two-word utterances and do not evidence the syntactical complexity that is clearly evident in comprehension. Nonetheless, most of the productive communications are novel and completely unexpected by the listener (Greenfield and Savage-Rumbaugh 1991). They are not imitations and are typically used to communicate things that those who utter them have not heard before (see Table 1.1 for examples of such communications.)

Table 1.1 Productive Communications of Bonobos

Action-Action	Entity-Attribute
Bite chase	Food blackberry
Chase tickle	Surprise balloon
Slap keepaway	Comitative-Action
Action-Agent	That come
Hide Austin	Conjoined Entities
Keepaway you	Bread banana
Agent-Action	Ice TV
Liz hide	Conjoined Locations
Matata chase	Austin gibbon
Penny tickle	Staff office-grouproom
Action-Goal	Demonstrative-Entity
Chase banana	That apple
Chase Grouproom	That yogurt
Chase mushroom-trail	Entity-Demonstrative
Goal-Action	Apple that
Grouproom open	Water that
Surprise come	Effect-Negative cause
Trailer go	Bad mushroom-trail
Action-Instrument	Entity-Location
Chase ball	Austin play yard
Tickle ball	Play yard outdoors
Instrument-Action	Goal-Agent
Ball chase	Austin you
Water chase	Goal-Instrument
Action-Modifier	Childside key
Chase bad	Instrument-Object
Chase one	Can-opener milk
Modifier-Action	Knife kiwi
Bad chase	Location-Comitative
One hide	Grouproom Matata

(Table 1.1 continued)

Action-Object	Modifier-Location
Hide peanut	Childside play yard
Slap ball	Staff-office play yard
Object-Action	Nonexistence-Entity
Ball slap	No balloon
That keepaway	No coke
Action-Recipient	Possession-Entity
Give Kanzi	You burrito
Action-Volitional object	Recipient-Object
Hug surprise	Austin balloon
Volitional object-Action	Object-Recipient
Orange juice-hug	Egg Austin
Affirmation-Goal	Hotdog Austin
Yes Austin	Peanut Kanzi
Agent-Object	Transport-Location
You surprise	Vehicle trailer
Object-Agent	Two-Mode Paraphrase
Juice you	Chase chase
Play yard you	Bad bad

The ease with which bonobos come to understand and use a form of language not characteristic of their species suggests that their communication capacities in the wild may have been underestimated. Previous attempts to understand the communications of wild bonobos and chimpanzees have typically employed methodologies similar to those used with birds and mammals who have far smaller brains. Using such methodologies, researchers study only very near-term correlations between vocalizations and events. If we wish to understand whether apes are using symbols to represent foods, objects, and planned travel patterns, we cannot discover such information simply by computing the relationship between vocalizations and the immediate responses of conspecifics.

Communications regarding things that are not immediately present in space and time do not lend themselves to the current stochastic coding schemes. This is because if there is a benefit to the use of symbols in place of "emotive vocalizations," it is the ability of the symbols to communicate about things that are removed in time and space. For example, if an ape were to produce a vocalization that others could interpret as "I am heading toward bananas," the responses of the caller's compatriots could vary widely. Some might approach and travel with him, those who do not want bananas might travel elsewhere, others might meet the speaker at the banana site, others might try to get there first, others might wait till the speaker arrives to see if he finds the bananas ripe upon arrival, others might go off in a different direction specifically attempting to avoid the speaker, and so on. None of these responses would be captured by a coding scheme that recorded the behaviors *immediately* following the vocalization, since that would indicate nothing regarding the interpretation of the sound.

Certainly, among captive bonobos and chimpanzees it is clear that symbols can be used to announce intended actions and travel destination plans (Savage-Rumbaugh, Pate, Lawson, Smith, and Rosenbaum 1983; Savage-Rumbaugh, McDonald, Sevcik, Hopkins, Rubert 1986). By announcing travel goals, apes give evidence of a capacity to engage in simple planning and to communicate those plans. They are also able to specify some information about their itinerary by combining two symbols such as "pineapple (and) grapes" to indicate the order of their intended travel route (i.e., they are going to get grapes by way of the pineapple place).

It seems reasonable to conclude that wild chimpanzees and bonobos are unlikely to wander randomly around the forest hoping to come upon food. Rather, they set out with a specific destination in mind, and they may in addition be able to communicate that destination to others. Little is known about how chimpanzees or bonobos determine where to travel next, nor indeed how far in advance they may plan their route. Yet, being large animals who require a great deal of concentrated food, they cannot afford to expend considerable effort traveling to a location that will contain little if any food. The ability of chimpanzees and bonobos to survive in the field depends upon some means of determining when and where they will be able to find significant food resources. Indeed, Milton (1988) has speculated that the ability to remember specific locations of widely dispersed, ephemeral, and high-quality plant foods in tropical forests has been the driving force behind the evolution of primate intelligence.

By contrast with the bonobo and chimpanzee, the less frugivorous gorilla is more likely to find something edible in his vicinity regardless of where his group travels. If he waits long enough before returning to an area he has previously visited, the vegetation will again be plentiful. The gorilla needs only to remember where he has been recently. Fig trees, the preferred food source of chimpanzees, are a far less predictable food item. They must be checked regularly and the chimpanzees must be in their general vicinity when they are ripe, for the fruit will be quickly consumed by monkeys otherwise. Arriving at a food source just as it is ripe may be a difficult business when the individual cannot simply hang around and wait. Any information that could be gleaned from other apes, monkeys, or birds would be of considerable help in making travel plans (Ghiglieri 1984).

Regardless of our ignorance of the abilities of wild apes to make and communicate travel plans, we do know that none of the current ape study sites reveal any indication of the manufacture of stone tools. In addition, current opinion in anthropology is that language was a very late invention, appearing perhaps as recently as 40,000 years ago (Noble and Davidson 1991 and see Milo and Quiatt, Chapter 10 of this volume). If this view is correct, it would mean that stone tool technology preceded vocal language by as much as 2.5 million years. It would also imply that since modern chimpanzees and bonobos have not even begun to make stone tools, they are a long way from using language.

TOOLS BEFORE LANGUAGE:
OR THE RECENT-ORIGINS VIEW OF LANGUAGE

The perspective that tool use antedates language requires the assumption that the cognitive capacities required to construct tools are less complex than those required to fashion a simple language. This view is somewhat difficult to reconcile with the fact that children use language for some time before they are able to construct simple tools.

Assuredly, it is not to be assumed that ontogeny recapitulates phylogeny, nor that the development of children mirrors the development of the species. Nonetheless, toys involving elementary construction become popular with children at 2 to 3 years of age, well after basic semantics and syntax are in place. Prior to this time, toys that reflect simpler schemata such as shaking, pushing, carrying, hitting, throwing, storing, stripping, opening, inserting, and extracting predominate (Langer 1986). Without assuming that phylogeny recapitulates ontogeny, we should still recognize that the complex constructive schema that stone tool manufacturing requires can only be developed after simpler schemata have been laid down that will serve as the subcomponents of more complex constructive tasks.

All of the simpler schemata mentioned above (shaking, pushing, carrying, hitting, throwing, storing, stripping, opening, inserting, and extracting) would be useful to a forager, such as the modern-day ape, who obtained food directly or with the assistance of simple, quickly fashioned tools like termite sticks.

An Acheulean hand ax would require more complex schemata. To construct such a hand ax one must rotate the stone in one hand and then strike it with another stone. This requires a specialization of the activities of different hands and, in addition, a sequential coordination of three very different actions (holding, positioning, and striking). The brain must simultaneously keep in mind: (a) the end product, (b) the current state of the tool, (c) the relationship between these two, (d) the activities of the left hand, (e) the activities of the right hand, (f) the relationship between the movements of the hands, and (g) the purpose for which the tool is being constructed. In addition, the toolmaker must monitor the environment from time to time so as to be able to respond to any relevant external stimuli appropriately even in the midst of tool construction. Thus, eight individually complex but interrelated things must be monitored simultaneously in order to fashion an Acheulean hand ax. There are no tests of how many parallel tracks of thought either the human brain or the ape brain can process at one time, but presumably the requirements of creating an Acheulean hand ax are more than the chimpanzee or bonobo can handle.

The number of items that can be processed in parallel is critical for such toolmaking activities. For while it is conceivable that chimpanzees or bonobos could engage in any of the component tasks listed above, it is not likely that they could engage in all of them simultaneously, as they do not seem to be capable of continually processing that many parcels of information in parallel. Apes may be able to make a tool by stripping leaves, but this does not require the planning and coordination of a different action sequence for each hand while also imagining the future form and use of the tool. Present hunger dictates the desire for the tool, eliminating any need to imagine scenarios of future use for the tool. Thus, a chimpanzee might be able to plan its travel, even planning more than one location in advance, and it might be able to make tools. However, it would be incorrect to assume that by combining these skills, a chimpanzee or bonobo could make a hand ax for future use.

Combining such schemata is not merely a matter of serially adding together their individual components; it is matter of processing the information about both things at once so that the relationship between them becomes self-evident. That is, the chimpanzee would need to be able to recall, and plan for, the function of the hand ax while constructing it. If the chimpanzee utilized all available processing capacity to construct the hand ax, he could not, at the time of construction, imagine the use of the ax. Thus, a chimpanzee might learn to make a hand ax, and if one were lying about he might learn to use it when he needed it. However, to make one *in advance* of the need, while imagining the need, and then to keep track of that tool for many days until the need arises appears to be beyond the ape's capacity.

But a chimpanzee or a bonobo might be able to make simpler stone tools, ones that require bimanual construction but not such extended planning skills as those necessary to make an Acheulean hand ax. The bifacial choppers that precede the hand ax in the archeological record are essentially pieces of stone that have had chips knocked off of them. (They can be differentiated from nonworked rocks because toolmaking produces fractures of a character that does not occur in the natural breaking up of stone.) The construction of a bifacial chopper requires only a topological understanding of the shape of the base rock. Generally, if the rock is struck on the edge, particularly on the thinner edge, with an angled downward blow, a flake will result. Bifacial choppers occur in Oldowan assemblages. Studies of these assemblages suggest that the shape of the tool is controlled by that of the initial blank, not by a template in the maker's head (Toth 1985).

There currently exists some disagreement as to whether it was the chopper itself or the flakes knocked off of the chopper that represented the desired end product of the toolmaker, though it is generally agreed that either one could be used to process both plant and animal products (Toth 1985). In addition, there are data (from microscopic studies of the polished edges of flakes) to suggest that the flakes were used for just such purposes (Keeley and Toth, 1981). If the chopper itself was the tool that the maker aimed for, then it is possible that some predetermined form, albeit a very crude one, did exist in the mind of the maker. However, if the flakes were his goal, he needed only the knowledge that striking stones together was likely to produce the sharp-edged flakes he wanted.

Upon reviewing the archeological record, Wynn (1988) has concluded that the correlation between stone artifacts of varying forms of complexity and the increase in brain size is weak and cannot be used to support the argument that tool construction fueled the drive toward a larger brain. However, Wynn (1988) observed that one strong link can be discerned between the appearance of *Homo erectus* at 1.5 mya and the simultaneous appearance in the archeological record of hand axes and other tools that demonstrate symmetry of pattern. According to Wynn, the appearance of symmetry in the artifactual record signals the emergence of a notion of reversal that, though linked artifactually to the topological geometry of space, has broader implications. Wynn also noted that the archeological record provides evidence for the simultaneous appearance of a concept of measurement in the guise of stone balls and discoids that have constant diameters.

Are stone tools of the bifacial sort within the capacity of modern-day apes, as Wynn (1988) suggested? One bonobo, Kanzi, recently succeeded in producing stone flakes for the purpose of cutting a string. When the string was severed, the lid on a box was released, permitting access to a reward (Figure 1.1). The activity of producing flakes was demonstrated for Kanzi; however, Kanzi's behavior cannot be attributed to imitation. He clearly understood that he needed to make a small, sharp flake and that he could do this by striking a rock. Kanzi's understanding of these concepts was displayed when he elected, without demonstration, to produce chips by throwing stones against a hard surface. Although Kanzi could make chips by striking rocks together, he invented the method of tossing a stone onto a hard surface, which caused it to split open and form one or more chips. Kanzi's usage of such a technique without its prior demonstration revealed that he understood the principle of breaking the rock to achieve the production of a sharp edge. When the hard substrate was removed by placing the tool site outdoors on soft ground, Kanzi again spontaneously devised a tactic not previously shown to him. He placed one large stone on the ground where he planned to throw, then selected a second, smaller stone and threw it repeatedly against the larger stone until a flake of the type he desired was produced.

(a) Stone hammer and core.

Figure 1.1 Kanzi manufacturing and using a stone tool.

E. Sue Savage-Rumbaugh

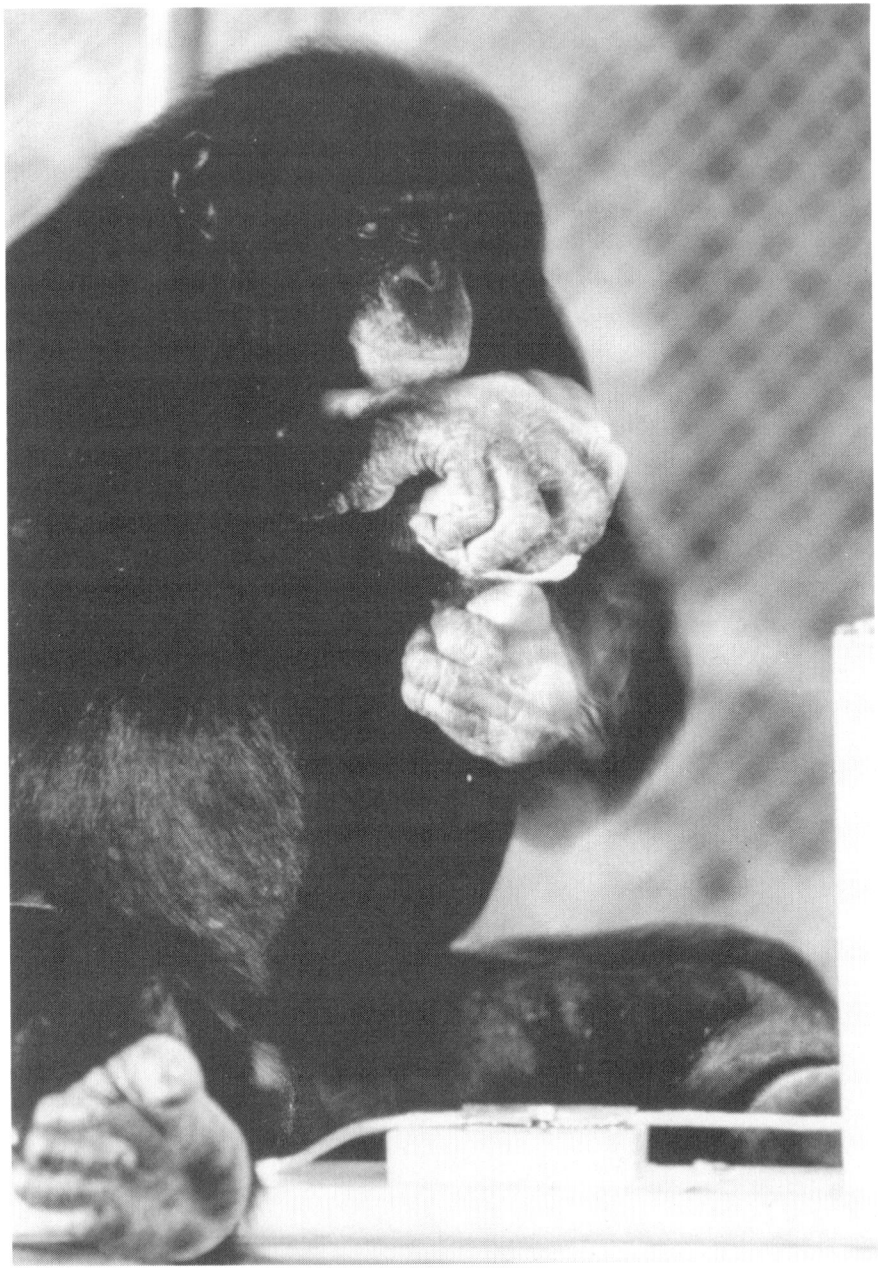

Figure 1.1(b) Striking off a flake.

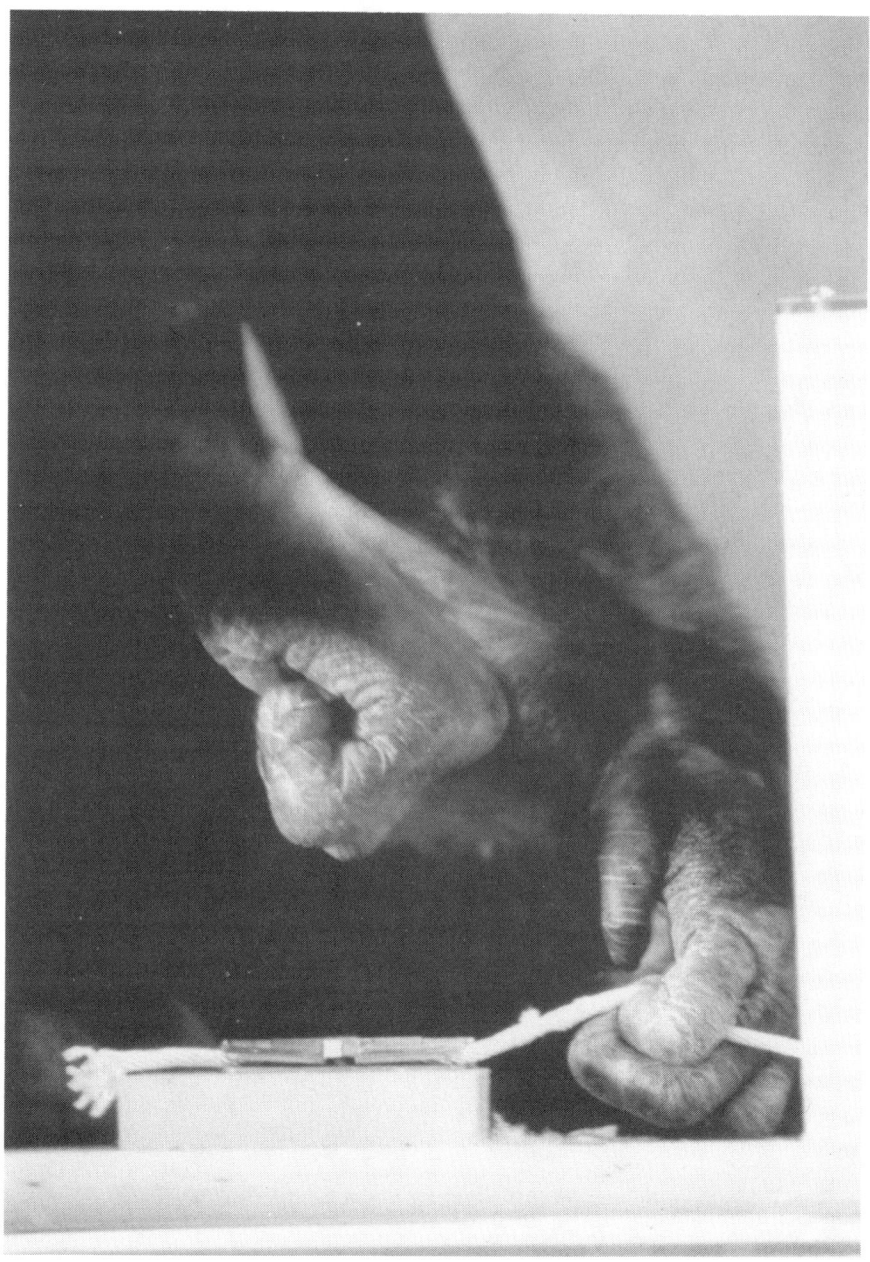

Figure 1.1(c) Using a flake to cut cord.

To date, Kanzi's tool-manufacturing ability has reached only that seen in the most primitive of the Oldowan tool assemblages. When he throws stones, instead of forming them using bimanual techniques, he produces cores that are difficult to differentiate from random strikes that might occur naturally. Hence, use of this spontaneously invented technique suggests that early hominids may have manufactured flakes for some time before recognizable bifacial choppers appear in the artifactual record.

Kanzi usually makes flakes only when the tool site is baited with a desirable item; however, such is not inevitably the case. He has, on occasion, produced tools simply when stones were made available and no reward was in sight or expected. Apparently, practice in tool construction is in some cases self-rewarding.

Kanzi's stone-chipping accomplishments give reason to believe that early hominids might well have made similar flakes even though there are currently no such tools associated with australopithecines. Such tools could have been constructed on the spot if material was available, as they require little planning compared to the Acheulean hand ax and can be made quite quickly. When a scavenged food source lay nearby to remind them of the need for a sharp flake, the hominids could quickly obtain one by striking two stones together or by throwing one stone on another.

But the strategy of making the tool on the spot would not have worked if sufficient natural stone could not be found near the area of a scavenged food source. If this were the case, it would become essential to carry rocks around as a matter of course, so that a sharp flake could be obtained as needed. Unlike wild chimpanzees, who gather their rocks on the way to their favorite nut-cracking site, scavengers must take their food where they find it. They could not always gather rocks once they had decided where they were going because it was not always possible to plan upon coming across a scavenged food resource. By contrast, chimpanzees can plan upon traveling to nut sites, as the seasons of plants and their locations in the forest are more predictable and stable than is the finding of animal remains.

For early hominids, then, it would become important to carry along the required stone material most of the time so that it would be ready for use if a scavenged animal was encountered far from proper stone. Boesch and Boesch (1984) have observed that chimpanzees will pick up a rock and carry it for several hundred meters on their way to a nut-cracking site. Such actions strongly suggest that chimpanzees know where they are going, what they intend to do, and the tools that they will need when they arrive. Nut-cracking tools are obtained only when the chimps are on their way to the nut-cracking site. They are apparently not carried about otherwise. However, hominid scavengers who did not know precisely where they were headed would need to develop the general understanding that at some future time they would end up in a situation where a tool would be required.

Potts (1984) has suggested that Zinjanthropus, as represented in Bed I at Olduvai Gorge (the FLK Zinj site; Leakey 1971), may have utilized certain sites as "stone caches" where tools needed to process animal bones were kept. According to this view, bones of scavenged animals were carried to the tool site and processed. Certainly, the ability to detach and carry part of a carcass to a predetermined site would imply considerable future planning. It would require the ability to carry rocks to the cache site initially in full knowledge of the fact that they needed to be gathered and cached. It would also require memory of this site and the carriage of bones from the site of kill to that of processing. In order to prepare the carcass for transport, some portions would need to be separated. Even preliminary separation would be far simpler and quicker with a sharp-edged stone than using the hands or teeth.

What kinds of circumstances might facilitate the development of the ability to think not only about current needs and plans but also potential future needs? What might foster the ability to remember to keep a tool at hand even though a current need for the tool was nonexistent? It is typically assumed that object carriage and bipedality go hand in hand since upright posture would free the hands for object carriage (Hewes 1961, Lovejoy 1981). While eliminating the need to use the forelimbs for locomotion would certainly make object carriage easier, it was not the freeing of the hands that was critical to the emergence of frequent object carriage; rather, it was the freeing of the mind.

Chimpanzees can and do carry even heavy objects relatively long distances although they are not bipeds. However, as noted above, they carry these objects only when they are thinking about using them. To carry an object that is not needed at that moment requires a mind free to attend to things such as imagined future needs. Such a mind would need to be free to remember to carry an object, even though the organism's current goal might have nothing to do with that object. Consequently, a mind free to disengage itself from the present situation, one free to remember to tell the hands to carry a tool even when it was thinking about something completely unrelated, was needed to guide the free hands of the biped.

BECAUSE ROCKS DO NOT CLING

There is only one type of persistent carrying activity that has been observed in apes: the carriage of infants by the mother (as well as the father, in lesser apes). However, infant carriage differs from object carriage because the infant supports much of its own weight by clinging. Even when unable to support its weight, it nonetheless grasps the mother's fur and cries out vociferously if its body is separated from hers by even a few inches. If an ape parent should forget to play its role properly with regard to infant carriage, it is quickly reminded to do so by the infant's screams.

A rock does not scream. If you forget to carry it, it remains quietly behind and you never know it is gone until the minute you need it and look to find that it is not there. Thus, the need to carry an object for long distances places a much greater burden on the memory of the bearer than does the carriage of an infant that clings and screams when the bearer forgets to do his or her part.

Not only do rocks or other inanimate objects require the bearer to remember that they are being carried, they require the bearer to remember this virtually all the time. In other words, there must essentially be a free attention-processing channel that attends to the rock, shifts its position, picks it up again if it was set down during a rest, and soon. All of this must be done even if there are frightening noises, fights, or any manner of other social and environmental stimuli that must also be processed. No matter what is going on, the bearer cannot forget the object when it is *not* needed or it will not be there when it *is* needed.

You might invent a means of strapping an object to yourself, but such a behavior is a complex act. It requires imagining the loss of the object and then looking for a second object to assure the presence of the first. It also requires the ability to wrap cordlike objects around your body in such a way that they will hold the rock as needed but can also be undone rapidly. Inventing such a knot is in many respects as difficult as constructing a stone tool, for knot construction requires the execution of a series of correctly sequenced actions, only the last of which results in any concrete and observable progress toward the goal. Thus, by the time intelligence had evolved sufficiently to permit hominids to strap things to themselves, they most certainly had been remembering to carry these same things with them for some time.

How did bipedal hominids come to be able to carry rocks that did not cling, that did not cry out if they were forgotten, and for which the bearer had no immediate need? It seems probable that such a skill was fostered by development of the capacity to carry infants who did not cling. Among primates, only human infants do not cling. Though infant carriage has been associated with bipedalism (Tanner 1981; Sinclair, Leakey, and Norton-Griffiths 1986), the real significance of this change in the fundamental pattern of primate infant transport has received insufficient attention among the scenarios of hominid evolution.

Whenever infant transport mechanisms are radically changed within a short time, a whole suite of other changes must rapidly accompany an alteration so basic to reproductive survival. In general, maternal caretaking patterns are remarkably similar and conservative in the order primates. Among New World and Old World monkeys, lesser apes and great apes, all infants cling throughout the duration of the nursing period. In some species individuals besides the mothers are permitted to carry infants, and in a few species the male participates in infant carriage and food sharing. However, these differences are relatively small compared to the absence of clinging behavior in human infants. Such a change must inevitably alter the fundamental nature of the parenting process.

For example, nonhuman primate infants are very thin at birth and during most of their infancy. They are in such close and constant contact with the mother's body that maintenance of body temperature under cool conditions is not problematic. By contrast, human infants possess an extensive amount of body fat, which helps maintain body temperature even when they are not in bodily contact with the mother. All other nonhuman primate infants are quiet all of the time, unless they are separated from their mothers. They have no vocal counterpart to the "fussy" cry human infants use to signal discomfort. An infant that is constantly on the mother's ventrum can nurse at will; it has no need to vocalize to indicate that it wishes to nurse. But an infant that is not in contact with the mother needs a vocal channel to get her attention and to signal its internal state. Primate infants lack the facial regard characterized by a mutual change in expression and a tendency on the infant's part to imitate expressions the mother produces while "en regard" (Meltzoff and Moore 1989). Infants that cling to the mother do not need facial signals to maintain her interest, nor can facial signals be seen if they are emitted.

Infants who cling to the mother must be able to support the weight of their own heads while they are clinging, even from birth. Human infants, in contrast, have heavy heads that must be supported when they are picked up. Unless the human mother carried the infant, providing support for its head, it would not be possible for brain size to increase much beyond that currently seen in ape infants; an infant with a large, heavy head would not be developed sufficiently to hold the head erect and to cling at birth. If an infant is carried with its head supported, the only remaining constraint on brain size at birth is the width of the maternal pelvis. In human beings the support provided by mothers after birth has permitted the brain size at birth to increase dramatically, making childbirth a difficult and somewhat risky venture. Finally, infants who cling to the mother must have the great toe rotated so that the feet can grasp and help the hands cling. Without such feet, most primate infants would have a difficult time staying with a mother who was rapidly attempting to escape a predator. However, infants who are supported no longer need feet that serve as an extra pair of hands for clinging. They are free to develop feet that enhance their locomotion as bipeds.

When an infant clings to the mother, the responsibility for maintaining bodily contact falls primarily upon the infant and only secondarily upon the mother. All nonhuman primate infants monitor the mother far more closely than do human infants. The moment that the mother startles or begins to move, they attach themselves to her. The mother's only job is usually to pause for a second to permit the infant to become securely attached and to provide intermittent support should it lose its grip or need a brief rest. By contrast, a human mother can startle and move away without the infant's noticing. Should a human mother plan a rapid movement, she must gather her infant in advance. She cannot simply react and rely upon the infant to make certain that it is with her. Monkey mothers provide almost no support to infants, ape mothers provide a modicum of support to very young infants, but human mothers (and fathers) must fully support an infant during all transport.

No one knows when human infants stopped clinging, but when they did, a great burden was placed upon the hominid parent who carried them. The burden was not simply the physical one of walking while carrying an infant. More importantly, once infants ceased to cling, it became the parent's job to monitor the activities of the infant rather than the infant's job to monitor the activities of the parent. Unlike young ape infants, human infants do not scream when they become separated a few inches from their mothers. Indeed, if human infants are full of milk and happy and placed on the warm ground, they quite readily go to sleep. Once the infant is asleep, it can be set down or cached while the parents go about their own business. However, the ability to place the infant on some substrate and tend to another task also requires the ability to remain cognizant of the infant's welfare. A mother who puts an infant down, even for a short period, must make certain that the infant is in a safe place and then must monitor the safety of the infant as well as her own safety. Things that could harm an infant are not the same as those that could harm an adult; consequently, the mother must simultaneously maintain two different awareness states, one for herself and one for her infant.

Moreover, unlike other species whose infants do not cling, such as canids or ungulates, human parents carry their infants much of the time. They do not leave them in dens or nests, returning mainly to feed them. Primate infants need frequent nursing and cannot survive if cached for long periods of time. Thus, the human parent must follow a rather unusual strategy, putting it down in a variety of locations and circumstances requiring different degrees of vigilance. Ape mothers never cache their infants. Indeed, among apes, only bonobo mothers permit other apes to carry their young. When an ape mother hears a startling sound and moves quickly, the infant responds simultaneously with the mother by quickly tightening its reflexive grip around her midsection. This grip makes it impossible to forget the infant, even if the mother were to be momentarily distracted by a predator.

Why would any primate that had developed a perfectly effective infant transport system abandon it for one that required extensive anatomical and behavioral-neurological compromises? The shift from infants that cling to infants that are carried has often been thought to go along with loss of hair. It has been assumed that because we humans lost our hair, our infants could no longer cling and we had to carry them. The explanation typically offered for our lack of body hair, however, is that it permitted cooling of the body during rapid, long-distance travel on the savanna, such as is required by hunting (Leakey and Lewin 1977). There are two problems with this explanation; the first is that although men are much more likely than women to be hunters, it is men who have retained the most hair. The second is that no other mammals who hunt in savannalike environments have adapted to these pressures by losing their hair. Why should such an adaptation have been effective only for hominids? There is another more telling aspect of the human anatomy that renders clinging difficult: our bipedal method of locomotion. Once we became bipeds, it was necessary to support our infants whenever we traveled for long distances on the ground, regardless of whether we had lost our hair or not. Ape infants cannot cling to a bipedal creature for long distances without some modicum of support. Gibbon and siamang infants can support themselves while their parents race across short distances bipedally. However, if the bipedal mother travels far, she has to use her arms to support the infant, for even modern ape infants cannot fully support their own weight for long periods of time when carried bipedally.

When an ape infant becomes so heavy that it is hard for mothers to lend support with their thighs and still travel at a reasonable speed, the mother begins to shift the infant to the dorsal position. On the mother's back, an ape infant no longer needs to be able to support its own weight (which by this point in development has increased substantially) for long distances by clinging. It essentially rides on the mother, clinging only to maintain its position, not to support its weight.

As the infant of a bipedal hominid grows heavy, it cannot move around to the mother's back to ease the burden of gravity because the problem of clinging to the back of a bipedal mother is equivalent to the problem of clinging to her ventrum. If the infant could not walk on its own by that time, it would have to be carried, probably on the shoulders. Here it could help steady itself by clinging to the hair on its mother's or father's head, much as an ape infant steadies itself by riding on the mother's back. Indeed, it is interesting to note that the only place our species has retained hair in sufficient quantities to permit an infant to benefit from clinging is on the head. Both males and female retain this hair in equal degrees, suggesting that perhaps one of its functions was to help infants steady themselves while being carried on the shoulders.

The constant selection pressure to cling tightly to the mother would begin to wane as mothers increased their support of infants. No longer would an infant who could not cling well be unlikely to survive to reproductive maturity. As the infant's clinging reflex began to wane, it would be possible to set the infant down for short periods to free the hands. Infant carriage places a great physical burden on the bearer, for it means that at least one hand has to be occupied much of the time. If both feet are occupied with transportation and one hand is busy supporting an infant, only one hand is left free for other activities. If two hands are needed for a bipedal task, the infant must be put down. The implications of putting an infant down also have not received sufficient attention in scenarios of hominid evolution. Infants who are put down must be picked up again before the parent travels very far. Infants cannot be stashed long, and while they are not being held they must be watched, for they are susceptible to predation or attack even by mammals far too small to threaten the parents. The attention and memory capacities required to monitor an infant that did not cling would come under rapid selection. Failure to remember or to appropriately monitor a sleeping infant could have disastrous consequences for the infant, and careless parents would leave few offspring. It is interesting in this connection that the Baka have a story about a mother who put her baby down while attempting to catch some fish; she forgot about the baby for a few moments, and it was stolen by a chimpanzee.

The need to separately and simultaneously monitor their own welfare and needs and those of their infants would have placed a great cognitive demand on early bipedal mothers whose infants were losing the ability to cling. In the typical primate mother-infant plan, whatever is a danger for the infant is also a danger for the mother. The mother need only concern herself with her own safety to make automatic decisions that are also in the best interests of the infant. However, a young infant that is not physically attached inhabits a somewhat different world from the mother's; the infant's safety is no longer synonymous with the mother's safety. While older offspring can monitor for themselves, very young infants cannot. During this period of time, the mother must attend closely to the infant's needs and must realize that they are separate from her own. This sort of bifurcation of the attentional domain occurs in apes, but the ability to assess the world from the attentional perspective of another seems to be a rather intermittent and difficult process. It requires an individual's ability to recognize that her own view of the world is not necessarily the same as that of her conspecifics, or the development of a "theory of mind" (Whiten 1991). The mother's need to process the information world of the infant separately from her own would operate as selection device for a larger brain, one that could process twice as much information while keeping information that is relevant to the infant's world separate from information that is relative to the mother's world.

A secondary by-product of infant carriage would foster the development of the mother's bifurcation of attention. When an infant is held bipedally in front of the mother, the eyes of the infant and the parent are in the same plane. When, in this position, the infant looks at an object, the mother can tell from the direction of the glance what it is that has captured her infant's attention. By monitoring its glances, the mother can form accurate inferences regarding the interests of her infant. Similarly, the infant can monitor the mother's direction of attention. Infants who cling underneath their mother's ventrum have little opportunity to monitor their mother's direction of gaze, leading to an inevitable lack of joint regard. By contrast, whatever the parent sees and talks about is immediately obvious to any infant being carried bipedally. Moreover, it becomes possible for the infant to communicate by glance or gesture because the eyes of the caretaker can easily monitor the infant. Thus, very early joint engagement of attention and joint regard can occur. It is out of such joint engagement that the concept of self arises and the value of language as a medium of communication becomes self-evident (Lock 1978).

When bonobo infants are reared by human caretakers, the effects of bipedal carriage and infant support can be clearly seen. The bonobo infant attempts to cling but needs frequent support. The more accustomed the infant becomes to support, the less it clings. When its attention is not devoted to clinging, other aspects of the environment become more salient to the infant. Bipedal carriage of the bonobo infant Kanzi sped up the emergence of joint regard and served as the foundation for the acquisition of intentional gestural and symbol use (Savage-Rumbaugh 1984). Kanzi was not raised by human caretakers alone but also by his bonobo mother, Matata. However, because bonobo mothers allow others to carry their infants, the human caretakers who worked with Matata were also permitted

Figure 1.2 Kanzi attemping to direct a gestural signal to his mother.

to carry Kanzi. The bipedal stance and infant monitoring typical of human caretakers permitted Kanzi to gain and utilize the attention of the caretakers in much the same manner as does a normal human child. As a result of this experience, Kanzi also attempted to direct gestural signals to his mother, but she often failed to respond because her attention was on what was happening in the environment around her rather than on Kanzi (Figure 1.2).

The tendency of human females to form long-term bonds with males may have its roots in the species commitment to bipedality. The need to carry an infant who does not cling could be expected to markedly reduce the foraging capacity of mothers. However, if a nearby male were to share the burden of infant carriage, the survival rate of the offspring could be expected to increase substantially. There seems little doubt that the need to carry an infant bipedally across relatively open areas would encourage cooperation between the sexes in a small-brained, 3-foot biped, particularly if there were already any tendency toward infant care among the males for evolution to exploit.

How and why would a small-brained ape make a radical shift from quadrupedal locomotion to bipedal locomotion if it entailed such a comprehensive reorganization of infant transport and infant monitoring systems? Lovejoy (1988) suggested that hominids became bipedal before they moved out of the forest environment. Supporting this view is the fact that modern lesser apes (gibbons and siamangs) locomote in a strictly bipedal manner when they travel terrestrially. These apes cannot travel very far bipedally without resting; nonetheless, they never travel quadrupedally on the ground. When female gibbons and siamangs brachiate, they support tightly clinging infants by pulling up their legs. Lesser apes employ their hindlimbs during arboreal locomotion only for pushing off of, or landing on, branches. As long as her infant can cling tightly at those moments, the mother can support it readily with her hindlimbs while brachiating with her forelimbs.

Gibbon and siamang locomotion on the ground is rare; they are not anatomically "adapted" bipeds, but neither are they "adapted" quadrupeds. Instead they, are specialized arborealists whose main mode of travel is brachiation. Nonetheless, their bipedal style of terrestrial locomotion implies that the common ancestor of both humans and the African apes may also have been a suspensory/biped who spent a great deal of time in the trees but who moved bipedally on the ground from one food patch to another when it became necessary.

The lesser ape's terrestrial mode of locomotion reveals that bipedality is a behavioral adaptation first and a skeletal adaptation second. The hip anatomy of modern lesser apes would not attest to the agility of their two-legged progress on the ground nor to their lack of arm use when terrestrial. Had none of the lesser apes survived, there would have existed little appreciation of the extent to which they move bipedally when on the ground. It is easy to imagine that similar suspensory apes of the Miocene could have become increasingly bipedal by spending more of their time obtaining food on the ground rather than in trees, without ever leaving the forest. The behavioral preparedness for such creatures to walk upright was most likely already in place for evolution to work with as the forests grew patchier. Those who could travel longer distances bipedally would have been the ones to survive as a new food patch was needed, when there were no trees between the old patch and the new one.

The monogamous social system of the modern lesser ape indicates that at least some Miocene suspensory/bipedal males may have participated in infant care and transport even prior to the appearance of bipedalism. Gibbon and siamang males bond with females and play an important role in rearing the offspring. Males share food with and carry older offspring, permitting the females to produce a second infant much earlier than if the burden of care for these older infants were to fall to the female alone. With a similar male support system already in place, Miocene female bipeds would be partially equipped to manage the burden of infant transport, as the distances they needed to travel increased during dry periods.

Given what we know about the bipedal characteristics of lesser apes, we may reasonably conclude that a bipedal/suspensory Miocene ape could have been ancestral to both hominids and the modern African apes. As the savanna grew patchier, some of these bipedal ancestors would have become confirmed in their adaptation to terrestrial bipedality and begun supporting their infants, possibly sharing this heavy burden between sexes. Those who could not cope with the increased demands of a bipedal way of life posed by infant carriage and monitoring would have become knucklewalkers, using their thighs rather than their hands to provide infant support.

But what advantage would a bipedal/suspensory mode of locomotion provide a common ancestor? Competition from many other apes and arboreal monkeys could make ground foraging a worthwhile activity. Additionally, on the ground are unique food resources, such as those accessible only in creeks or swamps and others (mushrooms, for example) that can be obtained only by digging in ground cover. Among modern apes, bonobos make frequent use of just such foods. Wet areas and the foods found in these regions are apparently critical to bonobo survival, as no bonobos are found in regions lacking wetlands (Malenky 1991). Moreover, whenever bonobos move in swampy areas, they do so bipedally, revealing the advantages of their bipedal/suspensory adaptation.

The problem of searching for food in murky water while standing bipedally also places a greater burden on representational processes. Although the forager can see the fruit on the limb ahead, it cannot see the submerged plants, fish, or snails hiding under the rocks nor the mushrooms that grow under the ground cover. In addition to fish and crustaceans, another item commonly found in streams and rivers would be rocks. Since aquatic prey tend to hide under and around rocks, any ape that utilized rivers and creeks as an additional dietary source would have become familiar with rocks and adept at moving them, if only to throw them aside as it searched for food. By similarly tossing rocks at any predators that approached on the ground, these early hominids could expand their terrestrial foraging time without having to stop and take refuge in the trees.

PATCHY FORESTS AND THE COMMITMENT TO TERRESTRIAL LOCOMOTION

There are multiple problems to be surmounted by suspensories/bipeds searching for food on ground. They must manage infants even while foraging bipedally in the swamps for food that is hard to see. Long-distance bipedal movement in swampy areas is very difficult, however, so apes that could utilize suspensory locomotion as well would have an advantage as long as the canopy of the forest housing the food resource was continuous. In fact, it is so difficult for confirmed bipedal creatures that have lost their suspensory capacity (i.e., humans) to travel in this environment that field researchers have yet to study what the bonobos do in wetlands because it is too hard to follow them there (Kuroda, personal communication.)

Hominid bipedalism, then, may not be the outcome of increasingly upright locomotion on the part of quadrupedal, terrestrial ape forebears. Instead, it may have evolved as the natural outcome of a more arboreal ancestor's transferring a suspensory mode of locomotion to the ground. Suspensory locomotion tends to place the organism in a vertical position most of the time. The skeletal changes that led to efficient suspensory locomotion in the trees would have predisposed the common ancestor of humans and African apes to an upright posture. Consequently, when such creatures foraged on the ground, they would have tended to assume a bipedal posture, even though their anatomy was not that of a confirmed biped. Computer simulations of the musculoskeletal activities of humans, gibbons, monkeys and chimpanzees tend to support this view (Yamazaki 1985). Looking at both kinematic indices and mechanical indices, Yamazaki (1985) concluded that "the bipedal walking of the gibbon is closest to man. . . . Consequently we conclude that a brachiation model is best for the pre-habitual bipedal walking of our ancestors in terms of a 'potential' for bipedalism" (p. 129).

Oreopithecus, a relatively large-brained (400 cc), siamang-sized ape that lived in swampy forests during the late Miocene, exhibited many of the characteristics of a common African ape–*Homo* ancestor (Straus 1949), including skeletal adaptations for bipedalism. *Oreopithecus* probably was able to suspend itself much in the fashion of siamangs but also displayed changes in the pelvis that presaged a full bipedal commitment but were suggestive of frequent bipedal locomotion. It had a relatively flat face with small canines and bicuspid lower premolars, typically hominid characteristics (Pilbeam 1972). Its teeth were also heavily enameled, suggesting that it consumed foods often associated with a gritty substrate. That an ape form such as *Oreopithecus* displayed suspensory/bipedal characteristics 8 to 9 mya strongly supports the view that bipedal locomotion among hominids need not have derived from quadrupedalism.

The extreme deforestation during the late Miocene must have demanded terrestrial travel between forest patches, placing great pressure on such apes to develop a mode of long-distance travel that did not require trees. With the onset of problems posed by the changing landscape of the later Miocene, a creature already committed to a bipedal/suspensory life-style would likely have become a better-equipped biped. Such a creature may to begin with have maintained some of its suspensory activities by holding onto trees when traveling bipedally, much as human infants go through a stage of holding onto objects while circumlocuting the perimeter of the room as they learn to walk.

Although it is often assumed that the savanna fostered bipedality, when bonobos walk up and down steep slopes, they often assume a bipedal stance and hold onto trees along the way, exhibiting a method of locomotion that displays a combination of suspensory and bipedal movements. They also assume a bipedal stance when carrying food, foraging in swamps, and digging.

A critical problem for a suspensory/biped attempting to traverse increasing distances would be how to support young infants who could not cling firmly enough to support their own weight for long periods of time. No longer could the mother's legs be drawn up to support the infant while the hands propelled them through the treetops. The choices for the mother would be limited. She could use her hands to support the infant or, by leaning down and walking on her knuckles, could use her legs. If her arms were already quite long, the knuckle-walking solution would prove relatively easy to adopt. If her arms were relatively short, however, she would have little choice but to use them to attempt to help her infant to hold on.

As long as the patchiness of the forest was not extensive, this common ancestor could travel bipedally from one patch to another by relying on the offspring's ability to cling. But as travel between food resources began to require that adults spend the majority of time in the upright position while moving, it would become increasingly difficult for infants to support their own weight the entire time. Mothers could pause, allowing the infant to rest, but if they did so they could not keep pace with other members of the group and would become vulnerable to predators in the open environment. The ability to travel long distances on the ground with infants must have become a critical factor in the effort to survive as the forest receded.

Some evolving species of apes could have become confirmed bipeds and others modified bipeds or knuckle walkers. By leaning over and enlisting the aid of their long arms during travel, mothers with very young infants that did not cling well could still keep up with the group. Later, as the infant grew too heavy to support itself, it could move onto the mother's back. Biomolecular evidence suggests that the strategy of knuckle walking was adopted independently at least twice, once by bonobos and chimpanzees and again by gorillas. Although these species appear similar and are often classed in the same genus, recent biomolecular evidence (Sibley and Ahlquist 1984, 1987) suggests that this is a case of convergent evolution, gorillas having split off from the line leading to man some 8 mya, whereas chimpanzees and bonobos branched off 4 to 5 mya. If the common ancestor of 8 to 9 mya was a suspensory biped, gorillas as well as chimpanzees and bonobos would have independently adopted knuckle-walking as the locomotor pattern of choice.

While it has been argued that such convergent evolution is unlikely, there is really little choice for a partially bipedal female whose infant cannot cling for the duration of the terrestrial distances that she must travel to survive. If she has no help, she will be forced either to hunch over and attempt to support the infant with her thighs or use her hands for support. Rumbaugh (1965) has shown that even squirrel monkeys will attempt to walk bipedally in order to carry an infant who does not cling. If the suspensory/bipedal form of locomotion were adopted by different Miocene ape lineages, which indeed seems probable, then each of these groups would be independently faced with the same transportation problem as the forest grew thinner. Because of the limited number of ready anatomical solutions, it would not be surprising that knuckle walking would be selected more than once.

The effect of forest patchiness on the common ancestor of humans and African apes is further supported by the observation that the restriction of the forest was less extensive in Asia, and there truly suspensory apes (siamang, gibbon, and orangutan) continued to evolve. None of the Asian apes displays either a knuckle-walking or a bipedal adaptation to terrestrial habitats. Gibbons move bipedally by leaping or jumping, whereas siamangs move by walking. Orangutans walk on the palms of their hands (like human infants who crawl) but cannot travel rapidly or for long distances in this fashion. Often they find quadrupedal locomotion so exhausting that they will somersault instead. The only means they have of traveling long distances without becoming excessively tired is in the trees. When they are required to travel in open areas between patches of trees, their rate of progression is so slow that they would be easy prey for any carnivore. Thus, both of the Asian apes have retained the primitive suspensory/bipedal locomotor pattern, with particular emphasis on suspensory travel.

ARE LARGE CANINES A DERIVED TRAIT?

If bipedalism is viewed as a primitive rather than a derived trait, it then becomes possible to reinterpret many of the current fossil finds that are viewed as hominid as ancestral to both modern apes and humans. This is particularly true if the hypertrophic development of the canines in apes is also seen as a derived rather than primitive trait.

While it is generally assumed that hominids are characterized by reduced diastema, small canines, and heavily enameled dentition, it is also the case that some known primitive forms such as *Oreopithecus* display similar characteristics. It seems equally plausible that the common hominid-ape ancestor had small canines and a dental arch not as strongly rectangular as that of modern apes. Canine development in modern-day apes is reliably correlated with the degree of intraspecific competition among males. Gorilla and orangutan males have the largest canines and display the lowest degree of male-male bonding and the greatest degree of competition for access to females. Bonobo males have the smallest canines and display the highest degree of male-male bonding, and the smallest degree of competition for females. Chimpanzee males fall in between gorillas and bonobos with regard to canine size, male-male bonding, and competition for females. Canine development in females is generally much reduced from that in males, and females generally display little competition for access to males. Moreover, the large differences across species in canine size among males has no counterpart among females, strongly suggesting that canine enlargement is a by-product of male competition rather than diet. This view is supported in that males rarely use their canines when feeding. However, they do use them in encounters with other males. The extreme muscular development around the heads of male gorillas and the cheek pads surrounding the faces of male orangutans uphold the theory that large canines are a secondary adaptation to male-male competition.

In species where female canines nearly equal those of the males in size, we would expect the absence of male-male competition if the canines of both sexes were small. Or, conversely, we might look for both male-male competition and female-female competition if the canines were large. Such is precisely the case among the lesser apes. The canines of both sexes are large, though those of males are only slightly larger. However, according to Frisch (1973), as regards sexual differentiation of canines, "the gibbon stands entirely apart from the other apes and can be compared only to *Homo*. Indeed some races of modern man, as the Javanese, exhibit a higher degree of sexual dimorphism than that observed in the gibbon" (pp. 81–82). Among gibbons, males defend the pair bond from other males, and females defend the pair bond from other females. Indeed, neither males nor females would likely remain monogamous if each mate did not prevent like-sexed individuals from gaining access to his or her territory and consequently his or her mate. If these correlations between dentition and social structure among modern-day apes can be extended in time to ancestral ape populations, it can be postulated that an ape with small canines such as *Oreopithecus* would be characterized by less intrasexual aggression than living apes.

It is suggested, then, that a bipedal/suspensory ape with small canines inhabited riverain forest and fed on fruits and leaves while foraging terrestrially as well as on fish, crustaceans, mushrooms, and other plants growing in the damp soil of the riverbanks and swamps. This ape was ancestral both to ourselves and to modern-day bonobos and chimpanzees. Because of their dependency on a riverain environment, these apes could not form pair-based independent territories as do modern-day gibbons and siamangs. Consequently, they developed strategies of intrasexual cooperation that permitted the pair-bonding characteristic of territorial primates to coexist as a substructure within the multimale, multifemale groups typical of most other primate species. Certainly, one ideal mechanism for a nonterritorial, group-living bonded pair would have been to increase the period of female receptivity. By being continually receptive, a female would be more likely to maintain a bond with a male even though there were other nearby females.

THE BONOBO: A MODEL FOR EARLY HOMINIDS

Of modern African apes, it has been suggested that the bonobo is the best living model for early australopithecines. This view is based on the numerous anatomical similarities with regard to craniofacial structure, dentition, and limb development (Zihlman, Cronin, Cramer, and Sarich 1978). Bonobos have a smaller chest and a smaller head that rests more vertically upon their spines than those of other apes. This makes it easier for them to assume a bipedal stance, and they do so more frequently than other apes. Their arms are somewhat shorter relative to their legs, which also aids in bipedal movement. However, it is not only the anatomical data but also the behavioral data that suggest that the bonobo is a model that deserves more attention. Recent data from field studies suggest that the bonobo of today manifests a life-style quite different from other apes and arguably similar to that which may have existed in the ancestral ape population prior to the emergence of either fully developed bipedalism or knucklewalking (Badrian and Badrian 1984; Badrian, Badrian, and Susman 1981; Horn 1980; Kano 1979, 1980, 1982a, 1982b, 1983; Kitamura 1983; Kuroda 1979, 1980; Mori 1983, 1984; Nishida 1982; Susman, Badrian, and Badrian 1980; de Waal 1989b; White 1989). Unlike all other apes, bonobos live in multimale, multifemale parties that are relatively stable for long periods. Also unlike all other great apes, bonobos demonstrate very close ties between adult males and adult females. Adults of both sexes groom each other frequently and share plant foods. All these traits make the social lives of bonobos much more similar than those of other apes to the hunter-gatherer tribes that live near by.

Even though strong sexual attractions exist between males and females, there is relatively little male-male competition or female-female competition among bonobos. Bonobos are notably more pacific than other great apes. Intrasexual competition is negotiated by two mechanisms, one physiological, the other behavioral. Like human females, bonobo females are sexually receptive throughout the majority of their cycle (Savage-Rumbaugh and Bakeman 1978; Savage-Rumbaugh and Wilkerson 1978; Thompson-Handler, Malenky, and Badrian 1984). Unlike human females, however, bonobo females exhibit anogenital swellings similar to those of the common chimpanzees. Bonobo females differ markedly from chimpanzee females, though, in that they retain a relatively large swelling throughout most of the cycle, thereby making intersexual copulation possible. Even when their swellings are at the minimum size, bonobo females still appear "swollen" by comparison with common chimpanzees, and they are still sexually receptive. Since copulation for either male chimpanzees or male bonobos is difficult unless a female is swollen, due to length and flexibility of the penis, the increased duration of swelling in the bonobo female appears to be a physiological adaptation to the need to extend sexual receptivity.

Intrasexual competition between females is mitigated through sexual activity, especially genitogenital, or GG, rubbing. Females engage in frequent ventroventral sexual encounters that appear to share many similarities with intrasexual encounters, including a climatic response. Such sexual bonding between females is frequently employed upon entering food patches and appears to function as an agreement to joint usage of the food patch (Kano 1989, White forthcoming).

Assuming that the prototypical ape ancestor was territorial, as are modern-day gibbons and siamangs, we may view the current diversity of social grouping patterns extant in modern apes and *Homo sapiens* as different solutions to a common problem: how to decrease intrasexual competition enough to permit group life. As the forest grew patchier in Africa, journeys between distant patches would be far less risky for apes traveling in groups than for those traveling only as bonded pairs with their offspring.

Gorillas apparently solved the problem of terrestrial foraging by forming groups with a single dominant male who controlled access to all females. This large male inhibits sexual advances of other males, who are generally his offspring and who are peripheralized. The

40

dominant male, being much larger than the females, also easily inhibits any intrasexual aggression among females. Minimal competition among females remains, however, in the form of attempting to maintain proximity to the adult male. The gorilla female, unlike other African apes, displays no external swelling, as there is no advantage to advertising her cyclical state to other males. If the dominant male controls sexual access to all the females of his group, he must remain with the group at all times to protect the females from approaches by other males. Consequently, females have no need to be continually receptive, nor do they need to advertise their sexual state. The penis of the male is small and copulation can be achieved without anogenital swelling on the part of the female (Fossey 1983).

Common chimpanzees also solved the problem by forming groups, but in their case the groups were not dominated by a single male. Consequently, there is less intrasexual competition among chimpanzee males, and they are far smaller than gorilla males. Intrasex competition among males is also mitigated by the need for males to cooperatively defend their territories against neighboring groups. When the number of males in a group becomes too low, the females are likely to transfer to other groups. Male chimpanzees are more tolerant than gorillas of sexual access by other males in their presence, and females have consequently developed swellings to advertise their sexual status to males. But chimpanzee males appear to revert to the ancestral pattern for at least short periods of time, insisting that a single female follow and depart from the group on what has been called a "safari." During this time, which typically lasts throughout the female's fertile phase, the accompanying male seeks to keep the female away from other group members and in so doing assure that he will have a chance to father offspring without competition from other males. The chimpanzee female is maximally swollen during the period of ovulation. The swelling gradually increases prior to ovulation and gradually declines following ovulation, thereby providing males with a distinct visual cue as to the time to insist upon a safari (Goodall 1965, 1968; Itani and Suzuki 1967; Suzuki 1969; Kano 1971; Nishida 1968, 1979; Nishida and Kawanaka 1972; Pusey 1979; Nishida, Hiraiwa-Hasegawa, Hasegawa, and Takahata 1985; Wrangham 1979).

Thus, the common chimpanzee seems to have solved the problem of overlaying group life upon the prototypical pattern exhibited by the lesser apes in a unique way. Both patterns are utilized but in different time frames. When females are maximally receptive, males attempt to revert to the prototypical pattern. However, when they are not accompanied by a receptive female, they form male-male coalitions and patrol the range. If males traveling as a group encounter a swollen female, they permit multimale access, thereby inhibiting sexual competition under these circumstances. The unusually fluid group structure exhibited by common chimpanzees both permits and results from their tendency to simultaneously maintain two different bonding strategies. Modern human hunter-gathers have opted for a variety of solutions to the same problem, though the most common one seems to be that of serial monogamy and the formation of nuclear families within a stable multimale, multifemale group (Woodburn 1980, 1982). This solution means that human females, like gorilla and lesser ape females, do not need to signal their sexual state to males. However, unlike lesser apes, human females live in groups in which other females may attract their mates. Consequently, human females, like bonobo females faced with a similar situation, have extended sexual receptivity to the duration of their cycle. In all known small groups of hunter-gatherers, human females are bonded to a single male during their childbearing years. Consequently, unlike common chimpanzee females, who participate in safaris during their receptive state, human females have no need to signal their state to males. Moreover, since they live in relatively stable multimale groups but are serially bonded to a single male, an overt signal such as an external anogenital swelling could prove disruptive to the male cooperation that has evolved amongst *Homo*. By remaining continually receptive and serially bonded to a single male, human females decrease intrasexual aggression among males and ensure the presence of a male while their offspring are maturing. Through this strategy, the human female benefits both by access to a single male who, as with siamangs, participates in provisioning of the offspring and by the presence of multiple males acting cooperatively to defend the group from large predators and from other potentially hostile groups.

The human solution, then, differs from the bonobo solution in that human females do not evidence anogenital swellings. However, in most other respects bonobo social structure and behavior appear to be more like that of modern hunter-gathers than do those of other ape species. (See Table 1.2 for a list of the similarities and differences among modern hunter-gatherers and modern apes.)

Table 1.2   Behavioral Characteristics of Modern Hunter-Gatherers Compared to Those of Modern Apes

Behavior	Modern Hunter-Gathers	Bonobos	Gorillas	Chimpanzees
Adults share plant foods	Yes	Yes	No	No
Males participate in child care	Yes	Yes	No	No
Males regularly share food with youngsters	Yes	Yes	No	No
Multimale Multifemale groups	Yes	Yes	No	No
Unrelated males cooperate to defend group	Yes	Yes	No	Yes
Females are receptive through the cycle	Yes	Yes	No	No
Intrasex sexuality	Yes	Yes	No	No
Forage in streams or swamps	Yes	Yes	No	No
Eat insects found in wet areas	Yes	Yes	No	No
Tool use	Yes	Yes	No	Yes
Exhibit hunting behavior	Yes	Yes	No	Yes
Construction of rain covers	Yes	Yes	No	No
Dig for subterranean foods	Yes	Yes	No	No
Variable copulatory patterns	Yes	Yes	No	No
Ventroventral copulation	Yes	Yes	No	No
Permit others to carry young infants	Yes	Yes	No	Not often
Length of time infants need support while carried	Up to 2 years	Up to 6 months	Up to 1 month	Up to 1 month
Strong intersex ties	Yes	Yes	No	No

In bonobo groups sexual activity appears to play a strong role in the bonding of group members to one another. It is common, occurring many times each day. A variety of different postures are associated with sexual interactions, ventroventral copulatory bouts being very frequent. During copulation there is often an exchange of intense gaze and vocalization. Sexual activity is associated with reunions, general excitement, food sharing, and reconciliation. Indeed, occasions that can be said to elicit hugging in *Homo sapiens* often correspond to occasions of sexual behavior in bonobos. And just as children participate in hugging in our society, bonobo offspring similarly participate in sexual interactions. Additionally, just as hugging is sometimes a group affair among *Homo sapiens*, so sexual behavior is a group affair among bonobos. And presumably, just as hugging tightens our social bonds, so does sexual activity serve this function among bonobos. Although bonobos are not monogamous, some selectivity is exercised in sexual interactions. Both males and females appear to be differentially attracted to particular partners. Unlike other apes, bonobo males often intersperse play with sexual activity.

While the primary function of sexual behavior is species propagation, in bonobos sexual patterns have become linked to food sharing to a significant degree. Sharing of plant foods is common among adult males and females, who often copulate while eating. The male-female sexual pattern is also expanded to include genitogenital rubbing between females. During these bouts one or both females apparently may experience orgasm.

Bonobo males play a much greater role in infant socialization than do other great ape males. Most male apes have only passing interest in their offspring or in the offspring of other males, but bonobo males appear to enjoy carrying infants, cuddling them, playing with them, and sharing food with them. Infants regularly leave their mothers at an early age to seek the company of favored males. Bonobo infants, unlike other ape infants, establish independent social relationships with other members of the group from the time they locomote independently, if not sooner. Though strongly attached to their mothers, bonobo infants are far less mother-centered or mother-oriented than are other ape infants. They enjoy being carried by individuals other than their mother, thus a mother is not solely dependent upon her own energies for infant transport.

Bonobos tend to be found along rivers, roads, villages, and in swamps. They are infrequently seen in primary forest, though this habitat is readily available. They do not hesitate to wade into swampy areas, where they apparently attempt to catch small fish. They are also avid diggers, searching for mushrooms and earthworms. While moving bipedally, they often carry food in both hands as well as in their mouths. If supplied with food such as sugarcane or bananas, bonobos tend to carry as much as possible away from the source. They then share the food in small groups. By contrast, chimpanzees tend to remain beside the food source, which is generally preempted by a dominant male. This male then determines who may eat and when.

SUMMARY

Current interpretations of hominid evolution fail to explain much of the extant fossil material because of two pervasive but possibly incorrect assumptions. The first is that quadrupedalism is a primitive mode of locomotion for apes, to which a corollary belief is that bipedal hominids consequently descended from quadrupedal apes as they moved onto the savanna. The second is that large canines (and other dental characteristics, such as a diastema and non-parabolic shape) are representative of the primitive dentition.

If, instead, these traits are assumed to be derived from a common ancestor, the fossil record is open to reinterpretation in a manner that matches the biomolecular dates for human-ape divergence much more accurately. It may also explain why there appear to be no ancestral apes. By assuming that both bipedalism and reduced canines appeared *after* the *Homo-Pan*-gorilla split, all fossils that show evidence for either bipedalism or reduced canines are assumed to be either ancestral to *Homo* or evolutionary dead ends. However, if teeth can get smaller, they can also get larger, and animals that are inefficient bipeds can become proficient knuckle walkers. Evolutionary adaptations work to fit every need, and consequently they do not always move in a single direction.

If the primitive condition is indeed a bipedal/suspensory ape, some of these creatures may have begun to spend increasing time on the ground obtaining foods from the swamps, streams, or small rivers. For each suspensory ape species, the choice between bipedalism and knuckle walking would have arisen during the late Miocene. Those that clung to the trees failed to survive. Apparently, three currently surviving forms selected knuckle walking. We perfected bipedalism and along with this choice made the required changes in the traditional primate mother-infant transport system. Those changes moved the burden of infant monitoring away from the infant and onto the parents. Monitoring an infant that did not always cling required an increased attention span and the mother's ability to differentiate what might be of danger to herself from what might be of danger to the infant.

For chimpanzee mothers, this is one and the same; since the infant is constantly clinging, the mother will probably notice anything that might happen to it because she will be undergoing the situation simultaneously. This is not so strongly the case for bonobo mothers, whose infants are likely to be carried by other members of the group. Similarly, bonobo mothers do not react as though threats to their infants are a threat to themselves, as do chimpanzee mothers (Savage-Rumbaugh 1984).

By developing the ability to carry and monitor infants, increasingly bipedal hominids expanded their ability to keep a large number of things in mind for parallel processing. The expansion of this capacity was utilized by members not carrying children as well. The increased attending skills and consequently enlarged brain permitted these hominids to expand greatly upon the elementary tool-use skills seen in modern-day apes because they could monitor the results of their actions more efficiently and because they were able to carry tools long distances. This tool-transport skill permitted them to take advantage of food resources that required tools regardless of whether or not rocks were available nearby. It also permitted them to equip themselves with a ready weapon that could be hurled to keep predators at bay and consequently to protect themselves more effectively as they moved between forest patches. If a potential victim has to look for a rock as a predator is attacking, it is generally too late. However, if a rock is ready at hand, it can make the difference between life and death.

Modern apes recognize the value of weapons for defense against predators and throw branches and rocks to threaten or scare other animals, but only when they happen to have them close by (Kortlandt and Kooij 1963). They do not carry rocks with them in order to be prepared for predators. On the savanna there are likely to be more predators than readily available branches or rocks.

The ability to throw and accurately strike a predator is not a trivial skill, and development of this capacity is likely to have paved the way for further elaboration of other cognitive skills. Throwing accurately at a distal target places planning demands upon the nervous system that are quite unlike those required to avoid or chase a target. The thrower must accurately predict the trajectory and plan and coordinate the precise muscle patterns necessary to achieve that trajectory. The ability to time this precision requires that the nervous system sequence many of these movements in advance and then execute all of them at once in the proper order and with the proper timing (Calvin 1983).

Throwing is often discussed in the context of hunting and is assumed to be primarily a male activity. However, the ability to throw an aimed missile at a predator probably long predated the ability to kill other creatures with such missiles. Females as well as males would need such a skill. Any bipedal ape that frequented streams could make the transition from moving rocks while foraging to throwing those same rocks at predators. Bipeds who frequented the shores of lakes would have needed to protect themselves from many other large animals who came to the lake's edge to drink. Rocks could also have proven to be useful tools in opening or smashing the shells of crustaceans, just as sea otters use them today. Early flake tools would have been most useful implements for removing scales from fish.

There is a significant difference between throwing by modern apes and by humans. Modern humans have the capacity to bend the wrist backward to almost a 90-degree angle and thus to snap the wrist as they release an object. It is this snap at release that gives the throw greater precision and accuracy than can be achieved by African apes. The adaptations for knuckling walking have included a stiffening and strengthening of the wrist. The African apes can abduct the wrist only a few degrees, not enough to give a snap to any thrown missile. Consequently, both the aim and the potential force of the throw is considerably diminished.

A similar wrist-snapping movement is critical to the fashioning of rock tools. In order to efficiently flake off large pieces of stone, the toolmaker must essentially engage in a "control throw" by snapping the wrist holding the hammer stone just before striking the core. When the bonobo Kanzi attempts to make stone tools, his inability to execute such a wrist-snapping motion greatly reduces the power he is able to translate from his arm movement into a blow on the cobble. It also affects the precision of his blow, causing him to miss the angle and even the stone if he attempts to make an arm movement of any length. He must therefore strike the stone with short, inefficient blows that require a great deal of strength, though not much of the strength is transmitted into the stone itself.

The inability of the ape to abduct the wrist suggests once again that the common ape-*Homo* ancestor was a biped who had not yet become a specialized knuckle walker. It also suggests that that evolving biped ancestor maintained a wrist free to abduct, just as it currently is in the orangutan. The ability to abduct the wrist permitted *Homo* to perfect both throwing and rock-striking skills and consequently to develop throwing as a much more effective predator defense system than apes could ever manage.

Scenarios of where we came from and who we are such as those offered here will always be open to critique, and none can be fully verified. Each can only be argued to be a little more or a little less likely. Nonetheless, the perspectives on our past that are currently in vogue impact upon our interpretations of our present actions as a species and influence those actions in a myriad of ways. Views of humans as hunters, creatures whose elegant intelligence was shaped by the chase and by social conniving against their fellow humans give us a very different sort of self-awareness than does the scenario I propose.

By ignoring the accepted dogma of what is a primitive trait and what is a derived trait, we allow a new perspective of the fossil record to emerge. Even if the perspective offered here should come up wanting in some of its aspects, it nevertheless may serve to direct critical attention to assumptions that are not recognized for what they are but are misconstrued as facts and so have continued to limit our interpretation of the fossil record and hence of ourselves.

NOTE

1. The preparation of this chapter and research described herein were funded by National Institute of Child Health and Human Development grant NICHD 06016, which supports the Georgia State University's Language Research Center, and by the Yerkes Regional Primate Research Center of Emory University. This research was also supported by the College of Arts and Sciences of Georgia State University. I am grateful to to Bill Calvin, Nick Toth, and Sydel Silverman for their input to many of the views expressed here, though their thoughts on these issues may be very different from mine. I also thank Duane Rumbaugh and Shelly Williams for many helpful comments. My greatest appreciation and respect is extended to Matata, Kanzi, Panbanisha, and Tamuli, the special bonobos who made this work possible.

PART II

Approaches to Comparison

Introduction

For any animal species, behavior is dependent in some degree on physical character — notably, in our own species, on the character of hindlimbs and pelvis, manual digits, and neural capacity and circuitry. If we could picture in detail just how these physical systems evolved in relation to one another in our forerunners, we could greatly clarify our understanding of the evolution of hominid culture.

But we still have much to learn about the anatomy of our early, prehominid ancestors prior to that series of speciations that produced the contemporary apes and humankind. And though we may anticipate, optimistically, that future fossil finds will fill out the record currently available of those ancestors' skeletal anatomy, still, fossilized bones in themselves can never give us an adequate picture of the whole organism, certainly not a moving picture of the living animal in action. It is likely that there always will remain room for great differences in interpretation and opinion as to the implications of particular skeletal features for muscle form and behavior function. Given our present knowledge, then, we must not be too hasty in assuming a terrestrial orientation and a confirmed hindlimb focus in the locomotion of our Miocene/Pliocene forerunners. As Savage-Rumbaugh has suggested (Chapter 1), where cognition and culture are concerned, we should consider carefully what we may owe to a late arboreal heritage.

While the character of individual behavior may not be written into the fossil record in complete and unambiguous detail, nevertheless a great deal *can* be inferred from physical remains. Cross-checking against the observable behavior of modern descendants or closely parallel forms, where these exist, can provide a measure of the "reasonableness" of inferences drawn.

Similarly, where fossil remains are incomplete, attention to regularities in physical form and in basic behavior across closely related surviving taxa can aid reconstruction of the lives of ancestors. This comparative approach is particularly important when it comes to reconstructing the social life of extinct animals, including early hominids, for both the fossil record and the archeological record until as late as 30,000 to 40,000 years ago shed little light on how individuals interacted and communicated (but see Milo and Quiatt, Chapter 10). Comparative studies of the behavior of the living primates is essential to understanding the evolutionary origins and development of hominid culture.

In recent decades long-term field studies across a broad spectrum of primate species have provided a rich foundation for evolutionary comparison. We are of course most closely related phylogenetically to the African apes and, in particular, to chimpanzees, and paleoanthropologists speculating about the lifeways of early hominids have drawn heavily on the findings of Jane Goodall and other students of chimpanzee behavior (see, e.g., Isaac 1978a, 1978b; Zihlman 1978; Zihlman and Tanner 1978). In recent years there has been a reaction to what some see as overreliance on data from a single species. Tooby and DeVore (1987), in a particularly influential paper, have cautioned against basing evolutionary reconstructions on "referential" models in which a single species such as chimpanzees or savanna baboons constitutes an analog for comparison. They recommend instead the use of "conceptual" models based on theoretical principles. Similarly, Wrangham (1987) has suggested that less emphasis be placed on similarities in human and chimpanzee capacities for certain dramatic activities (e.g., toolmaking or signing) that are absent from the inventories of most primate species, more on commonalities in behavior and social organization that extend across a broader range of primate species, especially where these can be linked to common environmental circumstances.

However, some have found Tooby and DeVore's distinction between "conceptual" and "referential" models artificial and misleading (Stanford and Allen 1991, Moore 1992). Certainly, we will not be content with comparisons that are mechanical and atheoretical, simply analogous, but, as Moore (1992) put it, "Nothing is gained by labelling a model 'conceptual' or 'referential'. Models are good, bad, or (most often) a bit of each" (p. 198). Nor can we grasp the essence of hominid culture by comparing our behavior with that of a single other species; nevertheless, chimpanzees and bonobos remain our most informative source for comparison. Indeed, one of the products of a wide-ranging comparison across primate species should be improved understanding of what behaviors to focus on and what questions to ask when we narrow the comparison to include only our closest relatives.

There are of course marked differences in behavior not only across species but across populations and across groups within contemporary ape/hominid species, for example, *Pan troglodytes*, *Pan paniscus*, and *Homo sapiens*. Consequently, we must be clear as to which populations and groups will prove most useful when it comes to making those comparisons on which we propose to base evolutionary reconstructions. In Chapter 2, Gordon W. Hewes recommends comparison of wild, unprovisioned, forest-dwelling chimpanzees with newly "discovered" contemporary human groups, groups reputed to utilize the simplest of technologies and foraging systems and to have been out of communication with other human beings for generations, perhaps for hundreds of years, prior to recent contact.

Says Hewes, "Comparisons of ape behavior and *generalized* human behavior are unlikely to be of much value. . . . We must find one or more human groups that lie at the absolute minimum of complexity." He finds these in the Mrabri of northern Thailand and the Tasaday of Mindanao, in the Philippines. He includes a third reference group, the aboriginal Tasmanians, whose technology McGrew (1987, 1992) has compared with that of chimpanzees, "chiefly to emphasize the far greater simplicity of Mrabri and Tasaday culture."

Such a comparison will be offensive only to those who defend the notion that all cultures are equal by every standard applied to them. Nothing Hewes can say is likely to placate critics grounded at this standpoint, so he doesn't waste time trying. However, we remind readers that while Hewes is intent on locating the most remote and technologically undeveloped human cultures he can find and comparing them with the cultures of similarly remote chimpanzees, he assumes biological "equality" within species, hence equal cultural potential, noting that all groups have "the potential (now increasingly realized) for rapid entry into modern complex societies." That, of course, includes the Mrabri and the Tasaday; it does not include chimpanzees, who in numerous ways and not least where language abilities are concerned, are biologically as well as culturally primitive in comparison with *Homo sapiens*.

In primates, behavioral and especially cultural divergences are strongly influenced by socioecological setting. Individual expression of behavioral possibilities that lie within a species' natural repertoire may be constrained or completely inhibited by features of the social environment, and the character of developmental learning is highly dependent on relationships established with kin and other close associates. Long-term studies of a number of chimpanzee and bonobo populations have provided a rich database for assessing, from a socioecological standpoint, cultural diversity in those close relatives to humankind; and videotaped records of behavior in nature, affording detailed analysis of complex sequences of social interaction and communication, turn out to be just what is needed to operationalize comparisons of how information is processed socially in different cultures, human and nonhuman.

The culture concept is such a powerful analytic tool that we can only wonder why we have insisted so long on limiting its use by definition to the study of human behavior. This anthropocentric habit has not furthered study of biocultural evolution. A species-bound analytic concept clearly cannot be much help in conducting evolutionary comparisons, and it has been awkward for anthropologists whose object of study was the evolution of culture and whose method of choice was that primary method of evolutionary biology, the systematic comparison of similarities and differences across (as within) species.

However, primatologists and anthropologists have recently begun to refer to *cultural* differences when comparing behavior across groups and populations within nonhuman primate species, especially with reference to chimpanzees and bonobos. In Chapter 3 Jo A. Myers Thompson reviews this literature, discusses the evidence for cultural diversity within species of monkeys and apes, and explores a specific instance of innovation and dissemination of learned behavior within and across captive groups of bonobos.

Comparisons of cultural diversity within and across species must take into account the ways in which behavior is constrained, in development and in expression, by (1) the physical makeup of individuals and (2) social circumstances associated with characteristic group size and group structure, population density, demographic trends, life history patterns, and intergroup and interspecies relations in a given habitat. These features are treated in the last two chapters of Part 2. In Chapter 4 Mary Ellen Morbeck discusses variation in physical attributes that — reflecting the influence of both habitat and cultural behaviors — may in turn influence future cultural diversity. Key physical variables include those relating to locomotion, object manipulation, and communication. Key social variables, reviewed by Vernon Reynolds in Chapter 5, relate to kinship, friendship, and reproductive associations. These are important to understanding the evolution of social cognition, information processing, and the manipulation of roles, consideration of which Reynolds applies to an illuminating comparison of mating and marriage in nonhuman and human primates, respectively.

In general, review chapters in Part 2 are intended to lay the groundwork for examination of critical transitions in the evolution of social cognition and symbolic behavior, role manipulation and pretense, tool use and manufacture, and language. Those topics are treated in greater detail in Parts 3 and 4.

2

The Baseline for Comparing Human and Nonhuman Primate Behavior

GORDON W. HEWES

ABSTRACT

Authors of comparative studies of modern human, fossil hominid, and pongid life-styles have frequently chosen human examples from recent or living hunter-gatherers as models for early Paleolithic counterparts. Often the San (Bushmen) and the now extinct aboriginal Tasmanians are represented as the "most primitive" and hence most like early humans. I propose two other groups as better representatives for comparison with chimpanzees: they are the Mrabri of northeastern Thailand and the Tasaday of Mindanao. Both are foragers rather than expert hunters (though the precise character and, indeed, the authenticity of Tasaday culture remains controversial). I present data in four systematic tabulations: chimpanzees, Tasmanians, Mrabri, Tasaday. The chief differences between the human groups and chimpanzees tend to be based on cultural behaviors related to language.

INTRODUCTION

The primate perspective on human culture has several angles. We can begin by examining culturelike phenomena in nonhuman primates, which has been one of the approaches well represented by Japanese primatologists in their pioneer studies of the Japanese macaque. Or we may focus on the pongids as being biologically closer to humans. Studies of chimpanzees, originally mostly in captivity, were strongly motivated by comparative human psychological possibilities. One of the first cultural anthropologists to take an explicit interest in such matters was A. L. Kroeber. Writing in 1928, he used W. Köhler's (1925/1959) data from the Teneriffe chimpanzee colony, with supplementary material on chimpanzees from Kohts (1921,1923) and Yerkes (1916a, 1916b, 1927, Yerkes and Learned 1925), and material on orangutans and gibbons from Furness (1916) and Boutan (1913, 1914). Kroeber did not follow up on his interest in the "subcultural beginnings" of culture in anthropoid apes, and for several decades few cultural anthropologists were willing to examine ape-human behavior from this perspective.

More acceptable have been ape-human comparisons, especially of toolmaking and tool-using behavior, in connection with efforts by physical anthropologists and prehistorians to reconstruct the lifestyles of early hominids. L.S:B. Leakey was influential in promoting such research, as in the highly productive career of Jane Goodall. Less obvious ways of combining nonhuman primate and human data include many studies of human, ape, and monkey infants, as well as of performances of retarded human subjects compared with those of nonhuman primates. Finally, direct comparisons have been made of aspects of human and ape skills, as in McGrew's investigation of Tasmanian and chimpanzee technology (1987, 1992) and Wynn and McGrew's paper on an ape's view of the Oldowan culture (1989).

Comparisons of ape behavior and *generalized* human behavior are unlikely to be of much value. The variety and complexity of human cultural life is so immense that the apes inevitably come out as hopelessly backward. Instead, we must find one or more human groups that lie at the absolute minimum of complexity yet manage to survive without direct or indirect outside support. Extremely impoverished slum-dwelling populations exist in many countries, often with drastically simplified cultural systems, but they would not persist without the more affluent surrounding societies that permit them to scavenge, beg, or otherwise parasitize host communities. Most often in ethnographic comparisons of existing human and either ape or early hominid modes of existence, relatively advanced hunter-gatherers are chosen to represent the extremely simple end of the world ethnographic continuum, such as the San (or Bushmen) of the Kalahari, the Pygmies of Zaire, or central Australian aboriginal tribes.

Where prehistoric cultures are concerned, whether we go back as far as the Oldowan or look at the more-developed Paleolithic cultures, perhaps 90 percent of the critical information about behavior of early humans has been lost — all but what stone tools can tell us. I have therefore chosen the cultures of the Mrabri of northern Thailand and the Tasaday of Mindanao in the Philippines to compare to the chimpanzees, retaining the aboriginal Tasmanians already used in McGrew's study chiefly to emphasize the far greater simplicity of Mrabri and Tasaday culture. I am fully aware of the acrimonious controversy about the ethnographic validity of the Tasaday, but it is not my purpose here to solve that thorny issue, except to state that I believe the Mrabri data, mainly collected over fifty years ago, reinforce the validity of the later Tasaday account, beginning in 1971–1972 and continuing to the present. An update of the situation of the Tasaday as of 1993 is given at the end of this chapter.

Briefly, I seek to show that the cultures of the Mrabri of Thailand and the Tasaday of the Philippines can provide information that contributes directly to our understanding of the differences and similarities in behavior and behavioral capacities of modern human beings, the earliest hominids, and existing chimpanzees. It is important that the differences between humans and apes not be so extreme as to make such comparisons practically meaningless.

An impressive worldwide ethnographic database is of course available, the admirably organized and indexed Human Relations Area Files (HRAF) archive. It is unfortunate that it does not include Mrabri or Tasaday. The aboriginal Tasmanians are present, although the data relating to them fall well below the rigorous standards later developed by HRAF for the bulk of its files. There are several other poorly documented foraging societies resembling the Mrabri and Tasaday, but their records are even more exiguous, or they possess some culture elements, such as seagoing canoes or advanced projectile weapons, that disqualify them for reasonable comparison with the technology of apes. Another good reason for selecting the Mrabri and Tasaday is that their habitats are much more like those of most chimpanzee groups in terms of forest cover and lack of access to seashore environments rich in saltwater marine food resources. We have no present way of knowing how the chimpanzee life-style might change under conditions of unimpeded utilization of littoral molluskan species.

Comparing even the least complex human society with that of chimpanzees is likely to offend some defenders of the notion that all cultures are exactly equal by whatever standards we may apply. Bicchieri (1990) found anything less than total equivalence to be typically "Eurocentric," although I am sure that several of the other great world civilizations also regard certain small-scale, marginal societies as inferior instances of human cultural achievement, at least in terms of technology or of scientific understanding of nature.

I should make it very clear that none of the "backward" human cultural groups that I have mentioned lack the potential (now increasingly realized) for rapid entry into modern complex societies, as I note in my update on the Tasaday. This is surely the primary difference between human beings and chimpanzees, whose exposure to the complexities of advanced human civilization still finds them, from a human behavioral standpoint, grossly retarded, most obviously with respect to language acquisition.

Language in fact constitutes the most glaring obstacle to any fair comparison of chimpanzee and human behavior. While our knowledge of Mrabri or Tasaday speech is fragmentary, it provides an efficient entry into Mrabri and Tasaday mental life compared to the inferences we must make regarding the thought processes of apes or other non-language-using animals.

We are lucky that the chimpanzees, Mrabri, and Tasaday have all been studied within the era of photography. Although photography had been invented by 1839, it came too late to record scenes of aboriginal Tasmanian daily life; only a few pictures were taken of the dwindling number of unmixed survivors (sixty-eight in 1839), all wearing European clothing. In contrast, a fairly good photographic record of the Mrabri exists, and for the Tasaday, the material is voluminous and of high quality.

The "paleoethnographic" contributions of Paleolithic archeology are well known, and gratefully received, but they are normally limited to the most durable remains of cultural activity. An attempt to reconstruct the rich sociocultural behavior of chimpanzees from their "archeological remains" would yield pitifully incomplete data, even where the chimpanzees used stone tools. Similarly, where cave archeology in the Philippines has revealed lithic implements not unlike those of the modern Tasaday, as at Tabon on the island of Palawan, little of the daily life seen among the Tasaday could have been predicted from the lithic evidence and cave stratigraphy at their cave home in Mindanao. Ethnobotanical information on the Tasaday is comparable to what is now known for various chimpanzee groups, aside from the information gap created by the chimpanzees' lack of language.

A legitimate question regarding the Mrabri, the Tasaday, and to some extent also the aboriginal Tasmanians is whether their recent or present level of primitivity (a term unavoidable in this context) can be explained as a survival of far earlier cultural conditions or instead as the result of retrogression, which in turn might be accounted for in several ways. This is not a matter I try to deal with here. Archeology might in these cases be of some assistance, as when it shows a cessation of the consumption of scaled fish by the Tasmanians about 2,500 years ago, despite the nutritional value of fish and their easy accessibility to the inhabitants of the island. Impressive as the evidence for extreme technological primitivity is for the Mrabri and Tasaday, it is not plausible that it represents a continuum from Pleistocene antiquity. Cultural regression has not been a popular topic in the social sciences, and when it is unavoidably obvious, it has usually been accepted only when it can be shown to be the result of oppression from other human societies. Endogenous explanations for backwardness are frowned upon in the present climate of opinion.

The Mrabri and Tasaday do not support themselves significantly by hunting; they may in fact be less active hunters than some chimpanzee groups, with a positive fear of larger animals that chimpanzees do not share. The numerous scenarios of humans as hunters do not fit early hominid life. Further, neither the Mrabri nor the Tasaday engage in any regular scavenging of meat from kills made by predators. Frogs, tadpoles, tiny freshwater crustaceans, insects, snails, and so on are not the staples we tend to associate with *Homo erectus* or the australopithecines. To be sure, shoreline sites of early humans do show considerable use of shellfish (as at Klasies River Mouth in Cape Province) as far back as 125,000 years. Our perceptions of much earlier hominid subsistence patterns probably continue to be biased by the paleoecology of the Rift Valley region.

What does come across strongly in the Mrabri and Tasaday accounts is the great importance of fire for them as compared to the fire-free chimpanzees. Far more than do other groups, the Mrabri seem to live in constant dread of voracious predators. Their fires provide almost their only security from tigers, although they certainly use fire for cooking. The Tasaday had little to fear from predatory mammals but nevertheless kept their cave fires burning continuously. This would also have discouraged bats from finding shelter in their cave homes, though this was not specifically mentioned. Smoke from fires also reduces the torment of flying insects, some of which are vectors of dangerous diseases, a fact of course quite unknown to the cavedwellers. Finally, the role of fire (and firelight) in warding off malevolent spirits, peculiarly prone to nighttime activity, must not be ignored. Once human groups began to conceptualize the prevalence of such nocturnal visitants, fire had another and very powerful function.

Although serious linguistic research has, as mentioned, been carried on in the field with Mrabri and Tasaday (and more specially the latter), firm identification of the external relationships of their language is yet to emerge. The Mrabri apparently speak or spoke a language related to other Southeast Asian languages or else used a version of neighboring languages in their contacts with sedentary folk. The Tasaday on contact spoke a version of Blit, related to Manobo in southern Mindanao, but there is some evidence not yet published that they also retained a distinctive language, possibly pre-Austronesian, not used with outsiders. If true, this would be the only known case of such a language in the Philippines. The vanished Tasmanian languages were used for fairly complex ceremonial or religious purposes and in myths and tales, although no corpus of aboriginal Tasmanian oral literature has survived. It is more difficult to determine what advantages the Mrabri or Tasaday, in their practically invariant daily routine, gained from their use of developed language. To be sure, the Mrabri used language in their very limited external trade when they could have engaged in silent exchange, and both Mrabri and Tasaday used language in their exogamous marriage negotiations. Significantly, a manual sign language, not yet investigated, was employed for communicating with two profoundly deaf Tasaday group members. The need for a complex language system (such as both the Mrabri and Tasaday possess) is no more evident than is the existence, in all of the Pongidae, of remarkable latent linguistic capabilities that have come to light only in the recent decades of ape language experiments. Savage-Rumbaugh's studies (1990) show an astonishingly high level of *receptive* understanding of spoken English in the bonobo (a communicative capacity difficult to account for on the basis of bonobo behavior in its natural forest

environment). Shaw (1989) reports analogous receptive understanding of spoken English in the common chimpanzee.

Differences in social behavior between chimpanzees and the
human sample discussed here are not striking, and the detailed study
of a large captive chimpanzee community at Arnhem (de Waal 1989a)
at times suggested a more intricate network of social behaviors than
seemed to prevail among either the Mrabri or the Tasaday. The
humans do use personal names, pronouns, and kinship terms, however superfluous these may be in the conditions of Mrabri or Tasaday
community life. And the apes are of course ignorant of paternity and
considerably more promiscuous sexually, although they tend to avoid
sustained incestuous relations. Males tend to dominate in both the
chimpanzee and human groups, with patrilocal residence and trans-
fers or migrations of females from group to group.

All are territorially defensive. Trade does not exist among chimpanzee groups, nor did it among the Tasaday, aside from marriage
gifts. Food sharing is practiced in all groups, but among chimpanzees
most commonly in the case of mothers and young offspring, except
when large amounts of food are available. Contrary to the notion of
Thomas Hobbes, it could not be fairly said of chimpanzees, Mrabri,
Tasaday, or Tasmanians that their lives were conspicuously "nasty,
brutish, and short." The "war of all against all" was clearly absent.
Life expectancy, though short by norms in some industrial countries,
was not dramatically less than among peasants in many village
societies with agriculture. Isolation and very small population size,
along with frequent movement of camping places in the case of the
Mrabri, may have limited the risk of certain contagious diseases. If
chimpanzees lived in very large population concentrations, their
vulnerability to infectious diseases might be higher.

If, as Steele contended (1989), cognitive demands increase with
local group size, the Mrabri and Tasaday, who fall below the levels
of group size in common chimpanzee local units, should suffer less
cognitive stress. The very recent upsurge in Tasaday population, from
twenty-six persons in 1971 to about seventy in only eighteen years,
must be engendering increased cognitive complexity if not some
degree of dissonance. In more familiar ethnographic groups, even
where local populations are extremely small, it is possible that the
oral tradition provides a surrogate "population" of mythic and
folkloric figures that may compensate for real local demographic

limitations. For the Mrabri and Tasaday, we have little evidence for the existence of such worlds of imaginary personages realized with any concreteness.

TASADAY UPDATE

The Tasaday were first reported to the outside world in July 1971 by a Manila newspaper and the Smithsonian Institution Center for Short-lived Phenomena. Within months, they were visited by a prominent Philippine politician, a famous U.S. aviator, anthropologists, and journalists, and their story was publicized worldwide. The urgency for fieldwork came in part from the threatened destruction of the Tasaday habitat by logging operations and the growing turmoil in Mindanao arising from armed rebellion against the Philippine government by both the Moro Islamic Liberation Front and the leftist People's Army. The Tasaday area was made into a special reserve, from which outside visitors were barred in 1974.

Following the overthrow of President Ferdinand Marcos, a Swiss journalist named Oswald Iten visited the Tasaday (in July 1986) and after a very short stay proclaimed them to be a fraud, supposedly confirming earlier suspicions voiced by anthropologists who had not actually seen them. Iten found them wearing cloth clothing, using metal tools, and living in palm-leaf huts rather than caves. It had been fifteen years since their first dramatic outside contact, and many Tasaday men had by then acquired Manobo wives. Supporters of the Tasaday as a legitimate isolated group then revisited them, further fieldwork was carried out, and the matter was also considered by a committee of the Philippine house of representatives (1987).

In October 1988 a small delegation of Tasaday arrived in Manila, where they lodged a lawsuit against their detractors, a lawsuit still deadlocked in summer 1990. In November 1988 a symposium held at the annual meeting of the American Anthropological Association to debate the Tasaday question reached no agreement. The Tasaday Community Care Foundation, organized in 1989, is sponsoring further research; a 284-page compilation of papers dealing with Tasaday culture published under the auspices of the foundation contained a very complete bibliography (Dandan 1989). I believe that the data now available strongly confirm the legitimacy of the Tasaday as a long-isolated cultural group with an extraordinarily primitive level of technology and refute the argument that they were "actors" hired for obscure political reasons to impersonate so-called Stone Age cave dwellers.

Table 2.1 Cultures. (Numbered sections correspond to the major categories in the HRAF Outline of Cultural Materials) Comparison of Chimpanzee with Mrabri, Tasaday, and Tasmanian

CHIMPANZEE

10. Identification and Orientation

PAN (includes Pan troglodytes and Pan paniscus) — refers mainly to chimpanzee groups studied at Gombe by Goodall and her colleagues, but includes data from other groups less intensively studied, from Tanzania to West Africa, and in Zaire (P. paniscus or bonobo), which may appear to some as methodologically unsound, although perhaps no more so than the pooling of ethnographic information on the Tasmanian aborigines possibly representing several culturally and linguistically distinct subgroups.

13. Geographical Setting

PAN communities exist over a wide belt of tropical Central Africa with many different local variations of habitat, from closed tropical and montane forests to forest savannas, and different topography and average temperature ranges. The best known groups in the Gombe forest area coexist with the baboons and other primate species, a variety of mammals, the largest of which in recent years have been bushbuck and bushpig, but until recently (or still very rarely) leopards, hyenas, hunting dogs, and hippopotamus; additionally a wide range of insects, some nutritionally important, reptiles (formerly, crocodiles along the adjacent lake, and snakes up to pythons), and birds. Water was readily available at Gombe, less abundant in some of the drier habitats. Significantly, few of the existing environments inhabited by modern chimpanzees closely resemble the prehistoric habitats of the earliest hominoids, notably in terms of coexistent other species, particularly large ungulates, but also in terms of seasonal or year-round semi-aridity.

15. Behavior, Personality

PAN psychology has been intensively studied, is wild and captive populations. Much systematic research has been devoted to chimpanzee cognitive capabilities, in contrast to what is known of the human groups here being compared.

Significant variations in local "cultural" behaviors have been noted, referred to by Goodall as "culturally acquired dialects," including differences in food resource utilization and tool-using habits. For the Gombe chimpanzees in particular, as well as for captive social groups (such as Arnhem), extremely detailed records have been kept on social interactions, dominance, agonistic, and affiliative patterns, emotional responses, etc. Compared to other great apes, Pan seem to be more readily aroused to high excitement and aggression, on up to lethally violent encounters with conspecifics. Investigators continue to find it useful to describe and analyze chimpanzee social behavior in frankly anthropomorphic terms, such as "Machiavellian" (de Waal 1989).

MRABRI

10. Identification and Orientation

The MRABRI ("Yumbri") are also known as Phi Tong Luang, Khou Pa, and from the Thai term translated as "Spirits of the Yellow Leaves." They are a Southeast Asian "Protomongoloid" people roaming over a tract of tropical forest in northern Thailand along the Thailand/Laotian border, sporadically reported since 1914, and ethnographically described in the 1930s (Bernatzik 1958), Weaver in 1956, and by expeditions of the Siam Society (1962, 1963). Although they nomadized in an area also occupied by scattered villages of sedentary rice farmers, with cultural features not strikingly different from much of the hill-country folk of Southeast Asia, the Mrabri lack agriculture and significant hunting, domestic animals other than dog, and presented a technological level of unprecedented simplicity or backwardness among living or recently extinct human populations.

13. Geographical Setting

The MRABRI habitat is generally a closed rainforest except for cleared agricultural land around non-Mrabri farming villages. Winter temperatures sometimes fall to about 5 degrees C. There is a definite dry season. Unlike the Pan habitats, very numerous bamboo species occur in great abundance (bamboo is scarce in tropical Africa), along with rattan, any of several species of climbing palm. Some large mammals are found—the guar, a huge and sometimes dangerous bovid, as well as elephant (present in some African chimpanzee environments), and the tiger, along with small deer (suntjac), wild pig, and various monkeys, rodents, bats, etc. Termites comparable to those so important for chimpanzees are unimportant. Honeybees provide food and wax. Aquatic foods, such as the crabs and frogs utilized by the Tasadya, are apparently unimportant.

15. Behavior, Personality

The MRABRI are described as excessively shy and fearful (in considerable contrast to Pan). Few if any instances of extremely aggressive behavior were noted. The ethnographers, linguists, and explorers with whom they have come into contact uniformly describe them as unintelligent, weak in ability to handle abstract problems, and lacking in what would be considered normal human curiosity regarding strange equipment, etc., or outsiders. Their neighbors, the agricultural Meao, regarded them as cold and reserved. They received gifts with little enthusiasm. Although observers may have had a repressive effect, there was agreement that they spoke to one another very little, only talking when it seemed necessary. They treated their older people well, and were willing to offer food to strangers.

In 1963, shown photographs taken of themselves a year earlier, "eventually" came to recognize and identify well established in the case of Pan in activity. They refused to accept small mirrors, being afraid to look into them.

(Table 2.1 continued)

TASADAY

10. Identification and Orientation

The TASADAY live in a small area of mountain tropical rainforest in southern Mindanao, Philippines, and have been known since 1971–1972. Like the Mrabri, although they live fairly close to Philippine rice farmers to the north, they lack agriculture, significant hunting, have no domestic animals, and present an equally primitive or backward material culture. Owing to a combination of political and other ideological factors, they have been widely suspected of being some kind of hoax, despite intensive field studies by reputable ethnographers, linguists, and ethnobotanists, mainly in 1972, but with revisits within the past three years. I am including data on the Tasaday because it agrees in nearly all details with the material on the Mrabri of Thailand, where factors which have led to suspicions about the Tasaday have never existed.

13. Geographical Setting

The TASADAY, although only 60 km from the Celebes sea coast, live entirely inland in a rugged mountain area rising up to 1,300 m. There is abundant rainfall, and much of the animal food utilized comes from mountain streams. Temperatures rarely come close to freezing. No open terrain exists in the immediate area, although open farmlands occur only a few hours walk to the north. There are no large mammals (the largest are wild pigs and small deer) and no dangerous predator. Several caves and rockshelters exist in the area, within a total territory approximately 25 km² in extent. Two cultural groups similar to the Tasaday reported lived within about 25 km, but they have yet to be visited or described by outsiders.

15. Behavior, Personality

The TASADAY seemed quite shy and gentle, rarely exhibiting angry expressions or even using loud voices. They showed a strong avoidance of genital exposure (at least to strangers). The world-famous aviator Charles A. Lindbergh, who visited them, commented on their "lack of spirit of adventure."

TASMANIANS

10. Identification and Orientation

The TASMANIANS are in a sense "well known" in the world ethnographic literature for their cultural isolation and primitivity, and for the tragic extinction (except in a minor genetic sense, since a very small group of descendants survives without knowledge of any of the former Tasmanian languages, and cultural details). Anthropological documentation in contrast to what is known about the Mrabri or Tasaday is fragmentary and, aside from recent archaeology, was collected before the emergence of scientific anthropological research, and also before photography existed as a means of recording ethnographic details of every day life, What is known about the now extinct tasmanian culture of cultures is not even based on so-called "memory cultures" collected from elderly informants, as in much of native North America, after massive European contact.

13. Geographical Setting

The TASMANIANS formerly occupied much of the fairly mountainous island of Tasmania (67,889 km².) the southernmost part of the former Sunda Shelf, with its aberrant flora and fauna, long evolutionarily isolated during most of the Cenozoic.

Unlike the environments of chimpanzees, Mrabri, and Tasaday, Tasmania has a cool temperate climate, with snow in the mountains in winter, formerly heavily forested with eucalyptus species. The fauna lacked eutherian mammals aside from bats, and there were no predators really dangerous to humans. The coastal areas were rich in aquatic animals—mollusks, crustaceans, birds, along with seals, as well as fish. Scaly fish were not eaten by the aboriginal Tasmanians.

15. Behavior, Personality

The TASMANIANS, in contrast to both Mrabri and Tasaday, were said to have been very curious about strange outsiders. The historical record of their decades of interaction with Europeans in the early nineteenth century was marked by much extreme violence (often quite obviously retaliatory). Contact between them and European and other outside groups (Australian aborigines from the mainland) led in the nineteenth century to frequent interbreeding and the formation of small mestizo population, which is all that now remains.

(Table 2.1 continued)

CHIMPANZEE

16. Demography

Data on various groups of PAN, both common and bonobo, range from a dozen or so to local aggregations of sixty or more individuals. The group range for the bonobo at Wamba was 58 km² compared to 26 km² at Gombe. The total African population of chimpanzees probably still exceeds 100,000 but with many wide geographic gaps, spread over a very large area. Thanks to Goodall and her colleagues, more detailed demographic data exist for the Gombe population than for any comparable-sized human group, except for the virtual lack of reliable information regarding paternity.

Chimpanzee "settlement patterns" vary widely depending on local topography and vegetation. In no known cases is there permanent or semi-permanent occupation of a single site, nor of caves or rockshelters even on a temporary basis, although many other mammals regularly occupy such natural shelters. Typically, at least in areas with fairly continuous geographic distribution of chimpanzees, there were zones of spatial overlap or no-man's and around each local group.

17. Archaeology

Archaeology has revealed practically nothing so far about the recent past of CHIMPANZEES, to say nothing of more ancient periods. Either because the areas frequented by them have not been explored with a view to discovering skeletal (or lithic ?) evidence of their existence in any part of Africa, or their bodies have been thoroughly consumed by predators or scavengers, we do not possess recognizable remains of "ancient" members of the species Pan, to say nothing of artifacts attributable to them.

MRABRI

16. Demography

The MRABRI, like chimpanzees, use a camp or sleeping place only for a few days at the most, often only for a single night.

Bernatzik believed that the total Mrabri population might have amounted in his time (mid-1930s) to no more than a few hundred, split into many tiny groups as small as eleven individuals. Young in 1961 put the total far lower, at a maximum of perhaps 100 to 150, likewise divided into many very small nomadic grouplets. A symbiotic relationship, largely restricted to occasional exchange of forest for village products, exists with respect to sedentary agricultural folk in the region, mostly Meau. Observers saw very few individual Mrabri they believed to be older than 50 years. Children were remarkably few, indicative of very high infant mortality or low birthrate. Tigers caused many deaths, especially of women and children foraging in the forest. Others were said to have died from falls from trees, snakebite, honey bears, or were killed by villagers who regarded them as harmful spirits. In all accounts males were more numerous, but this may reflect hiding of females from outsiders. Mrabri groups did not live in close proximity, although they were apparently exogamous as well as monogamous. With the Lahu, some Mrabri/Lahu intermarriage was reported.

They did not occupy any permanent sites, such as caves or rockshelters (as in the case of the Tasaday, or the Vedda of Sri Lanka). There was no apparent pattern to their nomadic wanderings, other than reliance on the availability of immediately usable food resources.

17. Archaeology

No archaeology has been undertaken in the area of the MRABRI. Given their virtual non-use of stone implements, practically nothing would survive in any case from their highly temporary camping or sleeping sites except traces of their campfires. Their minimal consumption of large animals would also limit the identifiability of their occupancy, as would their casual method of disposing of the dead. Thailand of course contains many ancient prehistoric sites, some with very early pottery and agriculture. Finding sites related to the modern Mrabri, or people living in a similar manner, would be quite unlikely. A local non-Mrabri tradition, also found in one monastic document, claimed that they are remnants of former slaves of the royal house of Nan, a small Thai principality, who were prisoners of war from another area. This report does not seem plausible.

(Table 2.1 continued)

TASADAY	TASMANIANS

TASADAY

16. Demography

TASADAY population in 1972 amounted to a total of twenty-six persons, four married couples and their children. Their marriages were with members of two similar forest groups so far not investigated, and equally few in numbers.

17. Archaeology

Archaeology of the TASADAY has indeed been suggested as one way of establishing their ethnographic validity. A deep midden lies at the mouth of their cave, which might repay excavation. Similar ancient cave sites have been excavated elsewhere in the Philippines, in some cases going back to 10,000 or more years.

TASMANIANS

16. Demography

The TASMANIAN population for the island as a whole, in early European contact times (ca. 1800–1830) has been estimated to have been about 4,000 to 5,000, scattered over the entire island.

Local bands apparently ranged up to about forty to fifty members, in turn groupable into so-called "tribes" (unorganized populations speaking the same or similar languages), which engaged less often in mutual warfare, ranging up to 250 to 700 members.

Settlement patterns varied according to resources, with coastal areas permitting semi-permanent occupation of particular sites. There were extensive uninhabited inland areas of rugged mountains.

17. Archaeology

Knowledge of the prehistory of Tasmania has recently grown considerably, indicating human colonization of what was at one time a peninsular extension of the mainland of Victoria, land connections with which have existed more than once in the course of the Pleistocene. Several carefully excavated Tasmanian sites, some of them caves (as is Judds Canyon), have produced evidence of several millennia of occupation. The earliest Tasmanian settlement may be as old as 30,000 years. Much of the island was grassland rather than in forest at the end of the Pleistocene. Some major cultural changes occurred within the past 3,000 years or so, leading to abandonment of the eating of scaly fish and the making of bone tools. This has not been adequately explained.

(Table 2.1 continued)

CHIMPANZEE

19. Language And Communication

Language does not exist in any form acceptable to linguists in wild PAN, although captive chimpanzees have quite recently been shown to possess remarkable capabilities to learn (from human teachers and caretakers) both manual sign language and other visual sign-systems, and receptive knowledge of human spoken language (in English, so far), with the bonobo apparently more adept than the common chimpanzee. Similar protolinguistic capabilities under conditions of captivity have been found in the gorilla and the orang utan. The verbal limitations of the great apes may lie partly in their less than human brain size, although microcephalic humans are known to speak in many cases, in neurological circuitry not conductive to language acquisition, in speech production in particular or, in short, their lack of a "language acquisition device." Absence of language may be the single most striking difference between apes and human beings, although Goodall's suggestion that (1982: 572, in discussing language and communication in chimpanzees) humans "can discuss events that happened in the remote past and make complex contingency plans for a future that may be close or distant" is something that may rarely occur even in human groups such as the Mrabri or Tasaday.

Chimpanzees have a wide range of communicative facial expression, body language and hand gestures, in addition to a rich repertoire of vocalizations, responded to as if they indicated intentions on the part of others. Carrying of ape infants on the mother's back or belly may inhibit gaze and manual gesture between mothers and offspring, unlike the situation in human mother-infant pairs. Chimpanzee leaf-clipping is probably communicative, as seen in Mahale and Boussou. Brandishing branches or sticks, rock-throwing, etc., are also communicative behaviors commonly seen in chimpanzees.

MRABRI

19. Language And Communication

The MRABRI speak a language probably belonging to the widely distributed Mon-Khmer stock of Southeast Asia, but its documentation is very incomplete, despite some local linguistic research tape-recordings. Bernatzik mainly depended upon a local Thai interpreter. The Mrabri, at least the adult males, knew enough works of several neighboring languages such as Khmu, Lamet, Meau, etc., to manage their very limited trade relationships. Various vocabulary items have been borrowed from these and from Laotian. What is striking about language among the Mrabri (and Tasaday) is how little its use seems to contribute to their technology or social organization, beyond facilitating their modest trade with sedentary villagers.

What little has been discovered about their "own" language, leaving aside what they know of Meao, Khmu, etc., or even a few English words acquired from recent investigators, is that it appears to be fairly restricted in lexicon (e.g., as in some other "primitive" languages, only two color words). Also, personal pronouns are said to be used less frequently than kinship terms, there are no stylized greeting or farewell forms, and there is an apparent disinclination to engage in animated speech among themselves, although this could be an effect of the presence of strangers. Bernatzik, Kraisri, and other observers agree that the Mrabri seemed incapable of responding to abstract or hypothetical questions. They did not appear able to count above twenty, and according to Bernatzik, really had no real comprehension of numbers beyond about two or three. As for non-verbal communication, manual gestures were not mentioned. To indicate that a particular forest path should not be used they might lay boughs across it. They had no graphic symbols or markings; simple line patterns engraved on musical instruments were not said to have any meaning.

No evidence for a body of oral literature or myths was reported, aside from some comments about dangerous tigers and malevolent spirits, or for word-games.

(Table 2.1 continued)

TASADAY

19. Language And Communication

The TASADAY, in contrast, speak a local Philippine dialect clearly related to Blit, one of the Mindanso languages to their north, and their discoverer, the native Philippine hunter Dafal, had no great difficulty in making himself understood in the brief period of years before they come to the attention of outside Philippine government authorities and then to the anthropological and journalistic world.

A trained linguist was able to make a study of the Tasaday language, including an 800-word lexicon. The Tasaday could name 207 different plants. There were three color terms and numbers up to five. The principal result of the language study was that the Tasaday could not have been simply a group of ordinary farm villagers recruited to impersonate a supposedly long-isolated people. The Tasaday vocabulary lacked the numerous common words of ancient Indio origin, disseminated in the Philippines over the past 1,500 years, along with terms borrowed in the last four or five centuries from Chinese and Spanish. It would have been very difficult to train non-literate southern Philippine villagers to identify which words were of Indio, Chinese, or Spanish derivation, and to omit them in the material they provided to the linguist. The materials suggest a linguistic separation from Manobo, etc., on the order of 571 to 755 years.

Two Tasaday individuals were profoundly deaf, and an ad hoc manual sign language was used by them and their fellow cave dwellers. Similar sign-languages for communicating with deaf community members have arisen in other isolated societies.

TASMANIANS

19. Language And Communication

The languages (probably two to five) spoken by the pre-modern TASMANIANS are not all extinct, and only scraps of their several vocabularies have been preserved. What little is known shows no connection with the languages of the Australian mainland. Like the Tasaday, they may have counted up to five. Regrettably, no texts or even complete sentences survive, although there is a suggestion of subject-verb-object word order. There is no information remaining regarding facial expressions or gestures, although the first Europeans who made contact with the Tasmanians managed rudimentary communication on a gestural basis.

(Table 2.1 continued)

CHIMPANZEE

22. Food Quest

CHIMPANZEE food habits reflect the high seasonal variability of most of the African environments in which they live. Marked local and regional dietary diversity within species is typical of primates in general (Chapman and Chapman 1990).

The Gombe food-list includes 184 plant species and the Lomako (Bonobo) group 114 species. A portion of the food now eaten by wild chimpanzees comes from the cultivated fields of neighboring human villagers, including sugarcane and pineapple. Most of the food consumed by the bonobo consists of tree fruit (estimated at 80%). Underground fungi (truffles) were also dug up by the bonobo. In West Africa, hard-shelled palm nuts constituted the major part of the diet.

Animal foods were consumed in far smaller amounts. Scavenging of carcasses left by regular predators is very rare; at Bombe only ten instances were seen in twenty-five years of observation, and these were mostly recent kills made by baboons (also noted for the bonobo). The mammals killed and eaten by chimpanzees were chiefly monkeys—colobus, ceropithecus, occasional prosimians (potto and galago), of ungulates bushpig, bushbuck, and duiker, rodents, including squirrels, along with mongoose, shrews, and hyrax. Group hunting was limited to some of the larger mammals. Very few birds were eaten, but birds' eggs were consumed when found. Invertebrates were eaten fairly often, especially termites "fished" from mounds, earthworms, various larvae, along with honey and nests (including larvae) of stingless bees. Young duiker and bushpit were more commonly eaten than adults. The bonobo of the Wamba area, though often in close social contact with the red colobus, did not kill or eat them (Ihobe 1990)

Group hunting was usually begun in silence, although at the kill some noisy behavior took place. Sharing of food was mostly limited to meat of larger prey (the largest bushbuck might weigh as much as 45 kg). Aquatic animals were not systematically obtained, though both land and water snails and freshwater shrimps were noted. Fish were rarely consumed, nor were snakes. Clay and termite earth was sometimes eaten, as well as tree-galls. Frogs, a standard food of the Tasaday were rarely consumed.

The omnivorousness of this menu is somewhat misleading, since the bulk of the diet in all areas was vegetal and seasonally restricted.

MRABRI

22. Food Quest

The MRABRI diet differed from that of the chimpanzee as might be expected from the geographic separations of their habitats. The Mrabri utilized more root and tubers, extracted with a digging stick. Like chimpanzees, they rarely gathered more food than they needed for a single day, though some might be left over for an early morning snack. Food was not stored, except for the very rare use of "smoked pork," simply left hanging over the fire at night. Plant foods included mostly locally edible items—fruits, berries, roots, shoots (especially bamboo shoots), and leaves. Broad leaves were also used as dishes.

Among the most important vegetal items in addition to bamboo shoots were wild yams, and the pith of wild sago palms. Unlike the chimpanzees, they did not rob the growing crops of neighboring farmers, from whom, however, they occasionally received rice in exchange for honey and beeswax, and woven handicrafts.

Nothing like the thorough analysis of diet made for some chimpanzee groups has been made, the Mrabri may have consumed little more animal protein. They "hunted" chiefly small mammals such as porcupines and lesser rodents, rabbits, rarely muntjac or barking deer, monkeys, and wild pigs. Larger animals were feared, such as the guar, tiger, honey bear, and elephant. Birds were utilized, though their eggs were relished when found. Lizards and turtles were taken, but snakes were avoided. Frogs were eaten also. Invertebrates eaten included freshwater crabs, snails, insect larvae, but not earthworms. Wild beehives were sought for honey and wax, which could be traded for occasional village products.

Male adults did what hunting was accomplished; the Mrabri were inept spearmen, and lacked the bow and arrow or blowgun. They were also inept butchers of the occasional large mammals they managed to kill. Tracking ability in hunting was minimal. Traps for small animals were known but not systematically set. Spears were for thrusting and piercing, not throwing. There is no mention of clubs as hunting weapons. The local male adult population in any one nomadic band is so small that large organized hunting of large animals was practically out of the question. The absence of accurate projectile weapons (such as the bow and arrow or blowgun) precluded much successful hunting of arboreal animals such as monkeys and birds.

Many plant foods were eaten raw, but cooking was done either in the coals of the fire, or by boiling in large bamboo segments, a common cooking practice in Southeast Asia. No utensils were used for eating.

Herbs and salt (obtained by trade) or salty soil, was used to flavor some foods, but foods were not salted to preserve them.

Water was drunk from bamboo segment containers or occasionally from folded leaf cups.

Women and children collected most of the plant and insect food.

74

(Table 2.1 continued)

TASADAY

22. Food Quest

The TASADAY diet was not strikingly different from that of the Mrabri. Detailed ethnobotanical records were made of the plants consumed by the Tasaday, as well as observations of the time required to collect them. Despite critics skeptical of the Tasaday as a genuine primitive group, qualified ethnographers claim that the local environment as exploited by the Tasaday is quite sufficient to support their very small population in reasonably good nutritional health. As with the chimpanzees and Mrabri, their dietary regime is extremely flexible. The entire subsistence area exploited was only a few kilometers in any direction. A difficult mountain hike of 10 to 15 km to the north takes one into rice-farming country. Daily food collecting rarely required members of the local group to go much further than 3km from their main cave dwelling.

The ethnobotanist who analyzed the Tasaday plant food-quest identified 120 edible species utilized by them (Yen and Nance 1977).

The plant food intake centered (when the group first was discovered) on a wild yam (*Dioscores sp.*), supplemented by a wild banana, fruit and flowers of wild ginger, palm fruits (so-called palm cabbage), bamboo shoots, and various additional fruits, shoots, roots, and edible leaves. The pith of a palm was extracted by a water-sluicing method, combined with a bamboo crushing device, apparently introduced by the Blit hunter Dafal in the late 1960s. The prepared pith, called natek, soon became a major food source and an easily stored stable. Rattan shoots were also eaten.

Animal foods were mainly obtained from mountain streams, chiefly frogs and tadpoles, freshwater crabs, and occasional fish. Although the forest contains wild pig, and a small deer, monkeys, and birds, these were caught rarely and without organized hunting. Before Dafal's arrival, they supposedly lacked knowledge of traps, including even fish-traps or baskets. Honey is not mentioned.

Cooking was usually directly in hot coals, occasionally by baking foods wrapped in leaves before being placed in the embers. Dafal introduced the smoking of wild pig flesh.

TASMANIANS

22. Food Quest

Although most of the details of aboriginal TASMANIAN diet have been lost or were never recorded, we see a very different picture. Although the pre-contact Tasmanian flora was not as rich in humanly edible wild plant resources, compared to tropical rainforests, roots, seeds, berries, fruits, and fungi were gathered, presumably to the maximum. Unfortunately, European colonists paid little attention to aboriginal plant collecting activities. Their hunting abilities were of greater concern, for the native Tasmanians soon took to killing sheep and cattle from isolated settlers' farms.

Adult males regularly hunted fair-sized mammals, using throwing spears and clubs, in organized as well as individual efforts. Their prey included kangaroo, wallaby, bandicoot, wombat, and long the coasts, fur seals and elephant seals (presumably clubbed when hauled out on shore). Stranded whales occasionally provided huge amounts of meat.

Some animals were trapped. Blinds were built to hide kangaroo hunters. Kangaroos were also hunted in fire-drives, in which they were killed by waiting spearsmen.

Not every local Tasmanian group made use of all these methods. They ate large quantities of muttonbirds (shearwaters and petrels) obtained from rookeries on cliffs and offshore rocks, which also supplied large quantities of birds eggs in season.

Women collected a variety of seafoods—crabs, oysters, and abalone, diving deeply for the latter, which had to be pried off underwater rocks. Women also climbed trees, previously notched, to catch opossums.

The use of scaly fish ceased, according to archaeological evidence, about 2000–1500 B.C., about the same time as the cessation of the making of bone tools. No good explanation of this has been presented. There was no scarcity of freshwater or saltwater fishes. Aquatic invertebrates continued to be utilized, as noted, along with ants and grubs.

Obviously the Tasmanian aborigines were far in advance not only of chimpanzees but of the Mrabri and Tasaday, with respect to obtaining animal foods, including the large ungulates introduced during the Colonial period, their hunting of which was one cause of the ruthless "Black War."

(Table 2.1 continued)

CHIMPANZEE

23. Domestic Animals
CHIMPANZEES keep no domestic animal, and no one has suggested that protohominids did either. Chimpanzees are known to interact at times in a friendly manner with baboons and other monkeys, but this is hardly indicative of a propensity to domesticate other species, any more than the "adoption" of a kitten by a gorilla in one of the language experiments in the anticipation of cat domestication by that pongid.

27. Narcotics, Stimulants
Narcotic and stimulant substances are not known to be used by PAN in the wild, at least not deliberately. In captivity, chimpanzees have often taken readily to tobacco smoking, and show no dislike of coffee or alcoholic beverages.

28. Clothing And Ornament
Clothing and Ornament alike were lacking among CHIMPANZEES, unlike the orang utan which occasionally drapes itself with leaves or vines. On the other hand, like other primates, the chimpanzees were very attentive to body care, not only from self- and conspecific grooming, but in the use of leaves as wipers to remove feces, urine, semen, mud, etc.

MRABRI

23. Domestic Animals
The MRABRI had dogs of small breed, used as a watchdog and as some protection while seeking food in the forest, if only because tigers are more likely to kill and make off with a dog accompanying a party of humans than to attack the human first. The dog was said to help in finding edible roots, and in digging them out of the ground, and in catching turtles and lizards. Dogs were not eaten. To some extent the dog was also a pet or companion for small children, who sometimes spoke to them.

27. Narcotics, Stimulants
The MRABRI when first observed by Europeans had already picked up occasional use of betel-nut chewing, probably from neighboring villagers, occasional use of opium, which they preferred to eat rather than smoke, and especially tobacco traded from their village neighbors, chewed or smoked in small pipes. Opium, grown widely in northern Thailand, was apparently regarded by the Mrabri as medicinal.

28. Clothing And Ornament
The MRABRI probably went entirely naked until recent contacts. By the 1930s they had adopted (at least when likely to be seen by outsiders) the wearing of very brief breech-clouts of cloth rags, obtained in trade for honey, wax and rattan ware, from their sedentary neighbors. Women wore somewhat more bits of old cloth. When asked why they wore these scanty, tattered garments, they answered "the others have clothes on," referring both to their sedentary neighbors and to the ethnographer and his party. Photographs taken in the 1960s show some Mrabri wearing more complete clothing obtained by trade. They used no footgear.
Grooming was mainly concerned with removing vermin from children's hair, and plucking out of beard hairs by adult males. Reportedly the Mrabri never bathe. Body ornamentation was very limited. Male ear-lobes were bored, but they did not wear ear-ornaments, aside from bits of wood or leaves. The ear-lobe piercing may have been learned from the Khamuk villagers. One report said the Mrabri were sometimes tattooed by sedentary villagers, who also sometimes cut the hair of Mrabri males, which was otherwise left unkempt and hanging to the shoulders. Women's faces were smeared with soot or mud, perhaps only when in danger of being seen by foreign visitors.
Unlike some other tropical forest dwellers, they did not make decorative use of feathers or flowers.

(Table 2.1 continued)

TASADAY

23. Domestic Animals
The TASADAY did not have dogs or any other domestic animals.

27. Narcotics, Stimulants
The TASADAY used the bui vine (an analogue of the pepper leaf use with betel) and the betel nut, chewed with lime—possibly a recent practice introduced by Dafal. The lime was obtained by burning snail shells.

28. Clothing And Ornament
The TASADAY had somewhat richer body-decoration. Their normal garb in the period from their discovery until the late 1980s was a leaf breechclout for males and a small leafy skirt for females, held in place in both cases with a vine cord. However, similarly informal and very scanty garb was worn by Bontoo women of northern luzon, when working in the fields in rainy weather, as recently as the early twentieth century.

In heavy rainstorms the Tasaday might use a broad leaf as a temporary umbrella. Bark cloth made from the bast fiber of a fig tree is mentioned, but its use as garment material is not explained further and is denied by the ethnobotanist as an aboriginal product. Rattan loops were inserted in pierced earlobes, and, after Dafal's arrival, brass ear-rings. Some individuals had simple tattooed patterns of X's and ovals, pricked in with thorns and an ash mixture.

The males plucked their beards, though not too completely. Men and women often tied their long hair up in ponytails with a piece of vine, to avoid getting their hair caught in undergrowth. The front teeth were filed to sharp points to make them look less animal-like. The Tasaday occasionally bathed in streams, and used a vine for scrubbing. A kind of shampoo was achieved by rubbing the hair with clay and sand.

TASMANIANS

23. Domestic Animals
The TASMANIANS did not have dogs or other domestic animals. They killed cattle and sheep belonging to the white colonists, but did not steal them to raise as livestock.

27. Narcotics, Stimulants
The TASMANIANS were able to obtain a mildly alcoholic drink from eucalyptus gum, naturally fermented in hollow trees.

28. Clothing And Ornament
The TASMANIANS were usually naked except in cold weather, when they might fling a kangaroo skin over their shoulders and backs; such skins were also used by women to assist in carrying infants. Men often let their hair grow into long ringlets, which were also mixed with red ochre and grease. Grease was also often rubbed on skin, with charcoal, clay, or ochre. Women usually wore their hair very shortly cropped.

Body ornament included keloid scarring, mainly on shoulders, with rows of dots and circles. Necklaces of small pierced blue iridescent snail shells were worn; strips of ochre-died kangaroo skin were also used as neck ornaments. Men might put feathers, flowers, or berries in their hair as ornament. Crude moccasin-like footgear were occasionally worn to protect the feet.

All in all, the Tasmanian aborigines devoted much more attention to personal ornament than the Mrabri or Tasaday.

(Table 2.1 continued)

CHIMPANZEE

31. Minerals
Use of mineral substances (other than of edible clay) is not known for CHIMPANZEES except for stone hammering tools and rocks hurled at intruders or predators.

34. Structures
The only structures made by CHIMPANZEES are nighttime nests, quickly improvised each evening in a suitable tree from nearby branches, twigs, and leaves. Technologically such temporary nests were far inferior to the nests built by a great many birds, some other mammals (notably the beaver), stickleback fishes, and many invertebrates. The structures built by bower-birds are architecturally sophisticated monuments compared to anything constructed by chimpanzees. Although caves and rockshelters exist within many areas inhabited by chimpanzees, these are not utilized (in many cases they would already be the homes of snakes, and would be vulnerable to leopard attacks).

MRABRI

31. Minerals
The MRABRI obtained salt occasionally from their village neighbors, and may have flavored some of their food with salty soil. Aside from the stone (flint?) used in strike-a-lights, presumably obtained form sedentary villagers, they made no other use of mineral substances, aside from mud or lay for smearing on the faces of women.

34. Structures
The MRABRI surpass the chimpanzees in artificial structures, although their nightly sleeping places are among the simplest human shelters. At times in clear weather they may merely spread leaves on flat ground without attempting to build any overhead protection. Ordinarily they assemble a windscreen of fan-palm-leaves, sometimes in of crude twill matting, usually next to a wall of growing bamboos, and propped against a horizontal bamboo pole. A rough low platform of bamboos is usually made as a mattress, with a sheet of bark as a sleeping pad. If tigers are in the vicinity, a more effective bamboo fence is set up, closed after everyone has checked in for the night. After a few days the leaves turn yellow, which led to their being described by the Thai as "Spirits of the Yellow Leaves" (Phi Tong Luang). Each nuclear family has its own shelter of fan palm leaves. Although caves exist in the region, the Mrabri did not use them as dwelling places.

Protection from marauding predators was chiefly achieved by the burning of several campfires. There was no attempt to make a clearing for the camping space.

(Table 2.1 continued)

TASADAY

31. Minerals
  The TASADAY used various stones for scrapers and pounders, picked up from adjacent stream-beds.

34. Structures
  The TASADAY did not have to build shelters for themselves, since they lived by choice in one or more hillside caves. There are several habitable caves in the area, occasionally also used for sleeping, but also for storage. Near the mouth of the main cave a rough line of boulders was observed, although it did not seem capable of repelling intruders. Thin bark sheets were used to lie on within the cave, usually on the shelflike projections. Several fires were maintained inside the caves. The storage referred to was mostly for firewood not needed for immediate consumption, and after the introduction of natek pith, also for that foodstuff.   The main cave was kept fairly clean with frequent sweepings, which led to the growth of a deep midden on the talus slope.
  Cave-dwelling prehistoric peoples lived in many other parts of the Philippine Islands, although not into recent centuries, as on Palawan and Luzon.

TASMANIANS

31. Minerals
  The TASMANIANS obtained red ochre from certain mines or quarries in the northern part of the island. Stone suitable for a variety of tools was widely available throughout Tasmania.

34. Structures
  TASMANIAN structures ranged from simple windbreaks of convenient plant materials to bee-hive-shaped huts covered with grass, bark, or turf, on the west coast. Such huts were up to 2 m, sometimes lined with feathers. In the drier parts of the east coast, shelters were simple, of strips of eucalyptus bark, forming slightly curving lean-tos against a rough framework of boughs, with a fire at the open side for warmth and cooking.

(Table 2.1 continued)

CHIMPANZEE

41. Tools

Tool making and tool using have heretofore been the principal concern in comparisons of CHIMPANZEE and human cultural behavior. Consequently, abundant data are now available on many aspects of chimpanzee tool-related activity. Moreover, some very interesting direct comparisons have now been published relating chimpanzee, recent human, and very early hominid tools, by McGrew, Wynn, and Boesch and Boesch, and others.

Chimpanzees use a variety of vegetal tool materials, and much less stone, in a technology devoted to extraction of foodstuffs from the environment. As much of this literature is familiar to primatologists, I do not need to present many details. Chimpanzees are now known, since the dramatic breakthrough report by Goodall in the early 1960s on termite fishing with twigs, to use implements for sponging out water, blood, brain tissue, etc., as wiping implements, fly whisks, clubs and switches, objects to brandish at attacking predators or conspecifics, missiles to hurl in like circumstances, and at prey animals, and probes to extract bone marrow (Boesch and Boesch 1990). The use of stones for pounding or cracking hard nuts was first reported by Savage and Wyman as early as the 1840s but not verified until quite recently. Sugiyama (1985) describes what appears to be the use of tools to make tools by West African chimpanzees, bringing them even closer to the behavior of human tool-makers. It is important to note one striking deficiency of chimpanzees as tool-users: so far they have not been seen to use vines, fibers, etc., for cordage purposes, for lashing, tying up loads, although many vegetal materials readily available in the wild could be used with little or no modification.

Even more significant than the lack of cordage or straps in chimpanzee technology is the lack of fire. Important as fire is for the cooking of foods, and for heating and lighting, by far the most critical advantage of fire seems to be use in repelling large predators from humans who sleep on or near ground-level, and whose shelters are not strong enough to resist their intrusion.

MRABRI

41. Tools

MRABRI tool use is considerably more varied and complex, although extraordinarily primitive by comparison with almost all other known human groups. To begin with, fire-using is an extremely important element in Mrabri culture. However, their knowledge of how to make fire may be relatively recent, since they use flint and pyrites to kindle, and even substitute a bit of iron obtained from neighboring villagers for flint. The fire-making kit, kept in a small bamboo segment, is considered a prize individual possession. No fire-fans or blowers were seen. Large fires were kept burning during the night, around the crude leaf shelters, in part to ward off tigers and elephants.

Cooking was by broiling or baking in the embers, or boiling in large bamboo internode containers. Torches were made of dried bamboo strips, although the Mrabri seldom went about after dark. Bamboo segments were also used, with plungers, for keeping embers hot for occasional forging of pieces of iron, a technique learned from the sedentary villagers from whom they also obtain iron scraps. In any case, this device was seen being used by a single individual.

Bamboo containers were also used for water, stoppered with wads of leaves. Bamboo poles were used in building shelters+mainly or only as a horizontal support against which to lean fan-palm leaves or screens. Among their few uses of wood other than bamboo was for digging-sticks. The total absence of pottery, a common element of sedentary Southeast Asian cultures, and known in Thailand for a great many millennia, was noted by all observers, although pots are a well-known encumbrance for nomads everywhere.

Bamboo was used for knife sheaths (they now have some iron or steel knives), and for mats, along with rattan. The Mrabri were familiar with cordage, string, rope, and knotting, and made rattan mats and baskets for exchange. They did not use nets, snares. or traps for hunting, Animal skins were not worked or otherwise significantly utilized. They make plaited mats and baskets of rattan, seemingly a technique borrowed from the Meau. A large fiber sling over the shoulder was used to carry trade items such as mats and baskets. The small local streams contain fish, but no fish-traps (usually of basketry) were employed. Sheets of bark were used as sleeping pads at times.

Lacking effective axes of stone or metal, the Mrabri did not attempt to fell trees, and in any case had no reason to cut down large trees.

The virtual absence of stone, total lack of pottery, and presence of tiny scraps of metal, suggests that Mrabri camp-sites would be practically unidentifiable archaeologically, aside from ash and charcoal in hearths, with scanty remains of animal bones. The Mrabri made no bone or shell implements, nor did they make use of porcupine quills.

The only poisons used, and then very rarely, for spear-tips, appear to be borrowed or obtained by trade from sedentary villagers. Other Southeast Asian forest hunter-gatherers are regular users of poisoned weapons.

Mrabri spears were usually without spearheads, though a few were fitted with a broken piece of steel. They made little practical use of their spears, which were for thrusting only. They did not employ man-killing weapons.

Given the otherwise universal use of the bow and arrow throughout Southeast Asia until the advent of firearms, its absence among the Mrabri is certainly surprising. They also lack the blowgun. Both would greatly enhance their ability to hunt arboreal animals.

The total technological trait-list of the Mrabri, leaving aside their recent use of metal items, traded cloth rags, and tobacco pipes, is the simplest or most primitive reported for any people in the modern world, and for much of the prehistoric past given the striking rarity of stone tools. Inhabitants of very broad alluvial plains, as in the Amazon Basin, make up for the unavailability of stone by regular use of animal teeth, just as some stoneless Pacific atoll groups were able to substitute thick shell, or sharks' teeth for what would have been fashioned from stone elsewhere.

(Table 2.1 continued)

TASADAY

41. Tools

The TASADAY tool inventory was comparably impoverished. The primary tool was a long (1.5 m or longer) digging stick, from which the bark had been stripped off with the teeth, then sharpened with a stone axe. Stone tools of exceedingly crude workmanship were made from river cobbles; some of the deficiencies in lithic technology can be blamed on the absence of stone really well suited for chipping. R. B. Fox (1976) observed that the small number of stone tools among the Tasaday were evidently sufficient to enable them to maintain their far more numerous tools of bamboo and wood, and that a similar dearth of stone scrapers, etc., had been seen by him in the Paleolithic assemblage in the Tabon caves of Palawan.

Stones were used for pounding, hammering, crushing, and scraping. For working vines or sharpening small sticks, etc., the Tasaday preferred to use their own teeth. Stones were shaped by rubbing or grinding rather than chipping for the most part. The standard scraper was very small (about 14 x 21 mm) and used to sharpen bamboo utensils. The scraper was not hafted, but rather crude stone axes were attached with rattan lacing. The total absence of hide working partly explains the dearth of small scrapers. Stone were not used as weapons.

As with the Mrabri, the main tools of the Tasaday were of bamboo, including bamboo knives. Thorns were used as piercing tools. Bone was not used for artifacts.

An improvised rattan or vine shoulder strap was used to help carrying firewood. Twine was made from fibres of a wild banana. Until Dafal introduced the use of large bamboo internodes for containers and for cooking, foods not simply warmed in embers were wrapped in leaves, and water was not kept in any kind of container.

Fire was made with a drill about 30 cm long; finder was dried moss or palm bark. Dafal brought in a flint and steel strike-a-light.

Spears were supposedly absent until introduced by Dafal, but even then not used significantly. Traps were also lacking before Dafal, including the basket fishtrap otherwise very widespread in the Philippines.

The bow and arrow, formerly used throughout the Philippines, was also unknown. It was the prime weapon of the Philippine Negritos.

TASMANIANS

41. Tools

TASMANIAN technology, as analyzed by W. H. Oswalt (1976) was ranked lowest in a worldwide comparison of thirty-six hunting, fishing, and food-gathering societies, using a carefully constructed scale of "techunits." In fact, the tool-making and tool-using skills of the aboriginal Tasmanians were considerably more advanced than those of the Mrabri and Tasaday, even given the significant loss of information about the Tasmanians since their culture became essentially extinct in the mid-nineteenth century. Crudely chipped stone tools called "tronatta" were used throughout the island. They were not hafted, despite the ready availability of strong eucalyptus gum used commonly for that purpose by mainland Australian aborigines. Flaking was only by percussion. Scrapers were handheld. A major food-collecting tool was the digging stick, also usable as a club to kill small animals.

There was no use of bone tools after about 2000 B.C., according to the archaeological evidence, except for perforators or awls. Hides were utilized, though probably merely scraped to remove subcutaneous fat, for which they had other uses. Abalone shell was sometimes made into razors for trimming the hair.

The typical projectile weapons of Australia, generally taking the place of the absent bow and arrow, such as the spearthrower (woomers) and boomerang, were not used in Tasmania. A throwing club or "waddy" was hurled with fair accuracy for considerable distances. Targets were sometimes set up for practice with such clubs, and with self-spears (no hafted spear-heads). Such spears were reliable for as much as 40 m, and used in both hunting and in warfare. There were no multiple-pronged spears, and because of the avoidance of fish, no fish-spears or fishhooks, fishing nets or traps, or use of fish poisons.

Small blue snail shells worn as necklaces were perforated by biting. Twined baskets and carrying nets were used to haul shellfish. No pottery was known (and was absent also in Australia). Kelp bucket-like containers were used for carrying water in coastal areas. Food handling tools were apparently limited to a wooden spatula. Food was not boiled, only broiled or baked in the embers.

The fire-drill may have been known, but this remains uncertain; it could have been brought in after outside contact, with mainland aborigines. Burning "slow matches" were carried about, to facilitate the rekindling for fires.

Edward B. Tylor observed of Tasmanian technology (he was writing about their stone tools), that it was "on a stage lower than that of the Quaternary Period," (1893).

(Table 2.1 continued)

CHIMPANZEE

42. Property
Property concepts are not known to exist among CHIMPANZEES in a sense other than that individuals resist or resent efforts to deprive them of food items or other objects in which they are interested, though they may tolerate removal of such items by juveniles.
Territorial "property" is considered under the heading War, 72, below.

43. Exchange
Nothing resembling exchange or trade appears among CHIMPANZEES.

48. Travel And Transport
CHIMPANZEES are known to carry stone (for tool purposes: for nut cracking or pounding) for up to 500 m. Termite-fishing twigs or sticks might be carried for a shorter distance but usually lengthy transport was unnecessary. Meat or entire small animal kills might be carried for considerable distances before being eaten or shared. Burdens were carried in the mouth , tripodally, using one hand or arm, or occasionally bipedally.
Infants were carried either on the mother's abdomen, or riding upon the mother's back. No carrying devices such as straps, cords, etc., were known.
Yearly wanderings in the Gombe area covered areas of around 14 km², following no clear-cut pattern. Old elephant trails were used as travel routes by the bonobos of the Lomako area. Various game trails criss-cross most forested or forest savanna areas occupied by chimpanzees. Streams were crossed at shallow fords, where chimpanzees preferred to walk bipedally.

MRABRI

42. Property
MRABRI notions about property have not been noted, and were probably not formalized or conceptualized. Local territory is "claimed," in the sense that permission has to be obtained by outside groups wishing to utilize it. However, no actual confrontations of this kind have been observed. Outsiders were seemingly welcomed unless they had sick persons with them. Individual artifacts such as knives, pipes, etc., appear to be regarded as private property. No property passed to heirs, since personal effects were placed with the dead body and subsequently avoided.

43. Exchange
The MRABRI engaged in limited trade with sedentary villagers, apparently without precise notions of relative values. Thus, they might accept a handful of salt for a basket or mat worked on for weeks, or expect to pay for a gun with a leaf-ful of honey. Silent trade, reported for some shy forest-dwelling groups like the Mrabri does not occur. The goods offered by the Mrabri consist of wild honey and beeswax, mats and baskets of rattan, and firewood. In return they may obtain salt, pieces of metal, scraps of old cloth, tobacco, opium (rarely), rice, meat (of domestic pigs), and alcohol. They could exist indefinitely without any of these goods. There was no special time for trading.

48. Travel And Transport
The MRABRI wandered about in the forest according to no discernible pattern, since food could be found almost anywhere.
There were no obvious forest paths. The easiest way through the forest was by wading along shallow streams. Streams were bridged (as they are widely in Southeast Asian forest areas by other peoples). To avoid getting lost in the forest, tree-stems or branches might be snapped or bent down to mark the way taken.
The Mrabri could not swim. No watercraft were used, although some rivers in the region would have been suitable for rafts or canoes.
Young children were carried on their mother's shoulder and arms. A shoulder strap of split bamboo was used when carrying bamboo node water cylinders. Heavier loads of firewood or food were tied together and slung over shoulder. Small items could be carried under twisted rattan belts.
The Mrabri were reported to be good at tree-climbing.

(Table 2.1 continued)

TASADAY

42. Property
TASADAY property conceptions are likewise un-reported.

TASMANIANS

42. Property
The TASMANIANS also are undocumented regarding notions of property, other than under territoriality (no. 72. War).

43. Exchange
The TASADAY apparently were so isolated that they did not engage in any known trade. Their practice of local group exogamy may have involved some exchanges, but this is not documented. No trade occurred with the settled farming villagers across the mountains to the north, until the arrival of the hunter/trader Dafal in the 1960s.

43. Exchange
TASMANIAN trading is not documented, although their acquisition of red ochre from distant northern quarries may imply some arrangements and reciprocal transfer of products.

48. Travel And Transport
The TASADAY stayed very close to their cave home base, seldom venturing more than a few kilometers away. The stream beds provided the easiest routes. Chunks of root an other foodstuffs were tied up with lengths of vine for transport. Leaf cones were used to carry tadpoles and crabs.

48. Travel And Transport
The TASMANIANS traveled over greater distances than the Mrabri and Tasaday. Trails on the west coast were kept open by frequent burning. To reach offshore rocks to collect nesting birds and eggs, to hunt seals and sealions, or gather mollusks and crustaceans, the Tasmanians make crude rafts or boats of bark or reeds, bound with bark strips. Women were generally excellent swimmers and divers.

(Table 2.1 continued)

CHIMPANZEE

49. Daily Routine
The daily routine of CHIMPANZEES did not follow any systematic program, aside from tendencies to rest around mid-day, and to sleep after nightfall near where they had last been feeding. Individuals slept apart (except mothers with very young offspring), or at different levels of the same tree, or in neighboring trees.

52. Recreation
Recreation was limited to rough and tumble play among the young, grooming, and tickling, among CHIMPANZEES. No formal games were played.

53. Fine Arts
Nothing corresponding to this grandiosely named category exists among CHIMPANZEES, except perhaps for the famous (and unique) "rain dance" described for the Gombe early in Goodall's fieldwork there. In captivity chimpanzees pay some attention to music performed by humans, and can be induced to produce non-representational drawing and paintings admired by some as approximations of some human abstract works of art. Captive chimpanzees can identify objects presented in two-dimensional pictures.

MRABRI

49. Daily Routine
The MRABRI arose at dawn, breakfasting on any leftover food that might be present, and either got ready to move their camp to another place, or start again to forage in the vicinity. The fires were revived. Individuals might lounge about for hours near the fires, smoking tobacco or simply idling. For toilet purposes, individuals would wander off into the nearby woods. Waste food, etc., was simply thrown beyond the shelter and fires.

52. Recreation
The MRABRI apparently had no games of any kind, and no competitive sports. Small children played with common objects in an unstructured fashion.

53. Fine Arts
In contrast, the MRABRI engaged in various kinds of musical activity, from "unrestrained dancing" to improvised songs, and the playing of both bamboo flutes and a vibrating instrument of bamboo with two prongs for twanging. Another instrument was made of a bamboo section with a raised bamboo strip, which could be struck to produce three different notes. Another bamboo tube was used as a pipe, not for music but as a warning signal. The 1963 expedition to the Mrabri of the Siam Society obtained some recordings; certain songs appeared to resemble those found elsewhere in northeastern Thailand and Laos.
No drawings were made; one ethnographer induced one individual to draw with a pencil, resulting in some circles on paper. There was no painting or sculptural carving. On some bamboo segments used as containers, simple twill-like patterns were scratched with a bit of traded metal.
Children were able to recognize photographs of people, trees, deer, etc. Such picture-recognition has been seen repeatedly seen in captive chimpanzees being tested in the language projects, as noted.

(Table 2.1 continued)

TASADAY

49. Daily Routine
    The TASADAY began the day with revival of their fires, and the women might gather more firewood. The cave was swept out, onto the cave-mouth midden. After a meal of leftover food, the group dispersed to collect more food. Men were the best tree-climbers. The men might venture further in search of wild yams, while women and children worked at the stream catching crabs and tadpoles. After nightfall, no one went very far away from the home cave, although there was little danger of predators, and people did not remain awake telling tales around their cave fires.

52. Recreation
    The TASADAY played while they bathed in streams; children slid off smooth rocks. A monkey skull was seen to be used as a kind of childs toy, and a child was seen to build a tower of stones about 40 cm high. Flower petals were placed on another's face in what seemed to be a kind of game.

53. Fine Arts
    The TASADAY sang fairly often, though few details are provided. The contents of the songs seem to have been improvised, relating to everyday matters.
    Their sole musical instrument was a mouth-harp of bamboo, but it was said to be another of Dafal's many innovations. In any case, only one individual was able to play it.
    The Tasaday made no drawings, paintings, or other depictions.

TASMANIANS

49. Daily Routine
    Little explicit information survives regarding the daily routine of the TASMANIANS. It was said that they slept in a sitting position around their fires.

52. Recreation
    No games or sports were noted for the TASMANIANS. The colonial contact situation was not conducive to observation of aboriginal Tasmanian recreational activities.

53. Fine Arts
    The TASMANIANS in contrast are said to have made crude drawings in charcoal of animals, people, bark boats, as well as geometric figures. That such behavior was ancient is seen in some prehistoric rock-paintings and carvings, notably at Mount Cameron West, with circle and grid motifs found also at mainland Australian sites.
    Dance enactments of hunting, shellfish and plant food gathering were reported.
    Music was said to be simple and monotonous, and the only instruments percussive: drumming with a pair of sticks or thumping on a roll of kangaroo skin.

(Table 2.1 continued)

CHIMPANZEE

62. Kinship And Social Organization

Kinship and social organizational behavior are very well documented for some of the CHIMPAN-ZEE groups, where special efforts over twenty-five years have been devoted to such recording details of observed interaction in a manner considerably exceeding what is available in most ethnographic descriptions of small-scale human societies. To be sure, the absence of language leaves out a major means for understanding what is going on, and the lack of information about paternity in the sense that it is known from verbal report in human groups drastically limits what can be determined about patrilineal relationships (if any).

There are no recognized regular group leaders, unlike the gorilla social system. Instead, chimpanzee social structure is exceedingly fluid, in what Goodall has termed a fusion and fission social situation. Temporary leadership appeared in short bursts of activity, nearly always on the part of males. While some observable differences exist, the bonobo and common chimpanzee are not strikingly dissimilar, although mixed foraging groups are more frequent in the bonobo. There are usually solitary females, in contrast to more strongly bonded male subgroupings. Males are more gregarious, spending more documentable time in each other's company than with females. Females spend more time in contact with their own offspring. Food sharing is mainly limited to meat from large animals and between mothers and their offspring.

Much of chimpanzee social-sexual behavior is governed by the striking and very visible rump swelling in cycling females.

Transfer of females from one neighboring group to another is fairly common, which would tend to lower the incidence of incestuous matings. Incest between mature males and their mothers was extremely rare or unobserved. Occasionally mother-juvenile son copulation occurred, as well as infrequent brother-sister copulation.

Local groups usually exhibit a definite territorial overlap with neighboring groups, with frequent patrolling of territories by adult males from both sides.

Consortships may last up to three months, in which adult males may display possessiveness and aggression toward other males seeking to copulate with an already bonded female. No consortships involve mothers and their own sons, or maternally related siblings.

MRABRI

62. Kinship And Social Organization

MRABRI society is very unlike that of chimpanzees, or for that matter of most known human groups, thanks to the very small size of local commensal groups and very sparse density of population. Local groups are patrilineal, in bands from one to three nuclear families. A typical group might consist of two brothers, with their wives and unmarried children, and perhaps a married son and his wife. Kinship terminologies appear to be very simple, with recognition of elder and younger siblings. The kinship terms appear to be used instead of personal names in ordinary discourse. Incest rules forbid marriage with siblings, uncle and niece, aunt and nephew, and first cousins marriages are rare and subject to some vague supernatural sanction. Marriages are monogamous (presumably chiefly because of very low population numbers) and locally exogamous. Widowers and widows may remarry. Widows and their children may rejoin their original families— i.e., in another camp group. Extramarital intercourse was frowned upon. Marriages were arranged by a youth's father, with some gifts of food, honey, rattan, etc., to the bride's family. There is no ceremony. Elopements might occur. Husbands could leave their wives, but wives were not free to separate from their husbands.

The sexual division of labor is not formalized, although only males are responsible for the infrequent hunting.

The leader of a local band is always male, but without any formal powers. The local group is aware of its own territory. No organized warfare or feuding occurs, and no weapons capability or violent engagement are known. In case of aggression from another Mrabri group, should it occur, one group or the other would remove itself from the vicinity.

(Table 2.1 continued)

TASADAY

62. Kinship And Social Organization

The TASADAY social group is the cave-dwelling unit, which is patrilineal and patrilocal, with wives being obtained from vaguely identified neighboring groups living in a similar manner, the Sanduk and Tasafeng.

Marriages are arranged by the couple's parents and there is a food gift, as among the Mrabri. There is a vague notion of common local patrilineal descent from a named ancestor named Fangul. There was no wedding ritual. Kinship terms appeared to be bilateral in type, but details on kinship were not easily compiled.

TASMANIANS

62. Kinship And Social Organization

TASMANIAN society was generally far more complex. Population size was much greater, with local groups containing family units up to eleven individuals. The band had a distinctive name and a recognized range, and a population of up to about fifty persons. The bands were exogamous, and band territories covered about 500 km², to a possible maximum of 700. Age-grading occurred. Older men might have more than one wife. There was no wedding ritual.

Leadership was vague and undefined. Aggregations of local bands constituted what Europeans have called "tribes," sharing a language, and with the same burial customs, as well as personal ornament style. Fighting usually was between groups of different so-called "tribes."

(Table 2.1 continued)

CHIMPANZEE

68. Offenses

CHIMPANZEES exhibited no patters of group opposition to "anti-social" behaviors, or punishments for miscreants. During the lengthy study at Gombe, cannibalism in which infants were killed and eaten occurred, mainly committed by a mother-daughter team. No adverse treatment of the killers was noted.

72. War

"War" is an inappropriate term for the conflicts occurring between some neighboring common CHIMPANZEE groups, where group territory is defended, silently patrolled by males, along the overlapping or "no man's land" usually existing between such groups. Females might take part in such patrols at times. When rival groups come close, both sides engaged in loud vocalization, hurling rocks or brandishing branches. One outcome of such violence is migration or transfer of some females from one group to another.

Goodall does not hesitate to discuss such chimpanzee "foreign relations" as a kind of evolutionary antecedent of human warfare.

72. Sickness

Special care of the sick was not characteristic of wild CHIMPANZEES, although closely related individuals might behave in a sympathetic manner. Non-relatives usually avoided sick or injured individuals. Wounds (on self) or on others might excite interest, and leaves might be used to dab wounds if they were not simply licked.

MRABRI

68. Offenses

MRABRI "law" is essentially non-existent, although little violent or aggressive behavior was reported. One of the most heinous offenses was theft of a fire-kit; confiscation of such a kit was a penalty for injuring a woman. It was reported that a youth who struck an older person in anger might be tied hand and foot and left for some days in the forest, where there was a real risk of being killed by a tiger. Bernatzik was told that in a case of murder, the killer would have to offer a cooked pig to be placed next to the corpse of the victim (pigs could be traded from sedentary villagers). Persistent bad behavior might lead to the group to expel a miscreant.

Suicide was not reported.

72. War

The MRABRI have such a low population density and so few contacts with neighbors of similar culture that nothing even like chimpanzee skirmishes takes place. The absence of effective lethal weapons for hunting precludes their taking arms against the far better armed sedentary villagers. Presumably if threatened by them, the Mrabri group would simply retreat into the forest.

72. Sickness

The MRABRI regarded illness as the work of malevolent spirits or ghosts, or of witchcraft of Lao or other sedentary villagers. Food offerings were made to relieve illness. The offerings were made by the ill person, mainly when he or she was recovering, and the offering was left some distance from camp. Fire was a regular part of treatment; the sick were moved close to fires, and ashes were placed on wounds. There was not any regular use of herbal remedies, and no known remedy for snakebite. Some use of herbs seems to have spread from the Lao.

There was no shaman or medicine man; in rare cases, the Mrabri might visit a village shaman for treatment such as object removal through suction. The avoidance of outside groups known to have sick persons with them might seem to indicate a rather sophisticated knowledge of contagious disease, but the practice of quarantine elsewhere in the world long predates any knowledge of the germ theory.

(Table 2.1 continued)

TASADAY	TASMANIANS
68. Offenses	**68. Offenses**
The TASADAY likewise reported no system of punishments for antisocial behavior.	The TASMANIANS punished marital infidelity by making spearwounds on the mans legs; adulterous women were said to be "cut to pieces with stones." It may be doubted that such drastic punishments were imposed. A "hated person" might be forced to dodge spears, though for what misdeeds is not stated.
72. War	**72. War**
The TASADAY, likewise, could mount no armed defense or attack owing to their lack of effective weapons. No territorial incursions seem to have occurred, at least until the past two decades during which considerable guerrilla warfare spread in southern Mindanao among sedentary agricultural peoples.	In contrast, the TASMANIANS are described as warlike, and continuous hostility seems to have prevailed, with long-standing feuds, chiefly over unauthorized territorial incursions. The Tasmanians had very effective man-killing weapons operative over considerable distances, chiefly spears but also clubs. Stones were hurled in battles. Small flat wooden shields were employed for defense in such encounters.
	The turbulent period of the "Black War" in which the aboriginal Tasmanians were gradually killed off by white settlers included several incidents of violent counterblows leading to settler fatalities.
72. Sickness	**72. Sickness**
TASADAY medicine was practically non-existent, with little knowledge of medicinal plants. There were no observed charms, medical treatments, or curing ceremonies, and no curers or shamans.	The TASMANIANS attributed illness to spirits, as well as to contagious black magic involving putting a bit of victims air mixed with fat in a fire. Spirit possession was reported. Shamans or medicine men existed and used a variety of remedies, including suction of disease-causing objects, herbal medicines, bandaging of wounds, swallowing ashes. There were supposed cures for snakebite.

(Table 2.1 continued)

CHIMPANZEE

76A. Religion
  Religion is not present among CHIMPANZEES. Koehler described the reaction of the chimpanzees in the Tenerife captive colony to a toy animal that was shown to them as akin to "religious awe." Comparable behavior has been seen in dogs. Such responses fall very far short of the minimal definition of religion. In wild chimpanzees some aspects of the famous "rain dance" appeared to Goodall as approaching some features of religious behavior in humans.

76b. Death
  Death observances or reactions were not interpreted as having any supernatural aspects among CHIMPANZEES. Although reactions to the bodies of dead conspecifics and dead animals of other species indicate their puzzlement or curiosity. Mothers with dead infants often carried them around for a while, seemingly unable to comprehend the non-responsiveness of their offspring. Chimpanzees were curious about dead animals, and might cautiously poke at them with sticks. Kortlandt's stuffed leopard, used in several experiments in West Africa, obviously excited great alarm, followed by a bold clubbing attack on a seemingly dead or oddly impaired enemy. There was no disposal of dead conspecifics, whose corpses would quickly become food for scavengers such as vultures and hyenas.

80. Numbers And Time
  CHIMPANZEES, lacking language, had no number system, although in captivity they can be trained to understand low numbers. They also lack any system of time-reckoning, although captive chimpanzees appear to be able to learn some simple time concepts.

MRABRI

76a. Religion
  MRABRI culture includes religious elements, notably the widespread belief in spirits (Tylor's minimal definition), and making of food offerings, and a ceremony, described by a non-specialist explorer (Weaver) in which spears were seen being ritually blessed or cleansed, with offerings of pig meat and pig blood and a ritual procession around the temporary frame of bamboo where the spears were displayed. This ceremony may have been rare or unique or simply a borrowing from the neighboring farmers. Bernatzik denied the Mrabri had any creation myths or other cosmic narratives, lucky or unlucky days (characteristic of Southeast Asian cultures), gods, cultic centers, sacred numbers, omens, masked dances, or sacred musical instruments.
  If another touched a man's spear, the toucher was supposed to offer an animal to appease the spear's spirit

76b. Death
  The MRABRI disposed of their dead as quickly as possible. At the place where the death took place, a small pile of sticks and branches was placed, and the body was covered with leaves. Relatives would kneel by the corpse and weep and then move on, leaving it in the forest. The dead were sometimes buried. A food sacrifice would be made but not eaten.
  The personal property of the dead individual would be left with the corpse.
  The soul left the body after three or four days, and did not itself die, and might be transformed into some non-human animal, such as a tiger. If the tiger itself were then to die, it would be reborn as another tiger. This belief might be considered a reincarnation theory, widespread in Southeast Asian village cultures.

80. Numbers And Time
  The MRABRI language has number-words, although they do not appear to handle them with understanding. They named day and night, and divided the daytime into sunrise, midday, and sundown, using the position of the sun to designate a future appointment. The stars were not individually named, but they had a collective term for them. The phases of the moon were not considered significant, nor did they recognize the year as a time unit; a person's age in years was not known. Overland distances were referred to as "so and so many day's walk."

(Table 2.1 continued)

TASADAY

76a. Religion
TASADAY religiosity was considerably less obvi-ous than that of the Mrabri, who after all were aware of the religious activities in the farming vil-lage cultures of the area. There were local spirits of a vague kind, associated with streams or rocks. There was an evil spirit or "vampire," the notion of which may have been introduced by Dafal. There was also a mythic ancestor, called Fangul, the first man on earth, and his two wives. Thun-der was referred to as "the big word." A bird warned the Tasaday of dangers (not specified). Something like a cult appeared to have developed around the person of Elizalde, the Philippine politi-cal figure who sponsored much of the initial work of the PANAMIN organization among the Tasaday, and whose commanding presence and ar-rival by helicopter was perceived to indicate super-natural powers.

76b. Death
The TASADAY left bodies of those who died away from the home cave in the forest, covered with leaves. If the death occurred at the cave, it was carried away and similarly covered. There was no ceremony and no grave marker. The dead were believed to reside in the tree tops.

80. Numbers And Time
The TASADAY paid little attention to time units, or to sun, moon, or stars. The moon was not often clearly seen, owing to the heavy forest canopy.

TASMANIANS

76A. Religion
The TASMANIANS, despite the very limited lin-guistic access to their beliefs, had a developed mythology with a first human being, a star god, who turned into a stone, a thunder demon, and a moon spirit, and what has been called a "hierar-chy" of supernatural beings living in caves, crev-ices, hollow trees, etc. The fragmentary information of Tasmanian supernaturalism which survives may well contain features spread to their remnants from the Australian aborigines who were occasionally brought to the island, or to gar-bled Christian beliefs acquired during their few decades of sporadic exposure to European culture.

76b. death
The TASMANIANS mostly cremated their dead, especially along the west coast. The resulting ash-mound was then covered with grass. The Peron ex-pedition of 1802 illustrated such a grave, with the marker consisting of a large cone of vegetation about 2.5 m high; it could also be made of bark. In central Tasmania there were some cairn burials. Along the east coast bodies might be placed up-right in hollow trees.
Female mourners plastered their heads with pipe clay and their faces with charcoal and grease. There was a taboo on uttering the name of a dead person. Some relatives wore amulets in small pouches, containing burned bones of the dead.

80. Numbers And Time
No information is available on TASMANIAN time concepts.

91

(Table 2.1 continued)

CHIMPANZEE

82. Cosmography
   CHIMPANZEES have no known cosmographic concepts.

84. Reproduction
   The reproductive behavior of CHIMPANZEES has been documented in considerable detail, especially at Gombe, to a level rarely achieved for human groups even in the most meticulous ethnographic accounts. Copulation was mostly dorsoventral for Gombe chimpanzees. In bonobo, ventral/ventral copulation was more common. Females become reproductive at about ten years of age, and gestation lasts eight months. The bonobo female appears to be sexually receptive for longer periods than the common chimpanzee female. Homosexual activity between paniscus females seems higher than in common chimpanzees.
   Chimpanzees exhibit no recognition of paternity or of the relation of copulation to reproduction. Incest avoidance may well rest on both genetic factors and the effects of mother-offspring socialization.

86. Infancy And Childhood
   CHIMPANZEE behavior with young offspring has been carefully studied. Mothers and adults generally were remarkably tolerant of rowdy behavior on the part of the young, and members of both sexes took great interest in very young babies. Lactation was prolonged, with birth frequency consequently low. Offspring whose mothers die are often adopted by other females. Juvenile chimpanzees engage in a great deal of rough and tumble play, wrestling, and tickling. Playing with inanimate objects is less common, compared to human young, but can include flowers, petals, sticks, etc. Extreme mistreatment of the young is infrequent, although the cannibalistic behavior of a mother-daughter pair at Bombe has been well documented (the victims were not close kin).

MRABRI

82. Cosmography
   The MRABRI believed the sun dies each night, and another sun is reborn for the next day. The sun was believed to be about 30 cm in diameter, the moon a little larger. Geographic knowledge was slight, with a paucity of place-names, even for large rivers.

84. Reproduction
   MRABRI sexual behavior was mostly hidden from the ethnographers and other visitors and usually seems to have occurred at night, and privately. Children learned about sex around age eight to ten. Girls married soon after their first menstruation. Young people of opposite sex eligible for marriage had little opportunity to meet; marriages were arranged by parents with suitable partners in other nomadic groups, usually living many kilometers away.
   Births took place near camp. Women might drink large amounts of cold water to hasten delivery, and the woman's abdomen could be stroked for the same purpose. The husband set up a post to which the woman would cling while delivering the baby in a squatting position.

86. Infancy And Childhood
   MRABRI infants and young children were treated indulgently, with no observed punishment for rowdy behavior. Mothers might attempt to constrain their offspring with warnings that a tiger might come to eat them, and children might be slapped for lying. Children did not often wander very far from camp, out of very plausible danger from the tigers. Group play among children of the same age group was rare because of the normally small numbers in any local population. There were no children's games, aside from a pebble-throwing game probably borrowed from neighboring Yao villagers.
   Toys might include bits of wood, flowers, leaves, etc., sometimes treated as if they stood for items like firewood or tubers, etc., or girls might imitate nursing a baby with a chunk of wood. Boys might play at hunting, catching small birds and insects. By age 8 a boy was considered able to do a man's work, such as collecting firewood. Children improvised songs as they played. Even very young children were permitted to smoke, as in most Southeast Asian cultures.

(Table 2.1 continued)

TASADAY

82. Cosmography

No data have been collected on TASADAY cosmographic knowledge. Even information about the two groups with which the Tasaday regularly intermarried was vague.

84. Reproduction

The TASADAY knew that babies might result after the man and woman "played together." There was some ribaldry about sex.

86. Infancy And Childhood

The TASADAY cut the umbilical cord with a bamboo knife and placed it in a tall tree to prevent a wild pig or some other animal from eating it. Children were treated indulgently by parents and not physically punished.

TASMANIANS

82. Cosmography

TASMANIAN knowledge of the geography of their "tribal" areas seems to have been fairly comprehensive, and information on the approximate boundaries of their subgroups has survived. Unlike the situation in the Australian mainland, however, very few native Tasmanian place-names have been adopted by the modern European population.

84. Reproduction

TASMANIAN sexual behavior is practically undocumented, aside from the rapes which were fairly frequent during the violent colonial contact period. The Victorian authorities who wrote about the Tasmanian aborigines were usually reticent about such matters.

86. Infancy And Childhood

Little information has survived regarding TASMANIAN child-care. There seems to have been a boys' initiation, at which time the shoulder scars were made, and there is reference also to a "fetish stone" in connection with such rites.

3

Cultural Diversity in the Behavior of Pan

JO A. MYERS THOMPSON

ABSTRACT

To see hominid culture in the true primate perspective, as the title of this book implies, we must consider the nonhuman primate portion of the continuum as a whole. Within that continuum the genus *Pan* offers the clearest examples of cultural diversity. The realization that there are group traditions in wild populations of chimpanzees provides a starting point for discussion of nonhuman primate culture. The body of comparative data has grown to a point where behavioral differences are evident across populations. In this chapter I look at evidence of nonhuman, nonlanguage culture for its own sake and not as a reconstruction of some evolutionary link between prehominid and early hominid cultures.

A. L. Kroeber, though he thought that nonhuman animals lacked culture, observed that the "higher" primates possessed "reactions and faculties closely akin to our own, and manifesting at least some measure of the basal psychic ingredients which enter into culture" (Kroeber 1928, p. 341). It was Kroeber's view that we had much to learn from other primates about the origins and nature of human culture and that we had made only a beginning at the experimental and observational studies, of the great apes in particular, that would instruct us. I hope to demonstrate that we are now well on our way.

INTRODUCTION

In summer 1960 Jane Goodall initiated the longest recorded study to date of a group of wild chimpanzees (*Pan troglodytes schweinfurthii*) in the Gombe Stream Game Reserve, Tanzania — now the Gombe Stream Research Center. Goodall's study, first reported in the *National Geographic*, captured the imagination of the public, and millions came to know the subjects of her research through magazine articles, scientific reports, books, films, and videos. One result of this concentrated interest was an image of *Pan* constructed in the image of those Gombe Stream chimpanzees: we attributed similar or identical labels of behavior to any organism that resembled them.

Many of the early reports focused on the discovery that this group of chimpanzees practiced a tradition of tool use and crude toolmaking in connection with subsistence. We saw in pictures and on film a chimpanzee deep in concentration, hunched intently as she worked to maneuver her tool (a modified blade of grass, leafy twig, or straw) into a hole she had manually cleared on the side of a termite hill. Her reward for successfully penetrating the termite mound was a tasty meal of the termites that attacked this intruding object. After repeated probing, the tool would become frazzled and worn. She would modify it until, perhaps, after repeated modification, its length was no longer sufficient to reach the depths of the mound. Still intent on her meal, she might discard this used-up tool and reach for another twig that she would alter to meet her needs. Her infant, playing nearby, would occasionally eye the activities. The infant might opportunistically taste the termites that dropped to the ground. Over time the infant would experiment with the discarded tools and perhaps attempt to fashion a tool on her own. As her strength and coordination developed, she, too, would become proficient at exploiting this food source that evidently tasted so good.

Since 1987, comparisons of a number of additional long-term studies have provided a more dynamic picture of chimpanzee behavior. Primary studies come from field sites at Gombe and Mahale in Tanzania (Goodall 1986; Nishida 1973, 1990; Nishida and Hiraiwa 1982), Budongo Forest in Uganda (Reynolds and Reynolds 1965; Sugiyama 1968, 1969), Kibale Forest in Uganda (Ghiglieri 1984), Boussou in Guinea (Sugiyama and Koman 1979, Sugiyama 1984), Tai Forest in southwestern Ivory Coat (Boesch and Boesch 1981, 1989), and Assirik in Senegal (McGrew 1974, McGrew et al. 1988). Comparison across groups has revealed variability in tool typology, techniques of tool use, tool transport, hunting, plant food selection, and gestural dialect (McGrew 1992). Some chimpanzee populations have been reported not to be tool users at all in the traditional sense first described. Research has demonstrated that wild, provisioned, and/or habituated groups of chimpanzees exhibit different traditions in different parts of Africa. Localized cultural diversity has been demonstrated in the use of insect or leaf tools, male-male social triads and coalitions, cooperation in hunting, clustering of mother core areas or ranges,[1] use of drinking sponges and other materials, interspecific relationships (Gibbons 1992), as well as species of prey eaten, lengths and diameters of tools, and perforation versus food acquisition tools (McGrew et al 1979) — nor does this complete the list of traditions that vary from group to group.

Data accruing at the several research sites provided opportunities for consolidation and for systematic comparison of behavior across chimpanzee populations. Primatologists long accustomed to exchanging views about the significance of differences observable in behavior from site to site began to realize that those differences, transmitted from one generation to the next, were interesting in their own right. Coordination of information and exchange of views on behavioral variability continued as before but with a dramatic shift in focus.

The recognition that it is culture-dependent modifications of behavior within groups that produce diversity across chimpanzee groups and populations — just as in human societies — was a central theme in a symposium hosted by the Chicago Academy of Sciences in December 1991 and titled "Understanding Chimpanzees: Diversity and Survival."[2] This *Pan* symposium, at which fieldworkers presented comparisons systematically derived from their studies of group and individual diversity in tool use and modification, social organization, and socioecology, may mark a watershed in the study of chimpanzee behavior. The increase in documented reports, from studies of new populations and from additional study of the more familiar groups such as Gombe and Mahale by an expanded number of researchers, has brought to light the behavioral variability of different long-term[3] sites. As the field of primatology matures, an expanding body of scientific knowledge provides the basis for systematic study and quantitative description of population diversity.

We have discovered that chimpanzees utilize a variety of techniques, depending on which group is observed, to harvest and process foods. The termite fishing described above is practiced by groups other than those at Gombe, but it is not universal in its distribution throughout the genus *Pan*. Other observed feeding customs include fishing or dipping for arboreal ants (Nishida 1973). In the continuing quest for efficient food acquisition, the termite sticks of yet another study site have been designed with a "brush" at one end, possibly a more effective means of catching termites (Sugiyama 1985). Meanwhile, chimpanzees of the Tai Forest largely ignore colonizing insects as a food source but utilize stones and anvils (Boesch and Boesch 1983) or wooden clubs (Boesch and Boesch 1984) to hammer open hard-skinned nuts. The common and notable feature of these and other group traditions is that the behaviors involved characterize particular populations and not others. This is the common ground on which a usefully comparative discussion of nonhuman primate culture might be raised.

THE CONCEPT OF NONHUMAN CULTURE IN REVIEW

Ascription of culture to nonhuman beings, particularly primates, is not new. Half a century ago Robert M. Yerkes (1943) observed that "the elements or makings of cultural exhibits are present and have often been noted and described" and, further, that changes he had reported in the behavior of his experimental subjects, "are *persistent and cumulative*, since they are constantly being passed along from individual to individual by imitative process and from one generation to the next by social tradition" (p. 52).

Descriptions of primate behavior were first framed in cultural terminology in the founding work of Imanishi (1952) and his students, notably Itani, Kawamura, Kawai, Mizuhara, and Sugiyama. In these pioneering observations of the behavior of Japanese macaques (with follow-up field experiments), the innovation and dissemination of a new behavior was taken as evidence for nonhuman culture (Kawamura 1959, Itani and Nishimura 1973, Itani 1974). Provisioning of free-ranging research subjects[4] provided a unique opportunity to record the origination and propagation of new behaviors. Research focused on identification of lineages, the importance of relationships, and the frequency of interindividual contact to the process of propagation. The behavioral innovations observed, and now long familiar to students of anthropology and psychology, include sweet-potato washing,[5] wheat-rinsing behavior (sometimes referred to as placer-mining),[6] bathing behavior in which the monkeys submerge varying portions of their bodies into the sea, and "give-me-some" behavior, that is, a begging gesture (Kawai 1965). None of these behaviors had been recorded prior to provisioning. Much of the significance of this work lay in the observation that behavior may be modified in the course of utilizing resources introduced into the environment, that the innovators tend to be young animals, and that transmission of innovations within the group begins with close associates. The work of these Japanese scientists was central to anthropological "consciousness-raising" where the nature of nonhuman culture is concerned.

In more recent years another behavioral innovation characteristic of Japanese macaques has been examined as evidence of nonhuman culture. Since 1975 stone handling in several different groups of monkeys (Hiraiwa 1975, Huffman 1984) has been studied as a self-rewarding behavioral acquisition that is transmitted over generations. Huffman has recorded the propagation and spread of this behavior at Arashiyama and Takasakiyama and has tracked changes in character once the tradition of stone handling was established in groups at those sites. Quiatt and Huffman have used stone handling as a springboard for speculation about the evolution of hominid tool use and tool manufacture, though a major difference between stone handling and stone tool use is that the former serves no instrumental function (Huffman and Quiatt 1986, Quiatt and Huffman 1993).

TOOL USE AS AN ARTIFACT OF SOCIAL LEARNING

Since the early reports of tool use and object modification by the Gombe chimpanzees, tool behavior has been at the heart of many debates over nonhuman primate culture. Tool manufacture — making tools to a pattern — has long been considered characteristic of human society and has been used by some as a defining criterion for culture. If animals other than ourselves make tools by design, it clearly calls into question the *tools* = *culture* equation. It is now widely known that chimpanzees use many and varied tools in the

99

acquisition of food. Forms of tool use reported in the wild in connection with behaviors other than food acquisition include aimed throwing or dropping of objects and use of objects in agonistic displays, tool use in social interactions, and tool use in removing foul-smelling matter from the body.

Evidence of habitual tool use may indicate a specieswide capacity for the behavior. Examples of tool use by great apes in nature are drawn almost entirely from the chimpanzee (*Pan troglodytes*). However, a number of studies have been conducted to record and measure use of tools by baboons, capuchin monkeys, vervet monkeys, and captive apes (Beck 1975, Cambefort 1981, Parker and Gibson 1977; for discussion, see subsequent chapters in this volume, in particular Ingold's Chapter 9). Many of the examples recorded have involved individual innovations that may not have become part of the group's behavioral repertoire. Yet feral capuchin (*Cebus apella*) groups have adopted feeding strategies that allow them to exploit palm nuts and small frogs, drinking water, termites, and ants, which other local primates do not utilize (Izawa 1979). The groups involved are found in La Macarena National Park of Colombia. Comparative analysis of other *Cebus apella* are necessary to determine if in fact these behaviors occur in some populations but not others. So far, no other *Cebus* has been reported to exhibit these patterns.

Individuals of one species appear not to learn behavioral innovations from observing individuals of another species even when the desired reward of food is mutual and opportunities for social learning exist over their lifetime. Beck (1974) cited comparative observations of chimpanzees and baboons living sympatrically at Gombe. During agonistic encounters the chimpanzees were known to throw objects at the baboons, but the baboons were not observed to throw anything at the chimpanzees. Likewise, the baboons would sit and observe the chimpanzees as they fished at termite mounds, often eating stray termites from the ground and being painfully bitten in the process. Yet the baboons were never observed to attempt to "fish" for the delicacy. As will be discussed later, social learning even within a species may be problematic.

Those individuals or groups that are able to utilize different feeding strategies through tool behavior can increase the number of food species exploited. That number is determined in part also by changes in food availability due to interspecies competition, regional habitat changes, or seasonal changes. Local traditions play an important role too, — not all chimpanzee populations use tools, and in some populations food items are available to groups at several sites but are exploited by only one or a few. However, confirmation of tool use clearly goes well beyond anecdote. The next question, then, is whether tool use is biologically, ecologically, or culturally determined.

LOOKING FOR A WAY TO EXPLAIN DIVERSITY

When analyzing behavioral differences specific to groups, we want to look for an explanation of the observed variation. This is important when anthropologists look to nonhuman primates for models to help understand early hominid behavior. All primates share the same evolutionary history in some portion. We assume that closely related contemporary species and, more especially, subspecies respond to roughly similar environmental pressure with similar adaptive strategies. Although genetic studies under way at Harvard University lend credence to the supposition that the biological distinctions between chimpanzee subspecies is greater than that between human groups, these genotypic differences do not entail morphological dissimilarities to the degree needed for a biological explanation of the variation in tool use. If the chimpanzee subspecies (*P.t. verus*, *P.t. troglodytes*, and *P.t. schweinfurthii*) exhibited phenotypic differences indicative of physical limitations or advantages in some and not others, and if we could link those differences to group differences in behaviors such as tool use, this genetic indicator might be grounds for declaring genetics as the responsible factor for group diversity. If differences were greater between subspecies than within subspecies in close geographic proximity, we might focus our attention there, but it has been widely recognized that populations of the same subspecies exhibit as much or more diversity within than across subspecies. Nevertheless, biological preparedness or the inherited propensity for functioning *is* important in affecting behavior.

In groups in which mothers rear their infants, both genes and environment are shared, so we cannot separate the effects. Organisms must live in an environment rich enough to support energy expenditure until successful exploitation allows the behavior to develop. The expression of genetic variation may depend on local conditions. Natural selection focuses on the costs and benefits of energy expenditure in resource exploitation. Differences in the quality of food in each particular environment may influence the character of wild chimpanzees' manufacture and use of tools for subsistence. Additionally, the environment must provide opportunities for a specific behavior. For instance, where rocks are used as a hammer and anvil, those rocks must first be present within carrying distance of feeding sites. Diversity therefore may be a result of environmental constraints unique to each location. Yet we know that palm nuts are available in most regions where chimpanzees live. Some chimpanzees exploit them as a food source and others do not. Those that do utilize this resource may access it in very different ways.

In some instances of behavioral innovation, genetics and ecology may not sufficiently explain local traditions. McGrew, Tutin, and Baldwin (1978) identified a behavior pattern they call the "grooming-hand-clasp." This gestural dialect, observed in chimpanzee groups of Mahale (McGrew, Tutin, and Baldwin 1978) and Kibale Forest (Ghiglieri 1984), was always connected with social grooming activities. The action is described as occurring when "each of the participants simultaneously extends an arm overhead and then either one clasps the other's wrist or hand, or both clasp each other's hand. Meanwhile, the other hand engages in social grooming of the other individual's underarm area revealed by the upraised limb, using typical finger movements. In doing this, the two chimpanzees sit facing each other on the ground in a symmetrical configuration" (McGrew, Tutin, and Baldwin 1978, p. 238). In all cases the holder was observed to be the groomer and was subordinate to the other participant. McGrew, Tutin, and Baldwin were unable to identify a unique ecological determinant to account for the selection pressure in the Kasoge group. In addressing the issue of this localized behavior, they designated social learning as the most probable source of behavioral propagation.

It seems likely that tool use and most other behaviors attributed to culture spring from some combination of biology, ecology, and culture. None of these elements functions independently of the others. It is their interaction that sets the stage for examination of the evolution of cultural behavior. But my goal here remains to look at the culture of *Pan* for its own sake.

BEHAVIORAL VARIATION ACROSS *PAN*

The memorable but dated image described earlier of the common chimpanzee, *Pan troglodytes*, subsumed that of the lesser-known species, *Pan paniscus*. The unique character of that relative, the bonobo,[7] was not intentionally ignored, but little was known about this ape until very recently, and still less information has been available in published form. A rough survey and census of wild populations was carried out in 1955, but no further data were reported until the early 1970s. In 1972 the Wamba Research Station was established in the forests of Zaire. This was the first long-term site dedicated to the study of the bonobo. Soon to follow, in 1979, was the formation of the Lomako Pygmy Chimpanzee Research Station. To date much of what we know about feral, free-ranging bonobos comes from observation of the provisioned groups at the Wamba field site. As with the chimpanzee, scientists must be cautious about the tendency to generalize to all groups from observations of one or a few local groups.

Comparison of chimpanzees with the lesser-studied bonobo accentuates the conspicuous behavioral differences to be found even between very closely related species within the genus *Pan*. Much published research has focused on contrasting these behavioral differences (Savage-Rumbaugh and Wilkerson 1978; Mori 1984; Savage-Rumbaugh 1984; Nishida and Hiraiwa-Hasegawa 1987; Susman 1987; de Waal 1988, 1989a; White and Wrangham 1988; Dahl, Nadler, and Collins 1991). The most apparent differences are found in vocalization, intergroup aggression, grooming, food sharing, and sociosexual behaviors. The differences have been attributed in part to habitat. The Zaire River acts as a geographic barrier to interaction between the species. Chimpanzees are ecological generalists found in environments as diverse as dry savanna to woodland. Bonobos are ecological specialists adapted to a specific environment in rainforest habitat. Differences may also be related to the availability of food resources and sociosexual ratios within groups. Bonobos have an extended diet that includes more plant foods and a wider range of invertebrates as supplements. To date bonobos have not been reported to use tools in the acquisition of subsistence. Female bonobos are sexually receptive for longer periods of the sexual cycle and sooner after they give birth. This may encourage a more equal

socionomic sex ratio (i.e., between reproductively active adult females and adult males) in bonobo groups.

The bonobo's geographic distribution is limited to the region of
forest along the left bank of the Zaire River. Significantly fewer
long-term studies, critical for establishing evidence of culture, have
been conducted in the wild, primarily because fewer populations have
been identified. In studying wild populations it is quite difficult to
track the propagation and dissemination of behavioral innovations.
The origin of the new behavior is unlikely to have been observed and
the transmission of the behavior must be tracked using quasi-experi-
mental design observations, difficult to apply in a natural setting
without provisioning the subjects of one's research. Thus, in order to
examine the complex behavior of culture, it may be beneficial to focus
on provisioned or captive populations.

STUDYING CULTURE IN CAPTIVITY

Knowledge of the chimpanzee (*Pan troglodytes*) in captivity
goes back perhaps as far as the 1600s. Behavioral study of this species
was limited and less adequate than contemporary efforts because life
for those early chimpanzees was bleak and short. In later years the
work of Yerkes (1943) and Köhler (1925/1959) made a decisive
impact on the future science of research on captive apes. However, in
the United States, some forty years after reports on the behavior of
the illustrious but short-lived Chim,[8] systematically collected data on
the behavior of captive bonobos were virtually nonexistent until the
arrival of the equally illustrious Kakowet[9] at the San Diego Zoo in
1960 (a notable year in a number of ways for *Pan*; see the discussion
of Goodall earlier in this chapter).

In reported behaviors for *Pan*, we tended to label as "species
specific" behaviors that appeared to be biologically determined and
phylogenetically very old and were expected to be revealed in every
population. Field studies have not yet cataloged the full variety of
behavior available to chimpanzees — complex individuals living in
similarly complex social groups. Captivity offers the potential to
identify predilections for particular behavior and to confirm cultural
determination in controlled conditions. Captive inbreeding between
chimpanzee subspecies has virtually eliminated any analysis of genetic
differentiation. The limited pool of information about particular

individuals over their lifetimes has discouraged comparative review of group behavior. Most important, though, manipulation of group demographics for breeding and other management purposes prohibits the formation of long-term social groups and thus tends to corrupt any study of captive-group traditions. Historical data on individuals is often poorly kept and inconsistent. Historical social situations and groupings are not always recorded. More often than not we do not have generations developing in the same environment.

In captivity the "wild" behaviors may be altered in their form, may be exaggerated, or may not have the opportunity to be expressed. Itani and Hazawa (1953) determined that the mother's active role in the mother-infant relationship plays a critical part in determining the infant's future interactions with its environment. Additionally, an infant's interest in sibling and playmate behavior furthers the process of behavioral transmission. However, in captivity the dynamics of the social group are often manipulated to maintain artificial harmony, and infants may not have the opportunity to interact with members of their family.

Captive environments may require more dependence on quick assimilation of social knowledge, something we might want to label "cultural selection." Captivity alters adaptive selection forces, for instance, by inhibiting locomotion and reducing the number of feeding strategies, but cultural development may be encouraged in strategies not directly related to foraging and feeding. Perhaps with the development of a technology related to food acquisition and food storage, our ancestors redirected their energies away from these primary activities, and broader cultural activities became more socially defining. In the wild, group traditions change very slowly. Captivity quickens processes of change.

Captive environments used to be cosmetically designed primarily to enhance human enjoyment. To a large extent we still have failed to focus on the variability of primates and tend to model the captive environment after the bygone image of *Pan*. In many zoos, where (for instance) ape housing facilities contain artificial termite mounds, social enrichment revolves around the assumption that cultural needs can be thus simply provided. We should not perpetuate such false assumptions, yet we must focus on captive research if for no other reason than that populations in nature are disappearing or becoming more "unnatural" as humans encroach.

A RECENT ARGUMENT AGAINST NONHUMAN CULTURE

Debate about the character of nonhuman primate tradition has intensified with the increase in evidence for culture through studies of population diversity. Some scientists contend that social learning, a central feature of nonhuman primate culture as defined by most anthropologists, is not representative of behavioral transmission in groups of nonhuman primates. A longitudinal investigation of observational learning by young chimpanzees (Tomasello 1990) supported the contention that culture is characteristic only of human life. As a psychologist, Tomasello was interested in the processes of social learning. His tests of observational learning (Tomasello, Davis-Dasilva, Camak, and Bard 1987; Tomasello, Gust, and Frost 1989) led him to conclude that nonhuman primate culture differs much more from human culture than other researchers have suggested. In Tomasello's view, monkeys and even chimpanzees do not reproduce behavior but attempt to reproduce the results of the behavior. If Tomasello is correct, transmission of newly acquired behaviors such as potato washing and wheat rinsing by Japanese macaques may require reinterpretation.

Chimpanzees appear to have an inherent propensity to manipulate objects in their environment. This presumably predisposes them toward including tool manipulation as part of their behavioral repertoire. Potentially, even without social assistance, chimpanzees are inclined to learn certain behaviors. Tomasello's argument is that chimpanzees do not imitate specific "strategies" in tool manipulation. Many scientists have pointed out that individuals who have the opportunity to manipulate objects in their environment are likely to learn the properties of the objects through exploration (ee, e.g., Beck 1980, Lancaster and Lancaster 1983). In any case, if manipulation of an object achieves a reward, such as food, the behavior will be reinforced.

With regard to social signaling or communications, Tomasello (1990) also reported that the social learning process does not involve cultural transmission. As is the case with tool use, Tomasello determined that population differences in primate communication must be attributed to different learning conditions. In populations with distinct signals, there is continuity of learning experiences available to individuals. A communicatory system unique to a population must, therefore, involve local conventionalization of population signals. Tomasello stated that "little nonhuman learning comes from observing and reproducing the behavior of conspecifics" (1990, p. 305). Therefore, his argument is against transmittable nonhuman culture. But there is also quite recent support for it, including some from my own research (Myers Thompson forthcoming).

A RECENT ARGUMENT FOR NONHUMAN CULTURE

In summer 1991 I compared all of the populations of *Pan paniscus* housed in U.S. zoological institutions to determine if there were in fact cultural variations among those groups. Previous studies had focused on social groups housed at the San Diego Zoo/Wild Animal Park (Patterson 1979, de Waal 1988), a center of bonobo breeding and dispersal to other captive settings. Combining my study with these made it possible to analyze temporal changes over the long term. Analyzing behaviors across groups over time and from one setting to another allowed me to put differences and similarities into a true cross-cultural perspective. In this analysis, I defined cultural behavior as behavior that:

1. is distinct between populations
2. is disseminated over generations
3. is socially determined or learned
4. is evidenced in many members of the group
5. is not a result of environmental conditions specific to one enclosure and not others
6. is not an item in the "natural" repertoire (this distinction is made to eliminate possibilities of conventionalization, such as with object manipulation)
7. is incorporated into the day-to-day social system as an action often observed in context
8. has an adaptive benefit
9. is an imitation of adult behavior and not simply a process of social shaping when an individual reproduces goal attainment (such as in food acquisition)

Further, by examining the records of social development for each individual, I was able to draw conclusions regarding the innovation or origin of the behavior and its transfer process.

De Waal (1988) described an apparent group-idiosyncratic behavior, hitherto unreported, that he labeled "clapping." De Waal defined this behavior as occurring when "two hands, two feet, or a hand and a foot are brought together one or several times, often resulting in audible clapping" (de Waal 1988, p. 222). His data revealed that only the seven first-generation bonobos in his study, all of whom were raised in the nursery, exhibited this behavior. De Waal observed that 85.5 percent of the clapping occurred during grooming sessions, with the groomee rarely clapping (de Waal 1988).

In 1991 I observed this behavior in two populations, both of which contained members of the original group. During the grooming bouts, an individual, typically the groomer, would clap once or twice during the activity. The behavior did not appear to be directed at anything or anyone but may have denoted pleasure or self-gratification in the task. There is clearly no short-term benefit as is the case with food acquisition. The groomees (individuals on the receiving end of the behavior) were not observed to clap or alter their behavior or position in a response. The clapping was incorporated on a day-to-day basis into the contextually unique dialect of gestural communication during grooming. Figure 3.1 shows a sequence in which foot-to-foot clapping occurred. This behavior, always contextual to grooming, was characteristic of and entirely distinctive to the two populations in which it took place. In other populations of my 1991 study, composed of members none of whom had been raised with or subsequently housed with clapping groomers, this activity did not occur, nor has it been reported in the wild.

To identify the origin of the behavior, investigation of social development had to include interviews of nursery caregivers. Figure 3.2 presents the genealogy and locus of social development for the U.S. population[10] of bonobos as of 1991. By locating the individuals who perform the behavior, we can trace its origin. This gesture seems logically to have its ultimate origin in human models. Yet the caregivers were unable to recall instances of the clapping behavior. It is interesting that this behavior was not observed to be practiced in play. Moreover, interaction with the general public involved activities such as knocking on the glass that served as a barrier between the human and nonhuman species. So, thus far, the human model or models for this behavior have not been determined. However, we do know that each individual experienced a somewhat different nursery life. This environment is not static; therefore, even though the indi-

(a)

Figure 3.1   Clapping sequence.

Figure 3.1(b)

Figure 3.1(c)

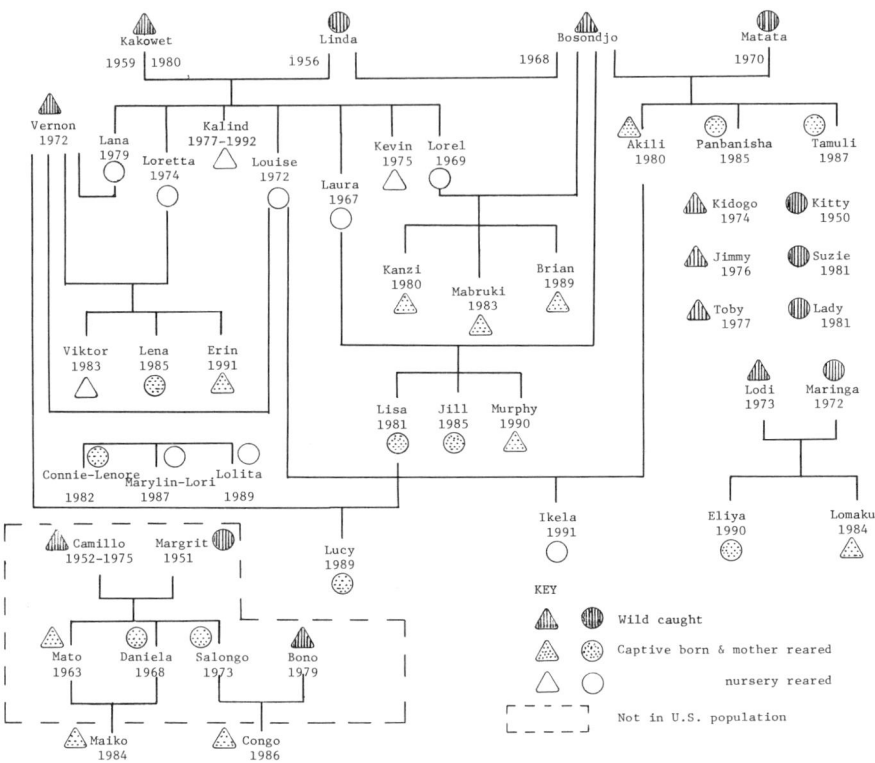

Figure 3.2   A genealogy of bonobos in captivity in the United States.

vidual caregivers were constant through time, the identical conditions could not be maintained from year to year.

Those individuals exhibiting the behavior in an interactive social environment were observed to participate in or be included in grooming bouts more often than those who did not. We may infer from this correlation that clapping has an adaptive feature. Apparently, reinforcement from others in the group ensured that the behavior was integrated into the regular communication system. However, the remarkable issue as far as evidence for culture is concerned is that the behavior clearly had been transmitted in invariant form to other individuals.

Clapping also has been transmitted across generations. All of the represented individuals performing the clap in 1983 continued consistently to display the behavior. Since the 1983 data were collected, Connie-Lenore and Lena were mother-reared by Louise and

Loretta, respectively. In 1983 Connie-Lenore, 2 years old at the time, was not reported to clap. However, she has since mastered the gesture. The other apes in the 1991 populations were either located in the group after their infancy or were 3 years old or younger at the time of data collection and were not observed to clap. So it seems that this gesture is socially learned during the first four years of life but is not exhibited until early juvenile ages.

DISCUSSION AND CONCLUSIONS

Returning to the definition of culture employed in this study, we can identify the criteria that support the inclusion of clapping behavior as an item of nonhuman primate culture. Specifically, of the bonobos under analysis in U.S. zoological facilities in 1991, two populations exhibited the clapping behavior in the grooming context. No other populations did. The behavior has disseminated over generations where mother-rearing is evident. Those individuals who have imitated their mothers' behavior had no siblings older than infants. The clapping gesture is learned in the infancy stages of development through social interaction but is not manifested until after infancy. Of the nine individuals in both facilities past infancy, six exhibited the behavior. Those that did not exhibit the behavior experienced their infancy and social development in another population, either wild or captive. Transmission and learning of the signal do not appear to be influenced by external environmental conditions (i.e., within different captive habitats).

Even in instances where individuals are transferred from one facility to another, the gesture continues and is relocated. To the best of my knowledge, this grooming communication has not been reported in wild populations. Therefore, it can be concluded that although feral bonobos might someday produce the gesture (they are morphologically capable), at present they do not. There is no mechanism for them to observe or learn this behavior.

Once clapping was introduced into day-to-day communication in a captive setting, the social system seems to have supported it as a functionally adaptive gesture. Clapping is evidently adaptive in that it enhances and increases positive social interaction, with accompanying benefits, for those who clap. Responses that are adaptive to the physical environment change as that environment changes. So this behavior seems to evidence increased adaptation to the immediate social environment. Finally, imitators of this behavior are likely to experience long-term gains from a snugger fit in the social interaction system, in addition to whatever short-term benefits may be associated with increases in grooming that derive from emitting the signal. This learned signal quite clearly involves transmission of behavioral information.

In looking for evidence of culture, we must each ascertain which aspects of cultural behavior interest us and define our terms from the standpoint of a particular, problem-oriented investigation. Each investigator looks for those factors in complex behavior that best describe culture within the limits thus imposed. We usually build on one another's definitions as we look for our own statement of meaning. But, as we have seen, culture is truly multifaceted. My own search for nonhuman, nonlanguage culture led me ultimately to an artificial environment in which the observed inventory of bonobo communicative signals had expanded to include an "artificial" gesture that proved culturally adaptive to its users — adaptive, arguably, in the sense that it facilitated increase in social knowledge acquired through grooming and associated activities.

The adaptiveness of clapping presumably helped ensure its retention over the long term, and the captive context of its propagation and dissemination afforded a unique opportunity for retrospective investigation. My research into the background and ongoing evolution of this innovation in gestural communication has convinced me that documenting the circumstances of tool use by chimpanzees in nature, though the first and most familiar exercise in exploring nonhuman primate culture, is by no means the only and perhaps not the most informative avenue toward understanding culture in evolution. Much is to be gained, too, from examination of phenomena such as the one I have described, especially if we are interested in culture as social knowledge.

In recent years, scientists representing a number of disciplines have begun to reevaluate the variability evident in social behavior, in comparative perspective and from a new standpoint of cultural learning theory. It seems to me that that reevaluation must increasingly take into consideration the cultural diversity that can be studied in captive populations. Captive studies, in addition to being valuable in their own right, are of particular importance in the case of species that are on the verge of extinction. For those species whose forest habitats are not in immediate danger of depletion or destruction, captive studies can still significantly reduce disturbances consequent on human intrusion. And, finally, information from captive studies of social behavior can be applied to the improvement of management practices.

NOTES

1. Clustering of mother core areas refers to the grouping or overlapping of those places that parous females regularly visit in a given period of time, such as a year.

2. Sponsored by the Wenner-Gren Foundation, WBEZ Radio, and the L.S.B. Leakey Foundation. The advisory committee included Jane Goodall, Toshisada Nishida, Richard Wrangham, Frans de Waal, Geza Teleki, Randall Susman, and Robert Fry.

3. *Long-term* is used here to represent studies of a local group of individuals for at least six months, and usually much longer.

4. The Koshima troop was first surveyed in 1948 by Drs. Imanishi, Itani, and Kawamura (Kawai 1965).

5. Kawai described the sweet-potato washing as a "behavior in which monkeys take the sweet potato to the edge of the water and wash off the sand on the potato by water" (Kawai 1965, p. 2).

6. Wheat-washing behavior is really not a washing action but an activity that separates grains of wheat from sand by depositing the mixture into water. The wheat grains float on the surface of the water and can then be eaten with less effort.

7. There is still debate concerning the correct generic name for *Pan paniscus*. Prior to assignment of species status, *Pan paniscus* was commonly referred to as the pygmy chimpanzee, a smaller version and one of the subspecies of *Pan*. The origin of the name *bonobo* remains in doubt. As do most scientists today, I choose to identify this species as the bonobo and *Pan troglodytes* as the chimpanzee. Whatever the derivation of *bonobo*, its use makes for clearer distinction of relationships at the level of informal usage, reducing the confusion introduced by the term *pygmy chimpanzee*.

8. In August 1923 Robert M. Yerkes purchased a bonobo (identified as a chimpanzee at the time) from a dealer in New York. He named this young male Prince Chim. Chim was estimated to be about 4 years old. Yerkes's tales about this creature marked the first recorded observations of bonobo behavior. Prince Chim died of pneumonia in July 1924 (Coolidge 1984).

9. George Pournelle, who was at the time the curator of mammals for the San Diego Zoo, captured a young male bonobo while on an African safari in 1960. This infant male, Kakowet, was estimated to be about 2 years old. During summer 1980 he died in captivity.

10. Only individuals either currently in the population or in direct lineage to current members are reported.

4

Object Manipulation, Gestures, Posture, and Locomotion

MARY ELLEN MORBECK

ABSTRACT

Object manipulation, locomotion, and posture interact with all aspects of a primate's life. These behaviors are based on species-defined body structure and physiological, biomechanical, and bioenergetic functions as each individual grows, matures, reproduces, and ages. In this chapter I illustrate the various roles played by structural-functional features and expressed behaviors of manipulation, posture, and locomotion in the lives of chimpanzees, our closest living, nonhuman primate relatives. Data are drawn from both anatomical investigations and long-term, multigenerational field studies. After describing forelimb and hand use in various social maintenance activities, I focus on how hand use, body movement, and posture contribute first to toolmaking and tool using and second to infant care. These complex behaviors of *Pan troglodytes* contribute to what it takes to survive, mate, and rear offspring.[1]

INTRODUCTION

Anthropologists view *Homo sapiens* as part of the natural world. A shared goal of contributors to this book is to explain the processes and products of human evolution. Comparative, functional, and evolutionary perspectives connect our research interests.

Attributes unique to our species such as habitual bipedalism, language, and culture are seen as built on a layered mosaic of biobehavioral features that characterize the primate continuum

(Hewes, Chapter 2). Hewes (1973b, 1978), in an outstanding contribution to the study of human evolution, emphasized the importance of knowing about nonhuman primates to understand the role of hand use and gestures in the origin of language. Contributors to this volume continue to link hand manipulative abilities, inherited and enhanced from our primate ancestors, to the evolution of both tool technology and language (Ingold, Chapter 9, and Savage-Rumbaugh, Chapter 1, of this volume). My research complements these studies since I focus on anatomy and behavior of individuals and the evolutionary roles of forelimb and hand use as part of object manipulation, locomotion, and posture in living and fossil primates (Morbeck 1975, 1979, 1983, 1991b; Zihlman et al. 1990).

Object manipulation, locomotion, and posture enter into all aspects of a primate's life. The ability to express these positional (i.e., locomotion and posture, Prost 1965) and manipulative behaviors are evolutionary adaptations. Positional and manipulative behaviors are based on species-defined body structures and physiological, biomechanical, and bioenergetic functions. The biological characters determine movement and weight-bearing abilities throughout an individual's life span. The anatomical and physiological features allow primates to engage in social maintenance and reproductive behaviors in a variety of environments. Thus, both biology and behavior contribute to an individual's survival throughout the life stages and reproductive outcome in adulthood.

In examining the dynamic interactions of manipulation, locomotion, and position with other aspects of chimpanzees' lives, I draw on anatomical investigations and long-term, multigenerational field studies of chimpanzees living in Gombe and in the Mahale Mountains National Park in Tanzania (Goodall 1986, Nishida et al. 1990, and cited references) and Tai National Park in the Republic of the Ivory Coast (Boesch and Boesch 1989, 1990; Boesch 1991b). I emphasize whole animals and their whole lives (Figure 4.1). From a functionally integrated, self-sustaining individual, we can move "down" to study body systems, cells, proteins, and molecules; "out" to investigate the dynamics of social groups and of populations in ecological communities; or "up" to consider aggregated individuals in populations, species, and higher taxa (Bates 1950, 1960; MacMahon et al. 1978; Grene 1987; Greenberg and Tobach 1988; Ho 1988; Thomson 1988, 1992; Kohn 1989; Liem 1990; Wake 1990; Morbeck 1991a and forthcoming; and cited references). In this chapter, then, I focus on

Figure 4.1 The "outside" view from behavior. Subadult male Gimbel uses a prepared stick to gain access to a termite mound in Gombe National Park, Tanzania. (Photograph by John H. Hoffman.)

life experiences and behaviors that are associated with manipulative, locomotor, and postural capabilities and expressed behaviors. I consider bones, muscles, and nerves; selected environmental interactions that involve hand use and positional behavior; and — for two well-known individuals from Gombe, Gilka and Madam Bee — their effects on individual contributions to the next generation.

My discussion of tool manufacture and use summarizes the anatomy and behavior of many individual chimpanzees. In contrast, for the second example of forelimb and hand use related to infant care, I focus on skeletal biology and observations of behaviors of the same individuals during their lives. This example shows how constraints on normal manipulative and locomotor abilities can compromise survival and reproductive success.

PRIMATE LIVES

Primates live in physical, biological, and social worlds. We manipulate these worlds with our hands. For example, forelimbs and hands contribute to traveling (occasionally even used by usually bipedal humans), feeding, communicating, and caring for youngsters. The evolutionary roles of object manipulation, locomotion, and posture can be interpreted using a new way to view multifaceted, multilayered life history characters (Morbeck forthcoming a).

Survival and time-based, reproductive features make up life history characters. Particular kinds of locomotion, posture, and manipulation; their biological (including cognitive) foundations; and associated social maintenance behaviors of feeding, social activity, communication, and predator avoidance are examples of survival characters. These features sustain life and promote biosocial health. They unfold as part of a life course of females and males determined by reproductive characters. Time-based, reproductive characters include age and maturation at birth, weaning, first reproduction, and death. They pattern the life stages of existence in utero, infancy, childhood, and adulthood and mark the transitions between them. Thus, life history characters of species, and of females and males within species, both describe the nature of and schedule growth, development, reproduction, and mortality, including when, how often, and what each sex can invest in reproduction (Zihlman et al. 1990; Morbeck 1991b, forthcoming a).

To characterize survival and time-based, reproductive life history characters in individuals and in species and populations in their many dimensions, I use four different but related perspectives. These include: (1) a central focus on whole organisms and whole lives, complemented by moving down and up through the biological hierarchy; (2) an empirical, natural history perspective that places known individuals in social and ecological contexts throughout their lives; (3) a perspective derived from more traditional life history studies of population biology that connect individual and population-level phenomena by emphasizing demographic variables and change through overlapping generations; and finally, (4) a phylogenetic view that emphasizes the present and past record of evolution of anatomical-behavioral complexes. My version of a life history approach highlights the role of the individual in evolution. It parallels, in part, broadly defined concepts of adaptation and biological roles (Mayr 1982, 1988; Bock and von Wahlert 1965; Bock 1980, 1989; Morbeck forthcoming a).

WHOLE ORGANISMS, WHOLE LIVES

Study of chimpanzees' bones and behaviors show how their upper limbs function as life history characters. Species-defined capabilities for behaviors that are dependent upon manipulation, locomotion, and posture contribute to many aspects of life. For example, they provide the basis for survival characters of quadrupedal walking and climbing, collecting and preparing food items, nest making, posture and resting, grooming and communication, and, less frequently, predator avoidance. (See Hunt 1991, 1992 for a review of locomotor behaviors at Gombe and Mahale. Also see Goodall 1986 and Boesch and Boesch 1989, 1990; Boesch and Boesch-Achermann 1991; Nishida et al. 1990; and cited references for discussion of feeding and social maintenance activities; and see Marchant and McGrew 1991 for review of hand use and cerebral laterality.)

Humans and chimpanzees share movement capabilities for many body postures and especially forelimb motions. Both species use forelimbs and hands as survival features to obtain food and to move around their environments. (Although not used for support, forelimbs are critical to the mechanics of human locomotion.) Both species also use forelimbs and hands to communicate with other individuals. We use visual and cognitive abilities to "read" upper limb locomotor behaviors, postures, and gestures that function as signals with particular meanings. Humans and chimpanzees are not the same in all anatomical features. The most significant differences between human sign language and communication gestures used by chimpanzees, however, is not in the minor dissimilarities of the musculoskeletal system. Size of the brain, its growth and development, organization, and function distinguish the roles that gestural communication can play in each species. This, for example, is shown by the brain function related to babbling as part of human sign language (Petitto and Marentette 1991). (See Lenneberg 1967 for an evolutionary approach to the study of the biology and behavior of language, Falk 1990 for ideas about the evolution of increasing brain size in hominids, and Snowdon 1990 for a review of nonhuman "language" capacities.)

During grooming, for example, chimpanzees exchange information about individual physiological states and social rank by using their hands. Grooming up and down the social hierarchy maintains harmony in the community. Body postures and contact facilitates communication about the social environment. At the same time, grooming promotes individual health through cleaning bodies of dead skin and insects. Forelimb movement and load-bearing abilities also play substantial roles in more dramatic forms of information exchange. Males use their forelimbs as part of displays, mating behaviors, and during territorial patrols and activities related to group protection. Arm raising, waving, hitting, swaying branches, brandishing and throwing detached branches, and throwing rocks are examples of how upper limb motions contribute to these behaviors.

POPULATIONS, DEMOGRAPHY, AND EVOLUTION

Long-term studies of the same individuals throughout their lives and of their populations through generational time are now available for nonhuman primates, including chimpanzees. These natural history studies allow particular anatomies and behaviors to be evaluated in terms of individual survival and reproductive outcome.

Differential mortality and reproduction of individuals (i.e., natural selection, an evolutionary mechanism that sorts variation) result from a myriad of possible biosocial factors that operate throughout an individual's life. Variation in manipulative and loco-motor skills affects who dies, who lives, how many offspring each produces, and how many of these live to reproduce. This shuffling of variation at the organismal level is measured in demographic variables of populations and, ultimately, as evolutionary change through generational time. (For examples from primate populations, see Dunbar 1988, Lee 1991 and cited references; for primates and other animals, also see Clutton-Brock 1988, Boyce 1988, and Bruton 1989.)

THE "INSIDE" VIEW FROM ANATOMY

Travel, feeding, escape from predators, social and grooming communication behaviors, mating activities, and infant care depend on movement and support capabilities of the musculoskeletal system. They are expressed during locomotion, posture, and manipulation by living individuals. Adopting the life history approach described above, we can move "down" into the skeletal system. Bones and teeth record survival and reproductive characters throughout an individual's life. Size, shape, structure, and composition of bones tell us what animals could do (species and individual features) and actually did do with their skeletons (individual life stories).

Skeletal morphology shows species capabilities for movement and support. It also documents the sequence and timing of manipulative, locomotor, and postural changes in females and males during the life stages (for example, state of ossification, growth in size, change in shape, or epiphyseal closure). The timing of these changes relates primarily to a species' pattern of growth, development, reproductive maturity, and aging, that is, reproductive life history characters. Expressions of these reproductive characters result in the skeletal morphology that underlies social maintenance survival features throughout the life stages (Morbeck forthcoming b).

Bones, therefore, reveal information about an individual's species (i.e., phylogeny), sex, and age. They also tell us about experiences of previous life stages, and, in skeletal and fossil series, about condition at the time of death. In addition, particular biological, physical, and, indirectly, psychosocial environmental factors also contribute to particular sizes, shapes, and structures of individuals' bones and thus to patterns of variation among individuals in populations (Morbeck 1991b, forthcoming b).

Chimpanzee forelimb anatomy and coordinated neural and visual abilities allow a wide range of hand placement and movements. During knuckle-walking locomotion and posture, weight is supported by the middle phalanges. Although chimpanzees carry objects in their hands while moving quadrupedally or tripedally, object use most often occurs in association in nonmoving postures. Objects, such as tools, usually are held across the palm with a "power" or "digitopalmar" grip that is characterized by strong flexion of long fingers. But some "precision" movements are possible since index and little finger metacarpal heads are slightly asymmetrical and allow limited rotation. And small objects can be gripped between the relatively short thumb as it is pressed against the side of the index finger (Lewis 1989).

The ways that chimpanzees can position their hands, however, as suggested above, are shared with other great apes and with humans. The anatomy that allows extensive shoulder mobility, full elbow extension with stability in the humero-ulnar joint, and rotation of forearm bones in the humero-radial and radio-ulnar joints, wrist flexibility, and strong, grasping hands (Schultz 1969, Washburn 1968, Morbeck and Zihlman 1988) apparently evolved as life history characters of a common hominoid ancestor.

Upper limb mobility and hand use of humans and apes overlie features of the infraorder Catarrhini and of the order Primates. They evolved in tandem with a colorful, stereoscopic way of seeing the world. For example, how the eyes, brain, and hands coordinate to interact with the environment are features of all Old World monkeys, apes, and humans. The order Primates, in turn, is characterized by biology that allows different movements of forelimbs and hindlimbs, hands and feet associated with climbing by grasping, and hand-to-mouth feeding behaviors (LeGros Clark 1959, Napier and Napier 1967, Cartmill 1974).

THE "OUTSIDE" VIEW FROM BEHAVIOR

Bones reveal movement potentials, but field observations confirm actual behaviors and how these affected survival and reproductive outcome. Manipulative, locomotor, and postural abilities, of course, are important to all aspects of expressed behaviors associated with survival activities. Furthermore, they directly underlie reproductive efforts. Like many other mammals, primate mothers protect and socialize their infants. But chimpanzees rely on their hands not only for traveling and getting food but for carrying and grooming their youngsters. This component of reproductive success usually is not emphasized in field reports.

Toolmaking and using and infant care are two examples drawn from the lives of free-ranging chimpanzees to describe some of the ways that forelimb structure and function, combined with cognitive abilities, contribute to solving problems of survival, mating, and rearing of offspring. First, tool use and manufacture by chimpanzees illustrate some aspects of the biological roles of capabilities for well-coordinated, systematic object manipulation associated with subsistence strategies. A second example highlights the role of forelimb and hand use by an infant toward facilitating its survival and infant care by the mother promoting both her own survival and successful reproduction.

TOOL USING AND TOOLMAKING

What is a tool? Use of objects for particular tasks, especially as related to food getting, is recorded among many animals, including primates. However, defining *tool, tool use*, and tool manufacture has been difficult (and controversial), since different species have different cognitive and anatomical capacities and exhibit different ranges of possible behaviors (Beck 1980). Goodall (1986 and earlier) and Boesch and Boesch (1990) use the same concept of a tool: "an object must be held in the hand (or foot or mouth) and used in such a way as to enable the operator to attain an immediate goal" (Goodall 1986, p. 536). Many kinds of animals, therefore, could be classified as tool users. But Goodall expands her definition by suggesting the importance of cognitive abilities. Chimpanzees can modify and "use an object as a tool to solve a completely novel problem" (Goodall 1986, p. 536). The point here is not to quibble about definitions. Instead, my purpose is to illustrate the kinds of roles that species and individual life history characters of forelimb anatomy and behavior as part of functionally integrated whole organisms can play during individual chimpanzee lives.

TOOLS AND CHIMPANZEES

Chimpanzees use and make tools to investigate and, to some extent, to control their world. They use and often modify leaves, grass, stems, vines, bark, leafy twigs and branches, small and large sticks, clubs, and rocks. (For reviews, see McGrew et al. 1979, 1992; Beck 1980; Nishida and Hiraiwa 1982; Goodall 1986; Boesch 1991a; Boesch and Boesch 1984, 1990; Tomasello 1990; and McGrew and Marchant 1992.) They use tools in a variety of ways: (a) to probe, insert, sponge, wipe, throw, hit, and pound; (b) to collect — from places where teeth and fingers can't reach — social insects, honey, remains of nut kernels, and soft tissue of mammal prey; (c) to clean their bodies (for instance, wiping off food, urine, feces, semen, and blood); (d) to interact with other chimps (for example, in friendly grooming behaviors as well as in less friendly aggression or dominance displays); and, finally, (e) to investigate potential danger and to protect themselves (for instance, against insects as well as large predators). Field observers note in particular that many important food items would be difficult or impossible to collect without tools. These include social insects, brains, eyes and bone marrow from animal prey, and kernels from hard-shelled nuts.

Chimpanzees apparently understand form-function relations among objects, their potential uses, and behavioral outcomes (Figure 4.1). They break or cut small and large sticks with their hands and teeth and shape probing tools by stripping leaves and bark with their hands. They often modify these tools again during use and sometimes recycle them. In addition, chimpanzees may carry a termiting tool or a nutcracking stone for hundreds of yards to use at a site they cannot see from their starting point, and, conversely, chimpanzees make repeated trips with nuts to cracking stations. Furthermore, they may use two or more kinds of tools to complete a task, for example, pounding with stone hammers to crack nuts and then removing the kernels with sticks. A rehabilitated chimpanzee on an island reserve in The Gambia used a four-component "tool-set" (Brewer and McGrew 1990) to extract honey from a bees' nest. She used, in sequence, what Brewer and McGrew described as chisels (one stout, one finer-edged) to break through the involucre surrounding the nest, a sharp-pointed bodkin to pierce the nest wall, and, finally, a flexible dipstick to obtain the honey that she had not been able to reach via the nest's flight entrance.

Boesch and Boesch (1990) also have suggested that at least as demonstrated by observed behaviors of chimpanzees in the Tai Forest, chimpanzees seem to have well-defined, preconceived ideas about tool size, shape, and potential use for particular tasks. They apparently know the technological procedures required to produce different preferred tool types.

Infants and juveniles observe and practice the skills of toolmaking and tool using. In the Tai Forest, for example, chimpanzees practice for about four years before they become efficient at nut cracking (Boesch and Boesch 1990). Mothers facilitate learning to crack different kinds of nuts with different hammer types for older infants with more mature brains and coordinated physical skills. For example, they may place appropriate hammers and nuts together. Mothers have also been seen to teach stone tool use actively. They show their infants how to position nuts for efficient pounding or how to orient and grip a hammer for the best hit (Boesch 1991b).

Data discussed here are based on observations of individual chimpanzees, in some cases throughout their entire lives. It would be interesting to know, first, the details of health, length of life, and reproductive history of these observed individuals and, second, whether the time and energy invested in tool use and manufacture — in particular by mothers teaching their offspring about tool behaviors — influence reproductive outcome in significant ways.

CHIMPANZEES AND INFANT CARE

The role of manipulation, locomotion, and posture as related directly to survival *and* reproductive outcome is illustrated by the skeletal biology and life stories of two Gombe chimpanzee females, Gilka and Madam Bee. Data from their skeletons are combined with field observations of life experiences. Skeletons of these and other Gombe chimpanzee individuals permit testing of hypotheses about how survival features are related to species and sex characters, life events and reproductive history, and, when integrated at the level of populations, evolutionary processes (Morbeck and Zihlman 1989, Zihlman et al. 1990).

During most of their adult lives, chimpanzee females, like other mammals, face the energetic challenges of pregnancy, lactation, and infant care. At the same time, females must stay alive and promote their own biosocial health. For chimpanzees, this involves forelimb weight bearing and movement in a variety of activities, for example, terrestrial and arboreal quadrupedalism and climbing while foraging, collecting and processing of food items, maintaining social relationships through grooming activities, and so on.

From the viewpoints of behavior of both a mother and her infant, upper limb mobility and hand manipulation are crucial to infant survival. Mothers feed and socialize their growing infants using their forelimbs to cradle and carry, groom, shelter, and play. In particular, they must hold newborn infants who have not yet developed full grasping abilities. Older, more-developed infants grip their mother's hair with strong digital flexors and fingers. When an infant is unable to grasp (e.g., as a result of disease or injury) or a mother's forelimb load-bearing, strength, or movement abilities are compromised the infant probably will die.

The individual life stories and reproductive outcomes of Gilka and Madam Bee show that when the nerves, muscles, and bones and potential forelimb and hand use are compromised, reproductive outcome also may be affected. These Gombe females had experienced paralytic poliomyelitis. Each lived with partial forelimb immobilization and loss of load bearing abilities for more than ten years prior to her death. The anatomy of their respective forelimb bones show similar physiological responses to diminished movement and load-bearing of the affected limb during life. The particular patterns of changes in bone size, shape, structure, and mineral content reveal individual differences (Zihlman et al. 1990, Morbeck et al. 1991).

Pronounced asymmetry of affected and unaffected forelimb joint and shaft size, shape, mineral content, and bone tissue distribution record the physiological responses to long-term unilateral forelimb disfunction at the level of the skeletal system and its component cells. For example, in contrast to normal Gombe chimpanzee limb bones, which show little bilateral variation, size and calcium content are reduced in the affected long bones of Gilka and Madam Bee (Figure 4.2 and Table 4.1). Secondary conditions are evident as well. For example, osteomyelitis caused by bacterial infection on Gilka's finger bones may be related to her dragging her hand during locomotion. A fracture in the ulna of Madam Bee's forelimb that was not affected directly by paralytic poliomyelitis never healed fully. This lack of union and possible pseudoarthosis at the fracture site may be related to overcompensation, especially in weight bearing, of the

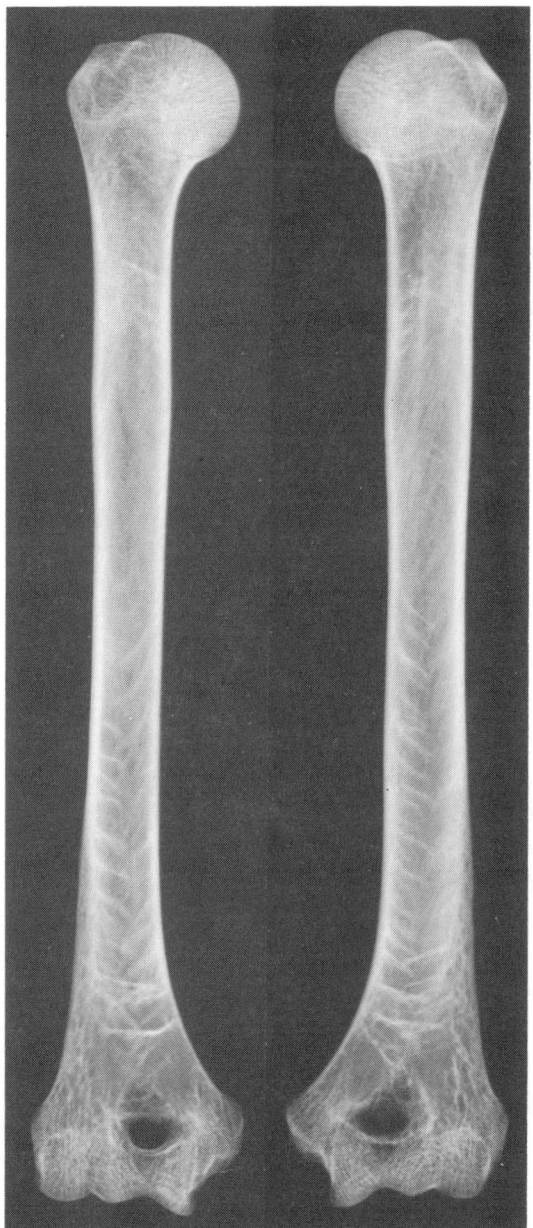

Figure 4.2(a) Gilka

Figure 4.2   The "inside" view from anatomy. X-rays of right and left humeri of Gilka (a), Madam Bee (b), and "Pallas" (c). Note that the right-left differences in the amount of bone are greater in Gilka and Madam Bee. These female chimpanzees experienced long-term, unilateral, partial forelimb paralyses due to paralytic poliomyelitis. "Pallas" was not affected by this disease; her humeri do not show pronounced asymmetry.

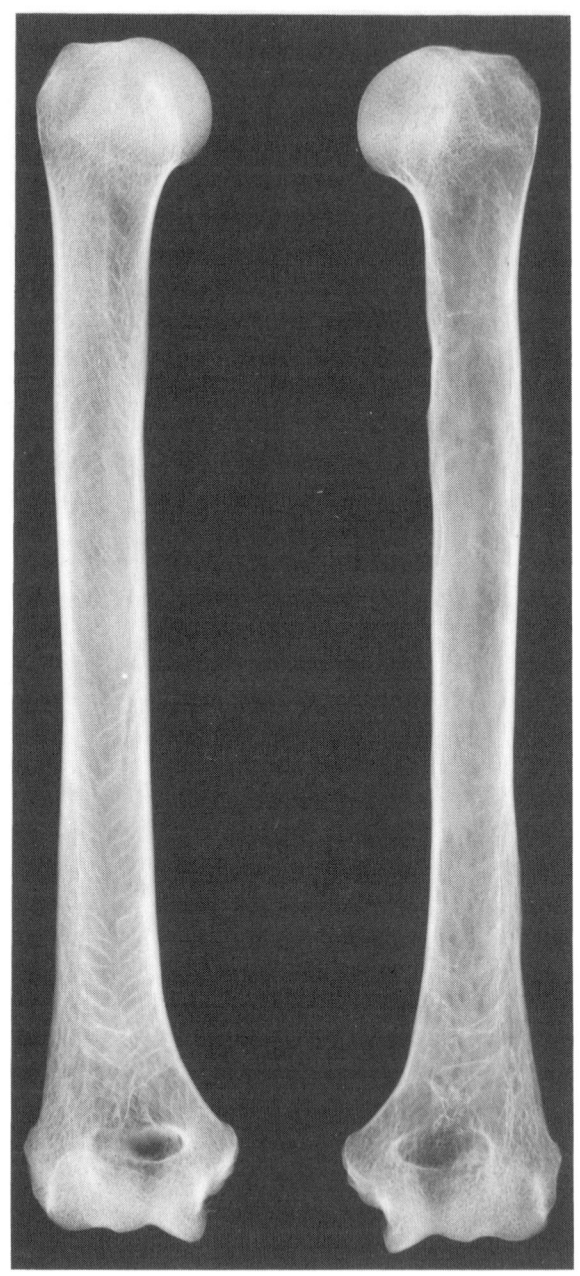

Figure 4.2(b) Madam Bee

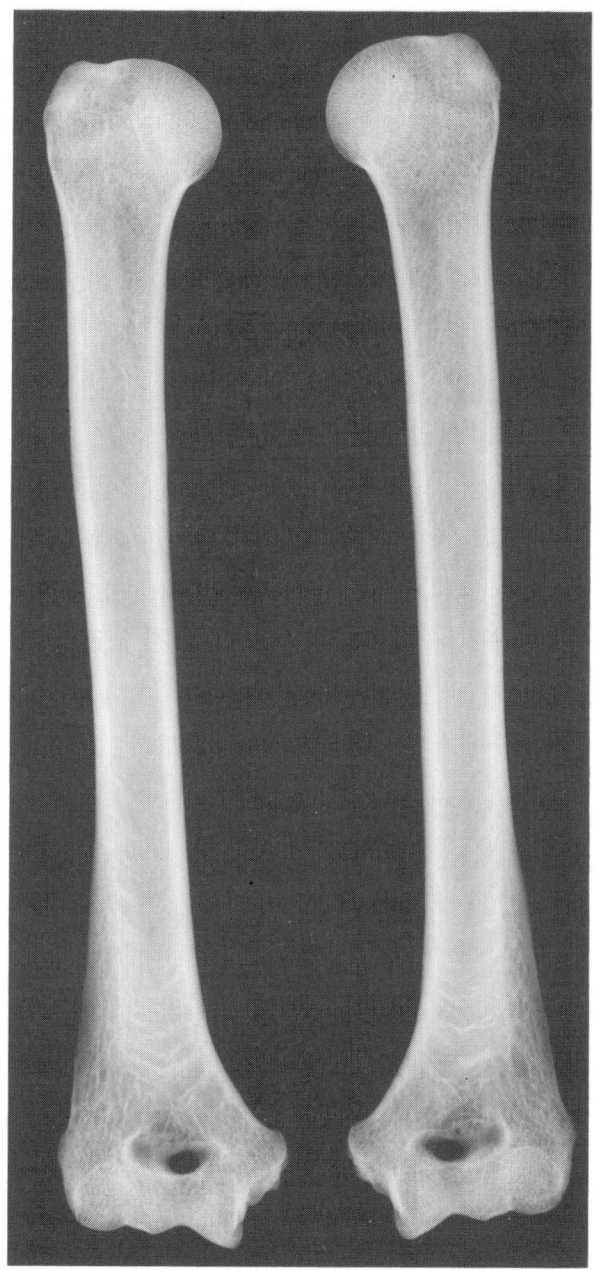

Figure 4.2(c) "Pallas"

Mary Ellen Morbeck

Table 4.1   Asymmetry of Bone Mineral Density (BMD) and Percent of Cortical Area (% CA) of Affected and Unaffeted Humeri of Gilka and Madam Bee

Gilka

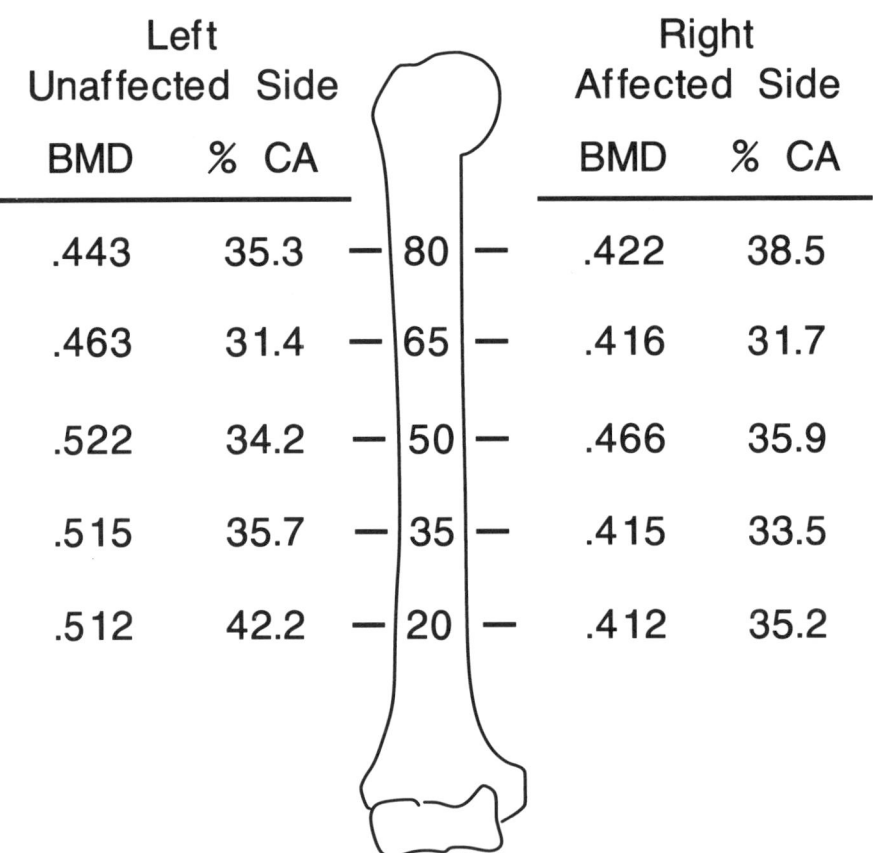

Left Unaffected Side			Right Affected Side	
BMD	% CA		BMD	% CA
.443	35.3	— 80 —	.422	38.5
.463	31.4	— 65 —	.416	31.7
.522	34.2	— 50 —	.466	35.9
.515	35.7	— 35 —	.415	33.5
.512	42.2	— 20 —	.412	35.2

BMD is assessed via single photon absorptiometry (SPA) for five equidistant sites along the shaft of each humerus. SPA measures bone mineral content (g/cm) and bone width (cm). BMD is bone mineral conent divided by bone width (g/cm^2). Since BMD is a two-dimensional measure, it provides an estimate of true bone density. The % CA is measured via computed tomographic (CT) scans at the same location. The % CA indicates the amount of cortical bone relative to the total cross-sectional area.

(Table 4.1 continued)

Madam Bee

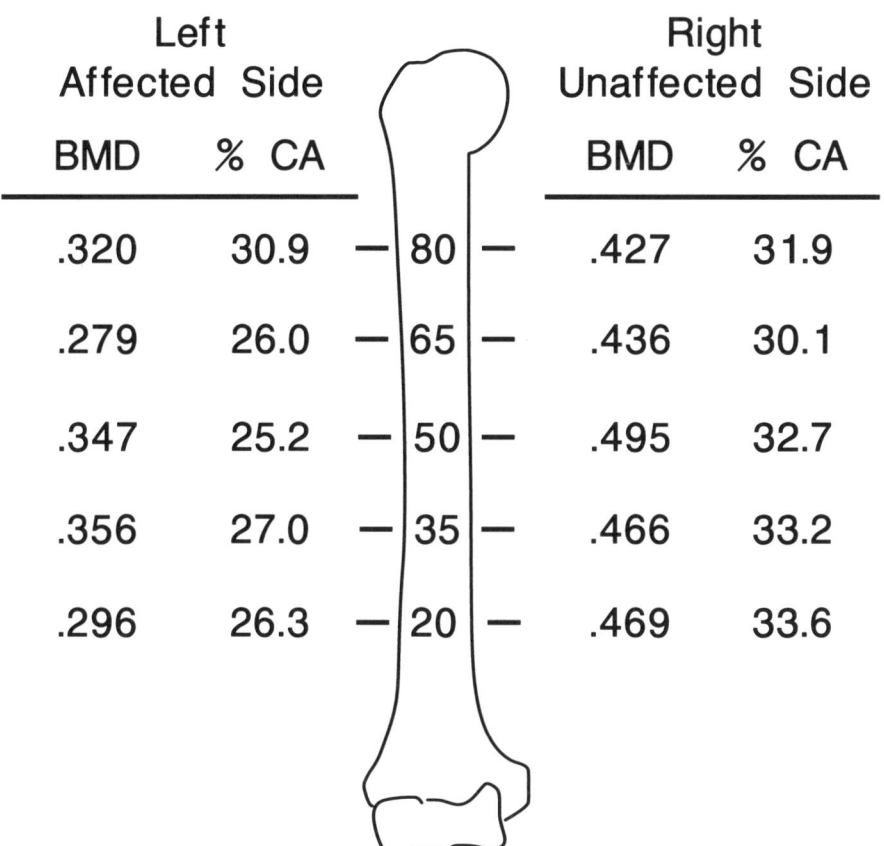

Left Affected Side				Right Unaffected Side	
BMD	% CA			BMD	% CA
.320	30.9	— 80 —		.427	31.9
.279	26.0	— 65 —		.436	30.1
.347	25.2	— 50 —		.495	32.7
.356	27.0	— 35 —		.466	33.2
.296	26.3	— 20 —		.469	33.6

"unaffected" forearm in response to disfunction of the "affected" limb (Jurmain 1989, Zihlman et al. 1990, Morbeck et al. 1991).

The bones, therefore, reveal the loss of upper limb weight-bearing and movement activities. Field observations flesh out the story of each individual's life. They illustrate how forelimb disfunction and its time of onset, severity, and duration relative to the life stages of each female had different effects on their respective lives, reproductive histories, and, ultimately, on the future of the Gombe population.

When Gilka contracted poliomyelitis, she was about 7 years old and had not yet reached reproductive maturity and adult morphology. She survived to maturity and became pregnant several times as a young adult. However, each of her offspring died when still a very young infant. Although the circumstances of the deaths differed, each was caused in part by Gilka's unilateral forelimb disfunction and the associated locomotor problems she encountered when collecting food and supporting or trying to defend an infant.

Madam Bee was about 19 years old and already had given birth to two daughters when her forelimb weight-bearing and movement abilities became constrained. Her daughters grew to maturity and one is known to have had surviving offspring. In contrast, both infants born after Madam Bee contracted paralytic poliomyelitis died. Madam Bee, like Gilka, lost her infants in part as a result of the influences of disease.

SUMMARY

The life history approach used here allows integration of natural history data studied from the perspective of biology of the individual and behavior of whole organisms with population biology and demographic data studied from the vantage point of individuals as part of breeding groups. Local population and individual case studies illustrate how life history characters — in this case those related to locomotion, posture, and manipulation — can be viewed in terms of (a) real animals and real lives; (b) the biological capabilities bounded by species characters that allow particular kinds of interactions with physical, biological, and social environments; and (c) the impact of individual contributions to population size, composition, and genetic variation.

More information is needed about the role of tool manufacture and use throughout the lives of individual chimpanzees and how these behaviors (with their biological foundations) affect survival and reproduction. As with the example of infant care, the importance of forelimb and hand use relative to health, life span, and reproductive outcome could be evaluated with detailed knowledge of a particular individual's benefits and risks of tool use (for example, better nutrition versus injuries sustained during nut cracking).

Bones, bodies, and behavior tell only part of the story of the potential evolutionary roles of manipulation, locomotion, and posture. However, the organism becomes the center of the evolutionary process when we study both biology and behavior in given environmental situations, viewing individuals as members of their local populations and noting survival and reproductive outcomes that can be measured in demographic variables of populations. This approach allows us to begin to track evolution-in-action through generational time.

NOTE

1. Thanks to D. Quiatt and J. Itani for their good work. I thank J. Goodall, A. Zihlman, A. Galloway, and D. R. Sumner for their contributions to this research project. I also appreciate the help of my friends and colleagues in the Department of Anthropology and the Desert Laboratory (University of Arizona). Ideas and data discussed here are based on projects and conferences supported by the Wenner-Gren Foundation and research supported by the L.S.B. Leakey Foundation and the University of Arizona Social and Behavioral Sciences Research Institute and Office of the Vice President for Research.

<u>5</u>

Kinship in Nonhuman and Human Primates

VERNON REYNOLDS

ABSTRACT

It is sometimes argued that kinship in humans is a cultural construct, while kinship in animals is purely biological. In this chapter I take issue with this fallacy and point out the many aspects nonhuman primate kinship and human kinship have in common. I describe the structure and functions of monkey lineages in a Darwinian framework and show that human lineages share many of the same characteristics. Evidence is available that monkeys, like humans, have a cognitive understanding of the kinship structure of their group and of the kin relationships between particular individuals. I suggest that the main departure from the primate model in human evolution came with the establishment of marriage systems, which had selective advantages in terms of enhanced lineage recruitment.

INTRODUCTION

The life process depends on information for its continuity. Information is encoded in a variety of ways on our planet. We can conveniently distinguish four levels of encoded information, drawing on the ideas of Plotkin and Odling-Smee (1981). Each of these levels integrates and extends the possibilities of the level below it. The information levels are as follows:

1. genetic (lowest level)
2. ontogenetic

3. learning
4. cultural (highest level)

This is not the place to go into the question of how level 1 translates into level 2 and so on. The translation processes between levels constitute some of the most complex phenomena in the life process, and the causal processes are only now becoming understood as a result of slow and painstaking research. In general, for the sake of clarity, I can just summarize: from the genetic to the ontogenetic level involves the organization of proteins into body and behavior; from the ontogenetic to the learning level involves the organization of neural hardware into knowledge stored in the brain; and from the learning to the cultural level involves the organization of knowledge in individual brains into shared information within a community. As stressed by Plotkin and Odling-Smee, there are feedback loops at all levels.

Kinship in humans involves shared information within a community and therefore belongs at level 4, culture. There is no known human community that does not have a kinship system dependent on a more or less consistent set of concepts about social relationships of the general kind father-son, mother-father, brother-sister, great-nephew–great-aunt, and so on. Relationships are reciprocal and involve rights and obligations on the part of those related to each other. Much of the core matter of social anthropology has been concerned with the elucidation of the different ways people in different societies conceive of the relationships among individuals in their community, the names they give these relationships, and the variety of rights and obligations among relatives (Morgan, Tylor, Malinowski, Radcliffe-Brown, Evans Pritchard, Fortes, Barnes). What makes the area of kinship particularly fascinating for those interested in the study of nonhuman primates is an observation made in the earliest studies of Japanese macaques in the 1950s: monkeys recognize a form of kinship, too. Today the fact that animal species recognize kinship comes as no surprise at all. Hamilton (1964) demonstrated the theoretical basis for kin recognition and kin selection in insects and, by extension (Hamilton 1975), to all other species, including humans. In such a perspective, there is continuity from bees and ants at one end of the spectrum to nonhuman primates and humans at the other, with blue jays, naked mole rats, and hyenas somewhere in the middle. All are concerned with a common process, the preferential treatment of close relatives; differences in organization of this process have to do with the extent to which it is organized at lower or higher levels. In insects there is very little evidence of level 4, whereas in humans there is very little evidence of level 1.

In this chapter I focus exclusively on kinship in nonhuman primates and in human beings. The argument, in line with the general theme put forward in Quiatt and Reynolds (1992), is for continuity. This may seem a sensible enough line from the perspective of a biologist, but historically, and indeed at the present time, it is widely seen as senseless by many social scientists, including some social anthropologists (e.g., Leach 1990). I do not here deal with the argument for discontinuity as such. Time, I think, will dispose of it, or perhaps show that it rests on misapprehensions of the nature of nonhuman primate behavior and a very one-sided view of human society. That is not to argue that all the differences between human and nonhuman kinship will somehow disappear, which would be absurd. The differences are real enough. However, I conceive of the differences as lying on a continuum rather than as wholly discrete. The continuum exists because of the continuity of life resulting from evolution. It is not the case that any human kinship system is the evolutionary product of any existing nonhuman primate kinship system, but it is the case that modern human kinship systems have their historical roots in the kinship systems of nonhuman primates of bygone times. To deny that is to deny human evolution.

The counterargument seems to hold that with the invention of culture, humans transformed their social lives out of all recognition, that all human social forms after this transformation are cultural constructs, built by human ingenuity and explicable by reference to indigenous logics and formal rules particular in time and place — in short, that cultures and kinship systems today have broken with their antecedent, nonhuman primate roots and it is folly to try to reconnect them. This overlooks quite a bit. It overlooks that kinship systems are not just structures in a void but have very real functions for individuals. It overlooks the existence, at the heart of all kinship systems, human and nonhuman, of a set of biologically significant relationships, between mother and offspring, brother and sister, between *genitrix* and *genitor*. These relationships are not somehow magicked away by culture; they are the stuff on which culture works.

In nonhuman primates such as macaques and baboons, kinship is recognized matrilineally, and matrilines are ranked relative to one another. Studies of ranked matrilines have shown that the existence of these structures is important in determining the eventual social rank of individuals (Kawai 1958, Datta 1986) and, furthermore, that matrilines rise and fall relative to one another over a process of years. Samuels, Silk, and Altmann (1987) described two periods of rapid change of female ranks after eleven years of stability among matrilines in a group of baboons. In January 1983 there were violent fights, several females in one matriline combining against particular females in other matrilines, as a result of which eighteen rank relationships were reversed. In the following August, three matrilines fell in status relative to the newly dominant one, involving downward shifts of status for nineteen females. The fights were not as violent as in the previous January. The reasons for these reversals were not clear, but the authors suggested that matriline size, health of individuals, and the number of sexually receptive females were the most important factors involved.

Such intermatriline fighting is unusual: stability is the norm. But the above case does indicate well the underlying competition among matrilines. Normally, matrilines appear to be able to maintain their relative status, and the mechanism of how they do so is well known, having first been described by Kawai (1958). It was not until some time after the explanation of kin altruism by Hamilton (1963), when the theory of kin selection was applied to the explanation of primate matrilines, that their significance became clear. At the level of inter-actions, stability among matrilines is accomplished because high-ranking mothers, by intervening on behalf of their offspring in disputes, are able to raise their own offspring's status relative to offspring of lower-ranking mothers. A second process, suggested by Datta (1986), is that young members of a higher lineage seek out opportunities to dominate older members of a lower lineage or, as opportunity presents itself, older members (including their own mother) within their own lineage. Younger monkeys do this by waiting until one or more relatives are nearby and then using these relatives as alliance partners to threaten an opponent. What we see in these cases is that a younger member of a dominant lineage enlists the support of kin against a nonkin rival. He or she is able to do this because of kin altruism. Kin are willing to support each other in agonistic situations, and this, we must assume, is a result of selection for behaviors that increase the inclusive fitness of one or other or both of those involved. In the present example, choosing the moment to

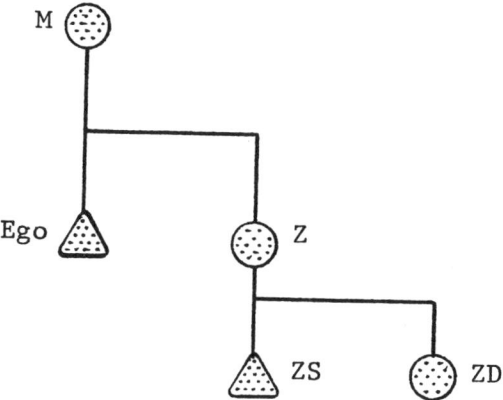

Figure 5.1 Generalized diagram of a three-generation monkey matriline. Key: M = mother, Z = sister, ZS = sister's son, ZD = sister's daughter. Horizontal lines link relatives by siblingship. Triangles represent males, circles females. Dotted symbols show members of the matriline.

act by the youngster, and support by the older kin, involve complex social learning factors. Fitness benefits most obviously accrue to the youngster, whose status is increased, but all members of a matriline stand to lose fitness if the matriline itself loses status relative to another matriline. Thus, fitness of matriline members is maintained by supporting kin and can even be said to be increased because the status of a matriline is marginally increased every time one of its members raises its status relative to one member of another matriline. Such marginal benefits may be very slight in themselves, but taken together over time they bring about the continued high status of a matriline, and this results in better access to food and preferred habitat space (and hence reproductive success) for its individual members.

Use of the word *matriline* or *matrilineal* for primates other than humans raises questions about the nature of the animal relationships under discussion, so that we may subsequently make a proper comparison with the relationships within human lineage systems. The structure of a monkey matriline, and the relationships in it, are shown in Figure 5.1.

In Figure 5.1 one of the animals is designated as Ego. This is an anthropological convention and is particularly important when members of a kinship system are designated by particular terms. It represents the individual to whom the other terms refer. Such terms (*mother*, *sister*, etc.) are applicable only in relation to a particular individual; in relation to another individual they change.

The key to Figure 5.1 states that horizontal lines above link sibling with sibling. What is implied by these lines? In this case, the answer is based on biological relatedness. Ego's mother is his biological mother, his sister is such because she has the same biological mother as he has, his sister's son is the natural son of his sister, and so on. The terms used are thus firmly rooted in biological relatedness, and if necessary the degree of relatedness can be calculated.

The terms used do not refer *just* to genetic relatedness, however. Indeed, in the course of everyday life in the monkey group, genetic considerations are irrelevant. What the human term *mother* presumably *means* in the monkey world of Ego (while Ego is an infant) is the individual who is warm, nurturant, supportive, and other such things. The monkey with whom Ego has a strong positive affectional bond is the one designated on the diagram as "mother." Likewise, the meaning for Ego of the other relatives designated is based on his personal experience of them during his life. By the same token, nonrelatives are monkeys with whom Ego has not had such interactions, and perhaps he may have had different, possibly stressful ones with them. The background experience of relatives and others is stored as information in memory, and in the course of time, as a result of complex learning, this experience is organized in categorical ways that enable monkeys to build up a comprehension of the relationships between other individuals in their group, as demonstrated by Cheney (1984).

The kinship diagram is thus a representation of monkey *social structure*, displaying certain basic *relationships*, arising out of particular *interactions* (to use Hinde's 1976 terminology). And further, it is a representation of what, in a different, neural way, a monkey knows about these particular animals and their relationships with him or her in its domain of social knowledge (Quiatt and Reynolds 1992).

The above points have been labored somewhat because it is of great importance that we understand what kinship in monkeys is, in order to avoid some confusions that arise when we come to make the comparison with human kinship. There are many implications of kinship terms when used of humans that do not apply to nonhumans. For instance, there are moral, legal, and other cultural rights and obligations pertaining to the roles of kin in human families, some of which are certainly absent in the monkey world.

Three things (at least) intrude into the human situation that are not so clearly present in the monkey one: first, the linguistic environment; second, complex technology; third, incorporation into a particular sociolegal environment. From an informational standpoint, all three involve increasing the variety and number of social stimuli by which we monitor and change environmental states. Human beings are bombarded by vastly more items of cultural information to be learned and integrated into social skills.

Are there any equivalents of these in nonhumans? Let us take the linguistic environment first. The maternal environment of monkeys is not a silent one. Cheney (1984) made an analysis of the ontogeny of vocalization in the vervet monkey, showing how early calls gradually change into the calls of adults. This was shown for the set of alarm calls, which in adults are specific to the type of predator but in young animals are unspecific and may even be mistaken. Cheney has described how, in their early alarm calls, vervet infants frequently make mistakes, calling when harmless species fly overhead. Their mothers, however, make no response to these harmless birds. Only when a predator is overhead does the mother react. Over the period of childhood, vervets narrow down the number of species to which they respond with alarm calls, until by adulthood they call only in response to predators.

The process by which this learning happens involves reinforcement from adults, especially the mother. We see here the development of categorical distinctions in primates parallel to, if not homologous with, the categorical distinctions incorporated into words; in both cases the outcome is an information-rich store of relevant knowledge in the minds of individuals.

Complex technology is most highly developed in humans, but its roots are certainly to be found in nonhuman primates. Young chimpanzees at Gombe learn to make and use termite sticks, and in West Africa we have the tradition of breaking nuts with large stone hammers. Both these technological traditions are transmitted mainly by females and by mothers to their offspring. We clearly must not exaggerate: the young primate is not *surrounded* by technology in the way the young human is. But a similar process goes on, albeit on a much smaller scale, and mothers are important in this process.

Third, we have the incorporation of the human infant into a particular sociolegal environment. In nonhuman primates, differences have been noted in the ways infants are incorporated into their community, both between species and between groups within a species. Langur monkey mothers pass their babies to other adult females for short periods within the first few days of life (Jay 1965). These females can thus get to know them, and they are incorporated into the social life of the group. In macaques the same process occurs, but later: mothers are more possessive and are reluctant to allow other females to interact with infants until they are older. Evidence of the way primates understand the details of the incorporation of others into their group comes from vervet monkeys. Seyfarth and Cheney have shown that at the age of 2 years, juvenile vervet monkeys are recognized by adult females of the group not just as individuals but as the offspring of particular females (Cheney 1984). They thus know something about the kinship structure of their group. We have also seen that the presence of dominant kin close at hand has an effect on the interactions of others in macaque groups, showing again a keen sense of how individuals are incorporated and located in society. These may not amount to the same thing as what we have called the sociolegal incorporation of the human infant into its group (by such rituals as baptism, registration, etc.), but they are the primate equivalents, and it is out of the interest in other group members and their place in society that our primate ancestors displayed that our present systems of naming, baptizing, registering, and so on have evolved.

PRIMATE PATRILINES

In most primate species it is males who move from group to group in search of mates. As a rule, juvenile males are more peripheral to group life than are females; eventually, they leave their natal group to find mates elsewhere.

In a few species, however, it is females who emigrate. In chimpanzees living at the Gombe Stream Research Center and at the nearby study site at Mahale, females move out of their home community and seek a mate elsewhere. As a result, the genetic lineage, or core, of these chimpanzee communities consists of the males. Probably the same organization exists in the Kibale Forest (Ghiglieri 1984) and possibly also in the Budongo Forest, Uganda. We do not,

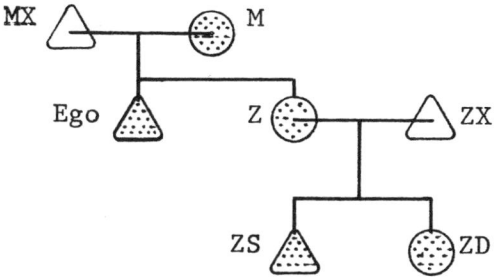

Figure 5.2   Olive baboon matriline. Key: as for Figure 5.1 with, additionally, MX = mother's mate, ZX = sister's mate, Lines joining symbols center to center represent mating relationships.

however, find ranked patrilines, because there is only one patriline in each chimpanzee community, judging from existing studies (Goodall 1986, Nishida 1979). The males in a chimpanzee community make up a very tight-knit group. They tend to move around together and share mates with each other, and they form the hunting groups that kill prey species such as monkeys, bush pigs, or small antelopes and share the meat after a kill.

Fatherhood is a complex topic. Nearly all primate species studied show evidence of male-infant relationships of a friendly kind, such as play. In some species, such as marmosets, males carry the young habitually, but this is uncommon. In most species males carry infants occasionally. Baboon males establish regular grooming relationships and associated ties with particular females who subsequently become preferred mating partners; in due course, when infants are born, the males defend them and take on protective roles toward them (Smuts 1985).

Figure 5.2 shows the structure of an olive baboon matriline in which the adult females have semipermanent consorts, who are also shown:

Males take less interest in infants whom they could not have sired, that is, who are born shortly after a relationship has been established with a female, and it has been suggested that care for an offspring by a male is very often consequent on a long-term relationship with its mother (Smuts 1985).

In some species paternity has become an issue of great importance. This is seen most clearly in the case of infanticide, for example, in langurs of the species *Presbytis entellus*. Sugiyama (1965) first described an attack by an adult male on a group of langurs in which he systematically killed all the infants. This phenomenon has been recorded many times by different observers (e.g., Mohnot 1971, Hrdy 1977, Vogel and Loch 1984; reviewed by Hausfater and Hrdy 1984, Struhsaker and Leland 1987). In areas where infanticide occurs, langurs live in one-male groups consisting of a male, his "harem" of females, and their young. He keeps out all other adult males. Such males form "bachelor bands" who live without access to females. However, from time to time such unmated males launch an attack on the one-male group and attempt to drive out the male. If they succeed, one of the attackers then drives out the others and becomes the breeding male, and it is he who kills the females' small infants. The result is that females resume estrus fairly rapidly, and the new leader is able to replace the offspring of another male with his own. As long as he can bring them to maturity (by resisting other takeover attempts), he has succeeded in transmitting his genes to the next generation.

This may not sound very much like the paternal care we mentioned in the case of baboons, but it is no different in principle. Just as the male baboon takes more care of an infant if he has had a long relationship with its mother (and it is therefore likely to be carrying his genes), so the langur male achieves the same objective in his brutal way. Both cases demonstrate that fatherhood, like motherhood, is primarily based on the fact of genetic relationship between parent and child, though what proximate factors are organizing the behavior is much less clear. But one thing is certain, namely, that in both cases, however different, the male concerned has to familiarize himself with the social situation he is entering into; he collects information and builds up an organized store of knowledge, which provides the essential basis for his subsequent actions.

In a few species of primates, such as gibbons, fatherhood is relatively uncomplicated (though even here this is not always the case — see Quiatt 1987). For simplicity's sake, let us take a straightforward situation. Gibbons are monogamous (perhaps serially monogamous), and thus the social group theoretically, and often in practice, consists of an adult male, an adult female, and their young. Both parents defend the young and take equal shares in doing so, primarily by emitting loud vocalizations at dawn and throughout the day, establishing ownership of their family territory (Carpenter 1940, Ellefson 1968). In addition the mated pair engages in vocal duetting, which, according to Brockelman (1984), serves to strengthen the bond between them. Paternity confidence is high in such a case, and the male does all that is necessary for the support of his female and their offspring. At puberty, however, the young male gradually withdraws to the periphery and subsequently leaves the group.

Complications may arise because of demographic changes caused by arrivals, departures, or deaths of males. Thus, Brockelman and Treesucon (1986; see also Quiatt 1987) noted that of three young males in a family group, the oldest two were in fact brothers of the current resident adult male, while the youngest was the female's offspring by that adult male's predecessor. Gibbon groups may not always be as regularly constituted as is sometimes assumed.

When group size is as small as it is in gibbons, with a mated pair and their offspring very often forming the social group, there is no scope for the development of matrilines or patrilines as functioning, intragroup entities. However, there is always the possibility that such lineages do exist as *dispersed* entities. If this were the case, then an unmated male might, for example, be able to join up with a mated brother on the latter's territory. At present, despite a few reports of adult males' attaching themselves to a mated pair (Chivers and Raemaekers 1980), we have insufficient information on the genealogical background of most hylobatid groups.

The question of dispersed lineages is potentially a very important one. It has been argued that on Cayo Santiago a subadult rhesus male leaving his natal group is likely to emigrate in the company of a brother or brothers, or end up in a group in which a previously emigrated brother has established residence (Drickamer and Vessey 1973, Meikle and Vessey 1981, Boelkins and Wilson 1972). The argument for brotherly influence on either timing or direction of migration remains unsettled; Colvin (1986) maintains that "there is no evidence that . . . [young males] emigrate in the company of their brothers," but if brothers do prove to exercise any influence, then we can see another very important function of matrilines: they serve to further the mating interests of males within the lineage (see Quiatt 1988). Whether this is also true for females in the case of chimpanzees is not known, but it would seem probable that a migrant female chimpanzee would find it easier to enter and settle in a community that had already accepted an older sister than in one where she had no prior contacts.

We have by now seen something of the nature of mother-, father-, brother-, and sisterhood in primates. We have shown how some kind of understanding exists between kin, that is, animals who would normally be genetically closely related to each other and who have been in close association with each other during the growing period. Such animals support each other in a number of ways against nonkin, and their special relationships are often continued late into life, or even until death, as in the case of chimpanzees where an ailing mother may be closely accompanied and groomed by an offspring.

At the psychological level, there is much in common between kinship in primates and kinship in humans. Continuity here is not hard to see. However, in the preceding discussion I have already spoken of social "structure." What I mean by that term is the way in which the (psychological) relationships between animals standing in the relation to each other of kin become, when considered together, the principle of organization of the society as a whole. The coherence of a rhesus monkey group on Cayo Santiago results from the sets of relationships within and between its matrilines. If and when the group splits, this usually happens between matrilines, one or more hiving off from the rest. This does not mean that matrilines are static; as I have mentioned, their strength waxes and wanes, and their rank relative to each other changes over time.

HUMAN COMPARISONS

Do the lineal structures of nonhuman primate groups bear comparison with their equivalents in human societies? I think that to some extent they do, more especially in smaller-scale societies.

In lineages everywhere, individuals are encouraged to maintain the lineage's rank, and members do what they can to keep up the status of fellow members. In patrilineal societies fathers can be most helpful and influential in a great variety of ways in maintaining the status of their offspring. Sociologists distinguish between "achieved" and "acquired" status. The latter, the status that derives from the position one is born into in society, is what we are now talking about. Brothers and sisters, too, can help one another to succeed by sharing information, social contacts, and resources. For instance, paralleling the case of the migrant rhesus monkey, migrants coming into a more affluent country from a poorer one in the hope of improving their financial position can be greatly helped to get started by the presence of one or more kin who have already settled in the new environment. As in primates, human lineages can operate when they are dispersed. Indeed, they can do better, for they can operate in this way even if the kin have never previously met one another. This is achieved through language use and in particular naming, to which I return later in the chapter.

MARRIAGE AND FATHERHOOD

We now need to make a schematic analysis of a human kinship system in order to make a more systematic comparison with the kinds of systems we have looked at in primates. We took as our paradigm case for primates the monkey matriline (Figure 5.1). If we were to consider this as a human family, we should rapidly notice a very odd thing about it, namely, that both the mother and her daughter are *single* mothers, that is, they are unmarried. This is quite a usual feature of families in Western society these days, but in world historical terms it is most unusual. More conventionally, in all societies the world over, and especially in societies in the less affluent parts of the world, childbearing is preceded or accompanied by the action of marriage, a characteristically human institution that legitimizes a partnership between a man and a woman in such a way that any children born to that woman, whoever the genetic father, obtain through that marriage a legitimate place in society as the heirs and descendants of one or the other or both of their legal parents. This is perhaps a curious definition of marriage, but it is nevertheless an accurate one and follows the definitions of other anthropologists (e.g., Mair 1972, p. 91). Marriage takes many different forms in different societies. The rights and obligations it confers on the husband and the wife differ widely from society to society. Marriage may be considered a long-term or a short-term arrangement. It may be exclusive, as in monogamy, or not, as in polygyny and polyandry. Where there is some emphasis on patriliny, the wife may change her name to the husband's, and her children may take that name, too, so that she and her children are lost to the lineage of her birth. In matrilineal systems, the children are likely to take the clan name of their mother and their mother's brother.

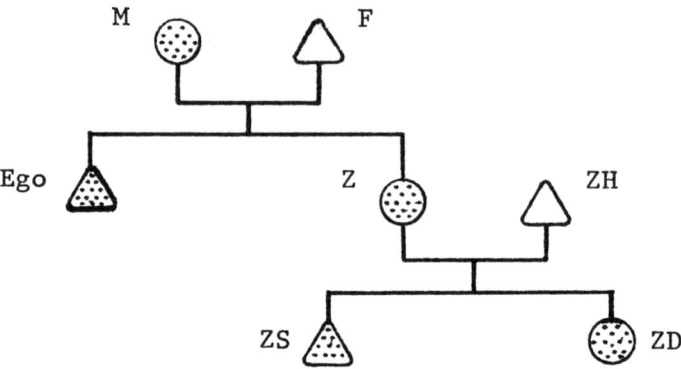

Figure 5.3 Three-generation human matrilineage. Key: as for Figure 5.1 with, additionally, F = father, ZH = sister's husband. Horizontal lines below link spouses.

Just as Figure 5.2 represented a three-generation matriline, Figure 5.3 represents a three-generation human lineage. To make the comparison as clear as possible, I have chosen to show a matrilineage.

Here, as before, the categories of individuals (M, F, ZH, etc.) are as seen from Ego's point of view. There are several departures from Figure 5.2. Note that parents are now designated in a different way. In Figure 5.2 the male partner of a mother was designated as her mate, and this was shown by joining the symbols for the female and male concerned by a line that ran center to center. In Figure 5.3, by contrast, the male partner of the individual labeled "mother" is called "father," and the male partner of the individual labelled "sister" is called the sister's "husband." These terms are precisely used here to designate affinal kin, kin through marriage, and as such mark a departure from anything found in the animal kingdom. Because these are marriage relationships and not mating relationships, the individuals concerned have been linked in the diagram by horizontal lines below, a standard convention in anthropology. Lest there be some confusion over the word *father* (which is an ambiguous term in ordinary usage), it is here used specifically to mean *pater* (the legal and jural status of father), not *genitor* (the biological father). This is not to deny that the *pater* may in many cases be the *genitor* but to emphasize the analytic distinction between the two.

Fortes (1983), like Wilson (1983), has pointed to the incorporation of the father into the family through the institution of marriage as the single most important innovation marking off human social evolution from anything found in the rest of the animal world. "Fatherhood," he wrote, "is a creation of [human] society . . . man's central rule-generating institution" (Fortes 1983, p. 20). Fatherhood is here identified as the father figure, the mother's husband in patrilineal societies and the mother's brother in matrilineal ones. In Fortes's analysis, like that of Fox (1980) and Ingold (1990a), human society is sustained by systems of rules that have juridical and moral force, are backed by authority, and are accepted and changed by consensus. The father, in Fortes's scheme, is the central disciplinary element in the process of rule transmission and enforcement. Fortes did not consider the argument for the equivalence of human and primate kinship to be valid, but he did not provide an in-depth analysis of the primate data of the kind provided here. His analysis of the emergence of human society is nevertheless very profound and worthy of serious attention, but it is characterized by a wider gulf between the human and the nonhuman than I accept.

I have noted above that it is often the case in primates that the male who is the current mate of the mother takes a protective interest in her offspring, and it is this interest that is safeguarded and formalized in humans by the marriage ceremony, in which he takes on specific obligations and accepts that his heirs will have specific rights. But before discussing the differences between mating relationships and marriage relationships, let us remind ourselves of one respect in which marriage and the incorporation of the father it achieves seem to have similar characteristics to the process of mate selection in animals. In many, perhaps most, human societies, marriage partners are not allowed to come from the spouse's natal lineage. There is here a close parallel with the situation in most animal societies, where *mating* occurs outside the natal group, one sex migrating for the purpose of seeking a mate. The explanation usually advanced for the evolution of out-mating in animals without regard to proximate explanations is the selective advantage resulting from the avoidance of inbreeding depression. Marriage is not the same thing as mating, a point to which I turn in the next section, but marriage very often does involve mating and may be the institution in which most mating occurs. Exogamy in humans thus shares certain features with out-mating in nonhumans.

FROM MATING TO MARRIAGE

I have noted that there is an important difference between mating (a sexual relationship) and marriage (a jural arrangement that usually contains a sexual relationship). This seems to mark a major divide between animals and humans, and we need to consider why this might be so. What are the factors that might have brought about the institution of marriage in humans? The term *marriage* includes a large variety of customs and is based on different ideas in different cultures. We do not intend to review these but rather to take some of the most striking features of marriage and try to draw out its distinctive traits when compared with primate mating patterns.

A useful starting point is the process of mate selection. Marriage involves among other things a mate selection process based on social knowledge. In some societies marriage is organized by the marrying pair, but in many this is not the case, and selection of a suitable spouse is in the first place done by the parents. Depending on the extent of parental control, economic and reproductive considerations will be foremost in the selection of a marriage partner. Where such control is great, future spouses have little or no physical contact before marriage and sometimes find they are incompatible afterwards. In our own society sexual compatibility is regarded as an important prerequisite for marriage. In arranged marriages the most important consideration for the parents is the social status of the family of a prospective spouse. In much of rural India, Hindu fathers with offspring who might make suitable spouses for each other meet and discuss the possibilities of a match, and a major consideration on both sides is the amount of dowry payment by the father of the bride to the family of the groom. Among African cattle herders, payments often go the other way, in the form of bridewealth paid by the father of the groom to the family of the bride. One of the factors deciding which way the payments go is the relative determination of the "value" of a bride as against a groom, and this depends on many factors, such as how much work the bride is expected to do for the family of the groom and whether marriage is uxorilocal or virilocal (Murdock 1949, Mair 1972, Radcliffe-Brown and Forde 1950).

For present purposes we need only note that marriage is a *corporate* process, operating between families or lineages rather than between individuals. This is one of the features that marks off marriage from mating. *Individual* men and women find mating partners in the community in which they live both before marriage and after marriage, and any given man or woman may have dozens or hundreds of mating partners in a lifetime. By contrast, the same individual may not marry at all or may do so once or a few times. Mating is ordinarily a part of marriage though not necessarily so, as in the case of the Nayars of southern India, where very young girls were ceremonially married to warriors but never had sex with them (Gough 1959; for a few other marginal cases, see Murdock 1949), or in the small number of cases in our own society and others in which a marriage takes place but is never consummated. These are interesting because there are rules in some societies about the legitimate dissolution of a marriage if one partner fails to consummate it by the act of sexual intercourse (Reynolds and Tanner 1983).

Marriage is therefore something very different from mating and not to be confused with it. Why then marriage? We have already argued that kinship systems are, in part, systems of knowledge. Affinal kin add to the network of people with whom one is involved in terms of specific rights and obligations, and in so doing they widen the field of relationships, adding to social knowledge. So much is this the case that in traditional village communities individuals may be related by descent or affinity to almost everyone else in the community, and in some inbred cases they may be related to each other in several ways.

This situation is in marked contrast with that which prevails in a monkey group, where matrilines, even though they can act together (e.g., in group defense), are more generally in a state of competition with each other and are ranked relative to each other. Recall the process by which an individual relies on matrikin to determine his or her status. The outcome of intralineage solidarity is to homogenize the status of lineage members, marking them off clearly as a group from the members of other lineages. In this situation, competition between lineages persists. By contrast, marriage reduces the competition between lineages and has often been (and still is) used to cement alliances between them.

We should not forget that competition between human lineages can be fierce. The institution of the feud is an example. In common parlance one talks of family feuds. Many English villages, for example, have families who will not talk to each other; each family is solidary, each revels in malevolent stories about the other, and marriage between the two is frowned on and consequently rare. But feuds can be much more devastating than this. In the early history of Iceland, as related in historical accounts such as Njal's saga (Magnusson and Palsson 1960), insults or bad behavior by a member of one family to a member of another could, if unchecked, lead to violent confrontations and revenge killings, reprisals being taken by kin of the losing party over lengthy periods of time. In this acephalous social system, kin reciprocity and support was the basis of social order.

Reverting to the situation in primates, kin solidarity is well attested. That formation of aggressive coalitions is determined in part by kinship has been especially well documented in macaque species, though we have evidence from langurs as well. Matriline conflicts in macaques have, for example, been described for *Macaca mulatta* (Kaplan 1977, 1978; Chapais 1983), *M. fuscata* (Kurland 1977), and *M. radiata* (Silk 1982). The evidence is overwhelming for macaques, and conclusions probably can be generalized to include most species in which females remain in their natal groups and take sides in quarrels.

There thus seems to be good evidence of lineage competition in both the monkey and the human cases. A number of social anthropologists have emphasized that marriage is a potent method for bringing about alliances between potentially inimical lineages, a point that Mair (1965) nicely expressed when she wrote that the people anthropologists study often say "we marry those with whom we fight" (p. 87).

GROUP SELECTION

If we conclude that humans developed the institution of marriage in order to reduce interlineage conflict, then we are moving in the direction of a group-selectionist model of human social evolution, and this requires justification. The problem is this: given that competition normally takes place between individuals and that kin solidarity does not challenge this because kin are genetically closer than nonkin, how can we explain interlineage solidarity when by definition different lineages are genetically different?[1] The group-selection answer is put in terms of group solidarity, that is, of competing groups, each group consisting of a number of lineages. Any group that has developed mechanisms between its lineages so that they are able to support each other will have advantages in fighting against groups in which lineages are opposed to each other. The latter will tend to fragment under pressure from larger, more solidary groups; will be driven from the best habitats; and their members will have reduced fitness as a consequence.

This is not unreasonable. It has been argued (e.g., Alexander 1987) that the explanation for the evolution of human institutions of many kinds is the pressure on whole groups of the practice of raiding, fighting, and warfare, which led to the evolution of mechanisms of group cohesion and group survival and brought about a reduction of individual and lineage competitiveness in the human case in favor of sacrifice and altruism for the group.

In fact, the extent of group action as against individual or lineage action is very much a response to socioecological constraints. At times when lineage competition is not of any consequence, individuals can be expected to behave in a self-interested way. When lineages come into competition with each other, we can expect to find lineage solidarity, and when groups come into conflict, we can expect lineage differences to be swamped in favor of corporate group action. These processes have all been documented in nonhuman primates. The last, for instance, was documented by Southwick, Beg, and Siddiqi (1965) at a temple site in north India. At this site there were a number of rhesus monkey groups. If two groups came into contact, severe intergroup fights occurred. Twenty-four such fights were seen in eighty-five days of observation. Individuals were frequently wounded in these fights.

The existence of group solidarity, such a prominent feature of human tribal groups, is thus on a continuum with the group solidarity seen in primates, even though primate groups are not endogamous in the way human tribes are. The rhesus example given above is not atypical but commonplace (e.g., Holloway 1974). Fighting between groups has been reported in a great many species. Normally, but not always, the context is competition for a valued territory or other scarce resource. The group conflict that occurred between the Kasakela and Kahama chimpanzees at Gombe, in which the males of the Kasakela group over a period of months exterminated the males and even a female of the Kahama group, was not clearly about the control of resources but may have had social origins, since the Kahama group was in fact a splinter group that had removed some females from the Kasakela group not long before the fighting started (Goodall 1986).

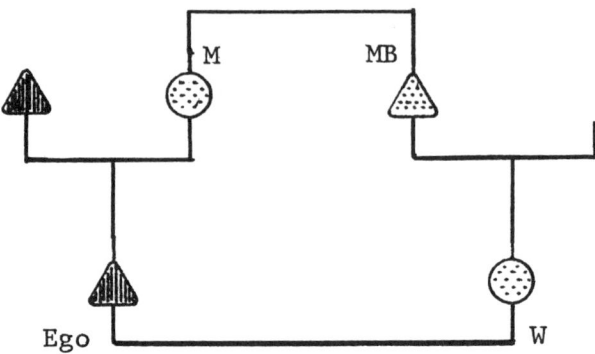

Figure 5.4 Matrilineal cross-cousin marriage. Key: Ego = the marrying male, W = the woman he marries, M = mother, MB = mother's brother. Lineages are patrilineal.

I have emphasized macaque or langur intergroup conflict as an example of groups acting together *as groups*, ignoring interlineage differences in favor of joint corporate action. Thus any argument that marriage arose as a way of uniting lineages in order to strengthen group action must take account that primates can act as a group without marriage links between kin lines. Group conflict is thus unlikely to provide more than a part of the explanation for the emergence of institutions linking up human lineages. Let us look at another kind of argument for the emergence of marriage, this time in terms of recruitment.

RECRUITMENT

Lévi-Strauss (1949) pointed out that lineages very often associate with particular other lineages, establishing mutual exchange of marriage partners, and he showed how the common institution of cross-cousin marriage could achieve this (see Figure 5.4)

The exchange portrayed in Figure 5.4 was, Lévi-Strauss assumed, following Emile Durkheim's much earlier insights about social solidarity, a primeval breakthrough in the organizing of society, associated with the elimination of incest by the introduction of systematic exogamy. The emphasis in that explanation is not so much on group selection in a context of intergroup hostilities as on the organization of intragroup mating partners into a sustainable pattern of long-term reciprocal relationships.

Something similar to this has in fact been reported in nonhuman primates. There is evidence of selective mating between particular matrilines and particular patrilines. This comes from a study by Smith and Small (1987), who studied mating preferences in three captive groups of rhesus monkeys at the California Primate Research Center. These groups were separately housed in field cages (0.2 ha), so that the normal pattern of male emigration seen in wild rhesus monkeys could not occur. As a result, each group contained both patrilines and matrilines. Smith and Small were able to show that females from particular matrilines chose males from particular patrilines as mates more frequently than chance would predict, and this was true in each of the three groups studied. The results were not an artifact of the age or sex constitution of the groups concerned but were the outcome of selective favoring of males from particular patrilines by females from particular matrilines.

In the wild, patrilines are not co-residential, owing to the pattern of male emigration at puberty. Females can thus only favor males from particular patrilines if they are able to detect patrilineality and if appropriate males are available, that is if they come into the female's group. Nevertheless, it has been suggested that matrilines on Cayo Santiago, where male emigration occurs, demonstrate a higher degree of genetic homogeneity than would be expected from random mating with incoming males (McMillan and Duggleby 1981). Thus it may be that even in free-ranging conditions females are selective about the lineage of the males they mate with. Perhaps, indeed, they prefer to mate with males from particular other *matrilines*. Either way (i.e., matriline or patriline preference) the result would be an increase in genetic homogeneity of the female's patriline over levels arising from random mating or mating with *any* male who happened to enter the group.

These nonhuman primate data are of interest for a number of reasons. The benefits to matrilines from this kind of selective mating as seen by Smith and Small (1987) are in terms of lineage solidarity arising out of increased kin selection, which is in turn a result of the increased genetic homogeneity of the lineage. An increase of genetic homogeneity of the same kind would inevitably result from the human institution of spouse exchange between particular lineages, whether matrilineal or patrilineal. The question arises, however, whether this resulting increase in genetic homogeneity would be reflected in an increase in cooperative, kin-selected behavior in the

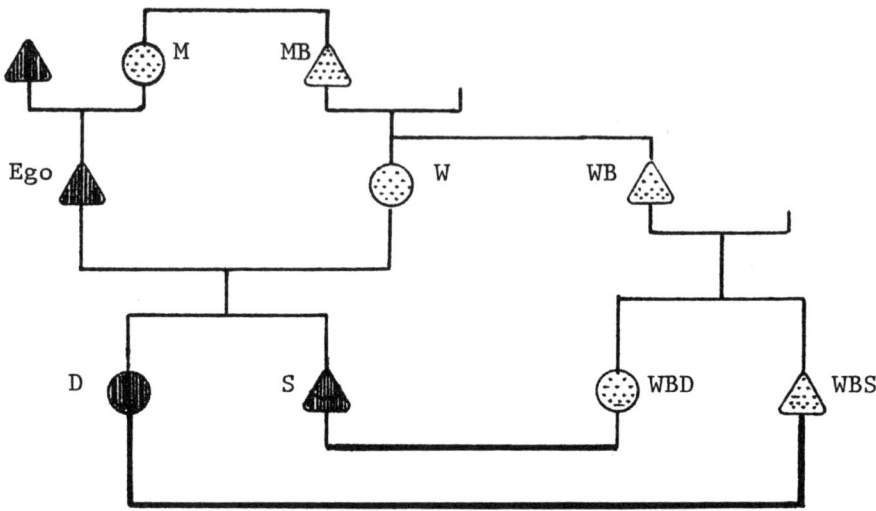

Figure 5.5 Wife exchange between two lineages. Key: as in earlier figures with, additionally: D = Ego's daughter, S = Ego's son, WB = Ego's wife's brother, WBD = Ego's wife's brother's daughter, WBS = Ego's wife's brother's son.

human case (such a question could also be asked of the primate case). In social insects there is no doubt that genetic homogeneity leads to cooperation, and we can speak of kin selection. In primates the situation is more complex because kinship behavior is a product of closeness during development rather than shared genes. Nevertheless, whatever its basis, kinship and lineality do seem to be rather fundamental features of social life in insects, primates, and humans, and it does seem that the preference for mating between particular primate lineages constitutes a parallel with the human situation in which certain lineages exchange spouses. Whatever the origins or causes of these situations, and even if they are rather different, there may be functions in terms of lineage solidarity to such arrangements, and a further function that we shall now explore, namely, continuity of lineage recruitment resulting from privileged access to particular mating or marriage partners.

The way in which a cross-cousin marriage system (ideally) works over the generations is shown in Figure 5.5.

The important thing to notice about Figure 5.5 is the bold lines connecting D and WBS and connecting S and WBD. These lines indicate the reciprocal nature of cross-cousin marriage in terms of the exchange of wives between lineages. In one and the same generation, the son of the lefthand lineage is marrying the daughter of the righthand lineage, whereas the former's sister is marrying the latter's brother. This exchange is sustainable down the generations. It can take the form of symmetrical, bilateral cross-cousin marriage. Or it can be asymmetrical, patrilateral cross-cousin marriage as illustrated above. Or it can be asymmetrical, matrilateral cross-cousin marriage. Whatever the particular circumstances — and these depend on the system of descent of the society as a whole — each lineage involved in cross-cousin marriage stands to gain from the arrangement, quite apart from (and possibly in opposition to) any gains to group solidarity. There is no reason why the group should gain as a whole. These two lineages may be a pair of aristocratic lineages that are keeping more than their fair share of the group's resources to themselves. They might in fact be impoverishing the whole of the rest of the group. We cannot know in any particular case unless we investigate the facts. The process of wife exchange between lineages can, however, provide a structural basis for social solidarity at the group level.

The exchange of women back and forth between a pair of lineages tends to be ideal rather than actual. Demographic factors at any given time may make it difficult to find an exactly appropriate partner for any given man, and so a well-known phenomenon comes into play, the fudging of kinship to allow the rules to be followed (Hiatt 1968, Chagnon 1988). Lévi-Strauss was aware of this when he wrote that rules have a lives of their own, in other words, they are at times honored in the breach rather than in the observance. This is a very interesting point for our analysis. When, for instance, a Yanomamo man is looking for a spouse and has no cross-cousin, though the rules demand one, then a man will consider women of appropriate or younger age who fall close to the desired relationship and, by devious twisting and reinterpretations of relationships between himself and the woman, will conclude that she is indeed a cross-cousin (Chagnon 1988).

In his account of the Yanomamo, Chagnon (1988) showed how detailed kinship knowledge was. Local groups were saturated with kinship links and ties, as well as oppositions and rivalries. Chagnon was interested in sociobiological interpretations, focusing on the way that "big men" could accumulate wives and thus increase their reproductive success by manipulating kinship links. From our perspective this is a case of manipulating information, restructuring social knowledge. Once a kinship system has come into being, individuals who can learn its rules and master its complexities are in a position to use it for their own ends. Our analysis of kinship systems as systems of knowledge lends itself to the idea that clever individuals (or parents, in the case of arranged marriages) will follow marriage rules that are of benefit to them but bend the rules if conformity threatens to leave them without a marriage partner.

If we are right in interpreting the evolution of rules of cross-cousin marriage as a solution to the problems of lineage recruitment and lineage survival, then such rule bending makes very good sense. Although the chosen spouse is not technically or biologically a cross-cousin, she is nominally one, and lineage satisfaction is assured on both sides; the game can go on. We are here looking at the history of social structure; we are seeing how the rules go on and the structure goes on, changing slowly over the generations as a result of progressive deviations from previous norms. Within this process, individual human beings achieve more or less reproductive success, but bending the rules is clearly a different process from that of cultural change. Bending the rules may or may not produce cultural change, and cultural change involves the transmission and survival of the marriage rules themselves. In providing each other with spouses, lineages achieve a reliable source of recruitment of future personnel, fudging is allowed for emergencies, and the lineages go on.

From the time that an arrangement such as cross-cousin marriage is instituted between two lineages, it is in the interests of each lineage to look after the welfare of the other, because each is responsible for providing the other with the wherewithal to continue its line. Cross-cousin marriage seen in this way is a product of biological evolution. It is the result of a cognitive structuring of society that has brought with it selective advantages to the members of the lineages engaging in it. Murdock (1949), in his survey of the world's cultures, found that "under unilinear descent . . . cross-cousins . . . are the commonest objects of preferred primary marriages" (p. 320). The arrangements ensure reproductive opportunity for most lineage members, even if they do not guarantee reproductive success. There will still be competition among individuals, especially in societies where polygyny is permitted, so that despite lineage solidarity and marriage links between lineages, some individuals will still fail to obtain a spouse, and some will be infertile. To this extent, the rules and structures of kinship systems and systems of affinity, although human constructs, themselves provide the setting in which competition, now conventionalized, takes place and in which differential reproductive success is achieved. It is in this sense that we can talk of humans living in a world of their own making, but also in one in which (at least until the advent of modern, affluent, contracepting societies) the normal processes of natural selection take place.

Figure 5.5 presents a two-lineage linkup. In societies with moiety organization, this may be all there is. Individuals belong to one of two clans, and the clans are exogamous. This kind of organization is not common, however. More often there are many lineages or clans in a society. Depending on how many there are, and precisely what kind of symmetrical or asymmetrical marriage links there are between them, the society may be closely knit as a unity, or there may be divisions such as those of class or caste. Models of interlinking marriage systems can be proposed and compared with actual situations. In a circular model, if lineage A exchanges spouses with lineage B, B with C, C with D, and so on, and in the end the last lineage, N, exchanges back with A, then a perfect form of social cohesion of the entire group will result. In real life, such systems are inevitably imperfect in operation.

NAMES

I have written so far about lineages and the importance of lineage recruitment without saying much about what human lineages actually do or what they are for. A lineage is a descent group; it has to do with the matter of descent down the generations. Let us look at the situation in our own society. Here, each descent group has a family name. This name is passed from father to children, that is, patrilineally. Property, however, is inherited from both parents equally. Thus, rules of inheritance are bilineal. Some property, notably rings and other jewelry, are often transmitted from grandmother to mother to daughter, that is, matrilineally. We can thus see from our own society that transmission of intangibles and of tangible objects is a complex matter, and social anthropologists have shown that the ways in which the rules governing how all the items — statuses, titles, honorific roles, rights, duties, and physical objects — are transmitted from generation to generation vary from society to society.

In our society and in many others, the name of the family is in some respects its "identity" and as such is the nearest thing to its real "essence." To besmirch a family name may be a punishable offense or one that can lead to feuding. The reason is surely that the name is a very powerful evocation, encompassing all the family members, past and present, at home and far away. In some respects, the name *is* the family. This intangible marker is the lineage indicator, and its mode of transmission marks off patrilineal from matrilineal descent groups. The extent to which it, or its equivalent, matters to the lineage members and to others is a measure of its importance, and of the importance of matters of lineage and descent in the society concerned. In our own case, its significance is somewhat less than in most societies in the less-developed parts of the world, where much of the economic life of the people is organized along kinship lines. In some societies a person is to a great extent defined by the kinship position he or she occupies, though this is a very complex issue (Carrithers 1985). In Iceland a person has two names, a first or individual name and a second name that states of whom he (she) is the son (daughter). Thus Magnus Magnusson, the well-known Icelandic author and TV personality, is Magnus, son of Magnus; had he a sister called Gudrun it is likely that she would be called Gudrun Magnusdottir, though she might later choose to take another name — her mother's last name, for instance, or her mother's first name plus *dottir*.

KINSHIP AND SOCIAL STRUCTURE

The study of kinship systems, whether in humans or nonhumans, is a study at the level of social structure. This is an important point because it introduces into the discussion supraindividual processes, going beyond single-lineage considerations to considerations of the whole group. Our understanding of the nature of human social structure is that it consists of institutions that are in turn the products of human inventiveness and are transmitted down the generations as systems of knowledge, the knowledge of individuals being subsets in the domain of social information available generally in the society.

The mechanics of ranking of individual members of monkey matrilines show how the outcome consists of a series of ranked matrilines in species such as macaques. Intramatriline cooperation and even altruism can be understood on the basis of the close genetic relationships among relatives in the lineage. By contrast, the mechanics of human lineages involve the actions of individuals toward each other, which are to some extent based on the rights and obligations vested in their roles. Nevertheless, there is a considerable degree of overlap at the biological level between primate and human lineages and the relationships within them; in the human case the biological relationships are overlaid with sets of jural rules. A preoccupation with kinship is characteristic of both primates and humans.

Moving on to a consideration of the exclusively human institution of marriage, this and the consequent incorporation of the father (*pater*) into the family have sometimes been seen as the most significant steps in human social evolution. There are similarities and differences between mating arrangements in primates and human marriage. Even though the latter creates a jural, not a biological, relationship, there are overlaps between the two in terms of mate selection and care of offspring. The main function that marriage seems to be performing is in the linking up of lineages, and its primary function at the level of social structure may well be the organization of an efficient system of recruitment to lineages. The corporate nature of the marriage process is noteworthy, so that in smaller societies the whole group can be unified by marriage links between lineages, reducing the dangers of group disintegration because of lineage rivalries. The main function of marriage, however, appears to be its advantages for lineage recruitment, and lineage competition rather than group unity is the basis for marriage systems, in particular cross-cousin marriage, which can be seen as a very effective mode of alliance between competing lineages.

How and why did marriage, now universal in humans, come about? The answer given by Lévi-Strauss (1949) was that it would eliminate incest and achieve a new level of social integration. These are not adequate biological explanations. Incest is already eliminated in all known species of mammals by a great variety of patterns of out-mating coupled with innate sexual avoidance of close kin. The improvement in social integration, while a self-sufficient answer in sociological terms, is no explanation in biological terms, since it lacks the necessary indication of how marriage would improve the reproductive success of marrying individuals or kin lines as against those that did not marry. If, however, it were the case that individuals and lineages that set up marriage arrangements between each other were then assured of a more regular supply of mates than individuals or lineages that failed to join the "marriage game," this would cause the spread of marriage.

CONCLUSIONS

Nonhuman society is very much the outcome of competitive strategies between individuals and, at a higher level, between lineages. For mating, an individual (normally but not always a male) migrates to another group, where he carefully selects a mating partner. The female he selects will be a member of a matrilineage, and the reproductive success he can expect will be related to the seniority of her lineage. An incoming male spends some time observing the status relationships between males and between females, along with the already established male-female partnerships in the new group, before he attempts to form a relationship with a female (Smuts 1985). This preoccupation with kinship and status is also very characteristic of human societies, especially traditional societies but also in Western ones. The status of the kin line into which a human being marries will to some greater or lesser extent determine her or his access to scarce resources. Much has been made in the past of the difference between marriage and mating, but as Reynolds (1991) has shown, there is an essential continuity. The particular advantage of some common kinds of marriage, and especially cross-cousin marriage, appears to be that it serves to enhance the reproductive success of the individuals in the marrying lineages over time. This advantage is posited here as a theoretical construct and needs empirical investigation. We also need to investigate why cross-cousin marriage is not a feature of some societies, and why indeed marriage itself is in decline in Westernized societies.

Marriage has been a focus in this chapter because it raises sharply the question of a possible discontinuity between human and nonhuman primates. I hope to have shown how marriage can be seen as a logical extension of nonhuman primate culture. Lineages themselves do not, I think, present a problem of discontinuity. The idea that they do is based on a series of misconceptions, of which the most serious is that humans themselves invented society in all its forms from scratch at some vanished point in history. That kind of thinking may have done for the ancient Greeks or medieval clerics, but it will not do for anthropologists, social or otherwise, at the present time.

NOTE

1. N. Allen (personal communication) has pointed out that for social anthropologists there is some ambiguity here in the word *kin*. Bilateral kin are closer to Ego genetically than nonkin. But lineages are unilineal. MZD belongs to the matrilineage, MBD does not, though they are genetically equidistant kin. Nevertheless, lineages defined collectively do have greater within-group genetic kinship than exists between lineages.

PART III
Cognition and Culture

Introduction

In Part 2 the contributors explored the problems of comparing our own behavior with that of our near relative, the chimpanzee; cultural diversity in chimpanzees and bonobos; the physical bases of ape and hominid behavior; and kinship in human and nonhuman primates. Each chapter provided an introduction to some aspect of the problem of comparing culture across primate species, comparison that we argued in our introduction to Part 2 is indispensable if we are to understand how hominid culture evolved.

In this section we focus on primate and human cognition — on self-knowledge and knowledge of the minds of others, on simulation and on dissimilation in the evolution of meaning, on the contribution of "ape language" studies to our understanding of human language origins, and on the question of how tool-using and toolmaking may have been related to language evolution.

We proposed earlier that culture be viewed as the mechanisms and processes by which information is socially mediated.[1] If culture is defined in this way, then discussion of *culture in evolution* should include and probably should begin with discussion of the evolution of cognition. Readers interested in the nuts and bolts of primate cognition from a psychological as well as a comparative evolutionary standpoint or in the neuroanatomical bases of individual (primate) cognition may wish to consult, respectively, Fobes and King (1982) and Martin (1990). A stimulating treatment of social cognition in vervet monkeys, with an unusually thorough and illuminating discussion of methods and procedures employed in the source study, a series of elegant, richly informative field experiments, can be found in Cheney and Seyfarth (1990). Cheney and Seyfarth's view of culture as *social knowledge* is in many respects similar to ours (see also Smuts et al. 1987). Finally, broader overviews of primate cognition in its relation to cultural evolution are presented by McGrew (1992) and Quiatt and Reynolds (1992).

The study of primate cognition in relation to cultural evolution has been greatly stimulated in recent years by attention to two problems: What do we mean when we say that behavior is intended? And what do we mean by *social intelligence* or *social knowledge* — and how is "social knowledge" different (if it *is* different) from "object knowledge," that is, knowledge about material structure and object relationships? These problems are connected in that they direct analytic attention to the kind of reflexive consciousness that human beings possess and to behavior that is self-conscious in the sense that an actor perceives herself or himself to *be* an actor capable of adapting roles to circumstances. Roles so perceived may be altered or manipulated in potentially rewarding social interactions *depending on the actor's expectations* of which behaviors are likely to be rewarded by goaltenders or cooperating companions.

Most if not all of us probably take it for granted that human behavior is at some level intentional. We may perceive our own actions and those of others as determined in part or in whole by some combination of genetic endowment, life history, social constraints, and learning; nevertheless, we usually have the sense of being in control (or, sometimes and disturbingly, very much out of control) of our own individual decisionmaking. Similarly, and from the standpoint of cultural evolution at least as important, we — or most of us — know better than to presume that we can manipulate other people like objects. We have discovered that however hard we work at coercing or seducing others into doing not just what we want them to do but what any reasonable man or woman *ought* to want to do, there still come times when some of the people with whom we interact change their minds and, having had second thoughts or simply on a whim, adopt a new agenda and new goals, leaving our own agenda, goals, and assiduous concern for their welfare entirely out of account.

Suppose we discover that a friend or business acquaintance with whom we thought we shared an agenda appears to have acquired interests that are at odds with our own. If the friend has put off informing us of these new interests, meanwhile continuing to interact with us in the manner to which we have become accustomed, and especially if we perceive that silence has been to the friend's advantage and not at all to our own, very likely we will feel deceived — and with good reason, for this is a definitional instance of deception.

The question of advantage need not be raised in every discussion of deception; after all, human beings (and, for all we know, other animals, too) practice many small deceptions out of tact or to ensure gratification, as, for instance, when planning a surprise birthday party. But to observers of animal behavior, advantage gained by deceivers is of interest both theoretically and from the standpoint of method, for the issue of deception by nonhuman animals is closely linked to that of animal "thinking" and the reflexive awareness that many cognitive scientists assume underlies intentional behavior.

Consider for a moment the many kinds of behavior that constitute animal mimicry, ranging from the color change that enables a chameleon to blend in with a background of like color to the broken-wing display by which certain birds distract intruders who threaten to discover their nestlings. The great majority of such behaviors involve responses that are clearly designed to conceal information from other animals, almost always prey or predators, yet do not appear to be under conscious control of the actor (think of the mantis, shaped by evolution, not meditation, to resemble a twig or leaf stem). It is convenient, therefore, to distinguish these stereotypic behaviors from the more labile routines or "tactical deceptions" (Byrne and Whiten 1985) by which animals of many species manipulate conspecifics.

Tactical deception has received much attention of late from students of animal communication (for reviews, see Byrne and Whiten 1988a and Whiten and Byrne 1988b), because responses by the target animal(s) to signals sent and subsequent disjunctions in the behavior of the signaling actor provide strong evidence in cases of deception that "true" intentions are masked by the actor's signals. Intentions of course are evidenced only in actions, and how conscious communicating animals may be of the discrepancy between the personal goals their signals are designed to achieve and the message receivers interpret is difficult to establish. In the examination of underlying intentions as nowhere else in the study of behavior does so much depend on the human observer's interpretation of events. Nevertheless, it appears that nonhuman animals do regulate their behavior so as to deceive one another, and animals of some species other than our own (chimpanzees, bonobos, orangutans, and probably gorillas) appear in tests of mirror self-recognition (reviewed by Gallup 1977, 1987, and Mitchell 1993c) to exhibit a reflexive awareness of themselves as behaving individuals. To the extent that animals of such species exercise that same reflexive self-awareness when monitoring their interactions with others, especially where there is good reason to infer that they attribute intentional states to those others (Byrne and Whiten 1992), it will be convenient in describing social behavior to introduce intentional terms and refer to, for example, "role *selection*" and "behavioral *performance*."

Deception, as we have noted, is important and may be crucial to observer verification of intention in animal behavior — tool behavior is another area in which it would seem that sequences of manifestly purposive behavior might be systematically examined from an "intentionalist stance" (Dennett 1983). But deception of course is not the only simulative activity in which animals engage. Simulation, whether or not it involves deception of a playmate, appears to be as important in the play of Old World monkeys and apes as in that of human children, and deception may be important to the construction and communication of meaning. In Chapters 6 and 7, respectively, Robert Mitchell and Jo Liska focus on these broader, perhaps more "positive" aspects of simulation, which they see as primary in the evolution of cognition and culture.

Mitchell's particular interest lies in how an animal's knowledge of resemblances in nature may be represented in imitation and pretend play, deception, and mirror recognition. Pretense and make-believe appear to be rooted in a capacity for recognizing resemblances and creating "particular forms of similarity." We are biologically prepared, says Mitchell, to notice perceptual similarity. Internal simulation of the external world, that is, a capacity for recognizing and *producing* resemblances, is a primate adaptation upon which the human mind is based. Mitchell develops this argument in detail, providing a cogent review of theory concerning art, pretense, play, and language as communication and/or metacommunication and drawing from different species (and, within species, from different age and sex perspectives) numerous examples to illustrate the ways in which normal development of psychological processes is contingent on the ability to recognize resemblance. He further shows how encompassing and deeply rooted in primate behavior is the human capacity for recognizing and producing resemblances, treating the evolution of that capacity and its variable development in species related to ours from a psychologist's standpoint and in comparative terms.

Liska, who has conducted extensive research in human communication and on the problem of language origins, takes a different tack. Liska, like Mitchell, is interested in the contributions of simulation and pretense to the evolution of meaning and, ultimately, language. However, she centers her attention on the cultural evolution of what she calls "syntactic symbols," which in her view are essential to a representation of "counterfactual relationships," that in turn "make imagination, planning, and speculation not only possible but communicable."

Syntactic symbols are the last (or latest) product in a cognitive evolutionary continuum that begins with symptoms, that is, symptomatic signs bearing "a natural, functional, or physiological relationship to their significates." Symptoms give rise to semblances of two sorts, ritual and iconic, both of which bear a looser relationship to their significates than do symptoms, and semblances in turn give rise to wholly arbitrary symbols, also of two sorts, proper and syntactic (see Figure 7.1 and accompanying text). Liska derives her conception of what she terms the "symbolicity continuum" from a set of generally accepted ideas as to how various sorts of signs and symbols might be usefully defined and distinguished from one another. She assembles these ideas — parts of theories — into a whole theory that has explanatory potential as well as classificatory utility. The specific evolutionary relationships that she hypothesizes need to be tested via cross-specific comparative studies and, where feasible, experimental investigations. Meanwhile, the symbolicity continuum provides us with an elegant model of prelinguistic culture, culminating in the appearance of syntactic symbols. It encourages us in our conviction that the right way to view the relation between culture and language is not to see culture as dependent on the evolution of symbols and language but to think of language as dependent on culture.

Mitchell's and Liska's concerns are parallel and complementary. Both are interested in the relation between sign and referent, in the difference between representation and symbolization, in the evolution of iconic representations, and in conventionality and arbitrariness as elements in nonhuman primate symbolization (e.g., in vervet monkey alarm calls). Both examine the evolution of cognition in its relation to the evolution of culture, that is, with attention to changes in the dynamic relation between social interaction and information processing in all of its aspects, including input, storage, and transfer. The main difference is that Mitchell puts a strong emphasis on the nature and act of *intraindividual representations* of external reality, whereas Liska focuses primarily on *interindividual communication*.

The intelligence that allows human beings not just to see the world through the eyes of others but to compare notes and talk about what they observe, organizing their observations and their talk (in effect, shared worlds) in particular cultural patterns — that intelligence appears to be well grounded in nonhuman primate behavior and, according to Humphrey (1976), in social behavior rather than object behavior. Whether or not social and object behavior differ significantly at the neural level remains to be established.

That neural reorganization is associated with human language ability appears to be established beyond doubt, with enough questions open to leave room for argument as to when and in what form, gestural or vocal, human beings first began to exercise that ability and to what extent it may be shared by our closest relatives. Over the years, beginning with the Gardners' methodologically innovative program of teaching American Sign Language (ASL) to the chimpanzee Washoe (Gardner and Gardner 1969), there have been numerous experimental investigations of the ability of bonobos, chimpanzees, gorillas, and orangutans to learn to comprehend and use, in interaction with their trainers, gestural and graphic sign systems of one sort or another. Such investigations are often lumped collectively under the label of "ape language" studies, which we hasten to point out is a misnomer in at least two ways: there is of course no language common to ape species, and all studies have involved teaching human-devised systems of task-oriented communication to ape subjects, not investigating natural systems of intraspecific communication. Nevertheless, the term is a convenient one and, as long as we keep such reservations firmly in mind, seems adequately descriptive. In Chapter 8 H. Lyn White Miles and Stephen E. Harper discuss the history of these experimental sign-centered programs of instruction (including Miles's own Project Chantek), review the ways in which conceptual and methodological problems have complicated research and interpretation of findings, and list the contributions of "ape language" studies to current understanding of the evolution of human language and culture. Their chapter, authoritative and broadly informative, takes us on a state-of-the-art survey that they have organized so as to inform us in particular of the many ways in which ape language studies can contribute (and indeed have contributed) to more reliable reconstructions of the communicative and underlying cognitive behavior of early hominids.

But what of "object intelligence"? In Chapter 9, the last chapter in Part 3, Tim Ingold reviews the arguments that link hominid technology with the rise of language. A social anthropologist who has published extensively on technology and human evolution, Ingold is perhaps the foremost explicator of connections between technological practices and social process from prehistoric times to the present.

Ingold observes that a causal "link between language and technology . . . has often enough been suggested but remains to be positively demonstrated." He examines the issue from six standpoints, adopting in each case a "potentially falsifiable" hypothesis for which "there corresponds a counterhypothesis that would deny the posited connection." The examination is rigorous and comprehensive; framed in plain terms, it is also fascinating to follow. In the end Ingold concludes that "we are still far from achieving a coherent synthesis," partly for want of data, partly because of the inherent difficulty of testing hypotheses about past behavior, but equally because the problem is one that we ourselves have constructed in our disagreement over what it is we mean by *technology* and *language*, the terms that we are trying to connect.

McGrew has suggested that while the tool kits of free-ranging chimpanzees and modern hunter-gatherers may differ in certain features, "the difference is far from wide, and the gap between [ancestral] hominid and pongid is narrow and bridgeable" (McGrew 1987, p.256, quoted by Ingold). To this observation Ingold adds the cautionary reminder, "However, the gap between the modern human and the chimpanzee is not." To understand the nature of human culture and how it differs from the culture(s) of other animals it is important to remember that other animals, including chimpanzees, employ neither a communication system as versatile and as capacious as our own in its processing features and storage routines nor a technology similarly versatile and capacious in functions like these. Yet the gap to which McGrew referred appears to have remained "narrow and bridgeable" (somehow, in the way that we conceptualize it — it is hard to think what else McGrew could have had in mind) for millions of years after the appearance of the earliest hominids. Then, between approximately 120,000 and 30,000 years ago (quite suddenly in evolutionary time) the gap widened. The sudden proliferation and distribution of technological variations that mark this acceleration in the evolution of human culture suggest that changes in language might have been involved. What those changes could have been and how they might have influenced the culture and biological evolution of hominid species at this juncture has been much debated; it is the subject of the chapter by Milo and Quiatt that constitutes the fourth and final part of our book.

NOTE

1. Culture also may refer, of course, to the artifactual product of that mediation, which in material form (as telephone lines, videotapes, the printed word) may be built onto or in vocal or gestural form fed back into the infrastructure that supports social mediation of information.

<u>6</u>

The Evolution of Primate Cognition: Simulation, Self-Knowledge, and Knowledge of Other Minds

ROBERT W. MITCHELL

ABSTRACT

Imitation, pretend play, deception, mirror self-recognition, and communication of nonnatural meaning are five psychologically interesting activities enacted (sometimes infrequently) by nonhuman primates. All five activities suggest a rich psychological interpretation, calling to mind planning, imagination, self-awareness, and knowledge of another's perspective. These activities share a common feature of simulations in that all are (at least apparently) based upon a primate's recognizing and utilizing some knowledge of resemblance. I detail how understanding of simulation leads to pretense, and how pretense is a basic component of self-awareness, mutual knowledge, and reflective self-awareness. First I present Walton's notion of pretense in make-believe, in which representations are things with the function of being props in games of make-believe. I then discuss Davis's theory of the origins of image making, Bateson's and Grice's theories of development of intentional communication from simulation, and Werner and Kaplan's theory of symbolization based on analogical processes. Next I discuss four hierarchically related types of simulative pretense: (1) self-simulation to influence others and recognition of simulation of nonself objects and activities by others; (2) simulation of others' activities dependent upon kinesthetic-visual matching, use of props for self, and recognition of self dependent upon kinesthetic-visual matching; (3) communicative simulation and use of props for others; and (4) creation and subsequent use of props external to or on the body as representations for others as well as self. The first type of simulative pretense is found in mammals and birds; the second in monkeys (to a limited degree), apes, and humans; the third in apes and humans; and the last in humans alone. Thinking of primate psychology in terms of simulation connects a variety of phenomena previously thought distinct and directs us to look at simulation of the external world as a primate adaptation upon which the human mind is based.

INTRODUCTION

According to Wittgenstein (1921/1961),

> We picture facts to ourselves. . . . A picture is a model of reality.
> . . . A picture is a fact. The fact that the elements of a picture are
> related to one another in a determinate way represents that things
> are related to one another in the same way. . . . Pictorial form is
> the possibility that things are related to one another in the same
> way as the elements of the picture. . . . *That* is how a picture is
> attached to reality; it reaches right out to it. . . . A logical picture
> of facts is a thought. (pp. 15, 19)

A simulation is something designed to resemble something else (see Mitchell 1987, 1991a, p. 75). In this chapter I examine the idea that with recognition and creation of particular forms of simulation comes a capacity for make-believe: that type of simulation in which the actor acts *as if* one thing is another. Very interesting psychological understandings are present in any "as if" simulation, specifically, the recognition of two distinct yet related interpretations of the same phenomenon and the experience of an interior "world" in some ways independent of the outside world. I suggest that the activity of symbolizing is dependent upon capacities for recognizing and creating various forms of resemblance (see Mitchell 1991a). I articulate the significance of recognizing and creating resemblance for the development of psychological processes. My aim is to show the widespread nature of the capacity for recognizing and producing resemblance in human and nonhuman primates. I present a framework that ties together the inception of psychological capacities for planning, self-representation, knowledge of other minds, and reflective self-awareness as a result of simulation.

I begin with the notion of representation. This term is used quite differently in various disciplines; I (initially) focus on three. In its most general sense, "representation includes both somatic and extrasomatic modes of acquiring, preserving, and manipulating knowledge" (Davis 1986, p. 193). For example, visual perception is a form of (somatic) representation, as are the "mental representations" or "internal schemas" of modern psychology and the "ideas" of past philosophers (Mandler 1983). Two more restrictive senses delineate types of extrasomatic representations, or artifacts. In the first, representations refer to images, where "an image refers to its object iconically or nonarbitrarily" (Davis 1986, p. 193). In the second, representations are "things possessing the social function of serving as props in games of make-believe" (Walton 1990, p. 69); that is, representations are (often) art objects such as music, films, novels, and paintings, as well as toys or other objects used in pretense. Although these two types of representation seem strange bedfellows, I show how the latter form of representation is generated by the development of the former.

I do not propose to provide a cohesive understanding of differing uses of "representation." Clearly, many art objects are not images, and many images are not art objects. Instead, I start with and elaborate upon Walton's notions of representation, make-believe, props, and fictionality, and try to show how these ideas are related to representation as image making. From there, I present Davis's model of the evolution of image making, the theories of Bateson and Grice to describe how nonhumans could use simulation to communicate, and the theory of Werner and Kaplan to detail how symbols are based upon simulation. I then describe the evolution of pretense via simulation in primates and detail the psychological consequences. As Davis (1986, p. 211) remarks, "We certainly need some . . . psychologically and evolutionarily plausible account of 'fictive acts of perceiving.' " I attempt to supply such an account based on simulation.

REPRESENTATION AS MAKE-BELIEVE

A child who imagines that the blobs of mud she fashions into pie shapes are really chocolate pies does not really believe that the mud pies are chocolate pies, but she *fictionally* believes that they are. She is engaging in "make-believe — the use of (external) props in imaginative activities" (Walton 1990, p. 67). For Walton, just as pretending is playing a game of make-believe, "appreciating paintings and novels is largely a matter of playing games of make-believe with them of the sort it is their function to be props in" (p. 53). The novel *Robinson Crusoe*, for example, is a prop that offers the reader a game of make-believe about the adventures of a man lost at sea who lands on an island. Van Gogh's *Starry Night* is a prop that offers the viewer a game of make-believe about a starry night. The viewer could, of course, engage in a game of make-believe about Van Gogh's painting such that *Starry Night* is seen as a representation of a man lost at sea; but the painting itself readily offers a more restricted game of make-believe — it "mandates imaginings" of a particular sort: those having to do with a night scene of strikingly wavy trees, whirling winds, and a fused sun and moon set atop a darkened town. In Walton's terminology, "*Representations* . . . are things possessing the social function of serving as props in games of make-believe. . . . A prop is something which . . . mandates imaginings. Propositions whose imaginings are mandated are *fictional*, and the fact that a given proposition is fictional is a *fictional truth*" (p. 69).

When imaginings about props are mandated by pretenders alone (as when, for example, children proclaim that tree stumps are bears), the props are not representations but are distinguished as "ad hoc props" because "there is nothing which appreciators of the stumps, qua appreciators of them, are obliged to imagine. The propositions fictional in the world of a work [of art] are those that are or would be fictional in worlds of games in which it is the work's function to serve as a prop. The stumps have no such function. So if there is a stump world, it is empty of fictional truths" (p. 61). Games of make-believe involve fictional truths or fictional worlds only when these worlds are prescribed or mandated by the props or by the establishment of rules about how to interpret the props (p. 40). A tree stump could be imagined to be a computer or an uprighted car or a soldier, none of which is mandated by the prop itself. But Defoe's *Robinson Crusoe* and Van Gogh's *Starry Night*, as props, mandate imaginings of a special sort, in that they "have their own fictional worlds, apart from the worlds of games played with them" (p. 61). Music also mandates imaginings. For example,

> in the case of much "expressive" music, it may be fictional not that one sees or hears or otherwise perceives external things but that one experiences or is aware of (one's own) feelings or emotions or sensations or sentiments or moods. The listener imagines experiencing excitement, passion, fervor, despair, conflict, feelings of exuberance, of striving, of determination, of well-being, of trepidation, of repose. Moreover . . . it may be fictional of one's actual awareness of auditory sensations (not one's perception of sounds) that it is an awareness of such feelings. In place of fictional perception of external objects we have fictional introspection or self-awareness. (Levinson 1990, pp. 335–336)

For whatever reason, the auditory experience of some music creates emotional experiences for listeners (Levinson 1990, pp. 324–325).

Experiences themselves, as in dreams, may be *internal* props of imaginings (Walton 1990, p. 49). In nondeliberate (spontaneous) daydreams or dreams, the imaginings are "governed by a blanket rule . . . that whatever is in fact imagined as part of the dream or daydream is to be imagined" (p. 44). Here the internal props prescribe imaginings of themselves, simply by being imagined. Fictionality derives from prescribed imaginings (no matter how prescribed) but can exist apart from any actual imaginings: "what is fictional in [a deliberate] daydream is what . . . participants in the daydream . . . are supposed to imagine; what they actually do imagine is irrelevant" (p. 44).

For Walton, make-believe is a "pervasive element of human experience," such that "works of art are neither the sole nor the primary instances of representation." As Walton said, "Make-believe may be crucially involved as well in certain religious practices, in the role of sports in our culture, in the institution of morality, in the postulation of 'theoretical entities' in science, and in other areas in which issues of metaphysical 'realism' are prominent" (p. 7). Make-believe is probably also necessary for understanding the other's point of view (Millikan 1984, p. 202; Gordon 1986). That fictive acts of perceiving are so generalized in humans as to incorporate all these forms of activity suggests that experiences of this sort are, currently anyway, based upon a distinct psychological capacity (although see

Walton 1990, p. 184, for what is involved in "divining fictional truths").

So, to recount succinctly: representations serve as (external or internal) props in games of make-believe in that they mandate imaginings. Although representations may refer to actual things (for example, Paris, Napoleon) in the real world, they very often do not. If one were to mandate imaginings about an actual object, verbal descriptions of it need not be true of it, and visual props need not resemble it (see Goodman 1968). A story about London may bear no similarity to modern-day London or Londons of the past; a stump representing a bear may not look at all like a bear. This fact has led numerous theorists to abjure any significance of resemblance for representation: one must avoid "the crude identification of the representational with the figurative" (Wollheim 1974, p. 28); one must "deny that *looking like a thing* could be a sufficient condition for *referring* to that thing" (Fodor 1981, p. 70). After all, "that a visual display happens to *resemble* some real object does not guarantee it *depicts* that object" (Davis 1987, p. 181). One is left with the conclusion, as Walton has shown, that resemblance cannot be the identifying feature of representations (Knowlton 1966, Goodman 1968).

Yet just because *all* representations are not figurative does not mean that figurativeness is not found in *some* representations. Although figurativeness cannot be used as an *absolute* guide to what is being depicted even by something that is highly realistic, it is still highly informative in most instances. The problem with worrying about sufficient conditions and guarantees for knowing what something is an image of (or even that it is an image) is that the experience of recognizing resemblance is often not based upon logical deduction or absolute, guaranteed criteria. By analogy, consider the way we come to understand a verbal representational work of art. Take, for example, Silverstein's (1964) poem "Slithergadee," in which a braggart briefly and repeatedly boasts that he will not be captured by Slithergadee but then abruptly ends his boast in midword. As Walton (1990, p. 162) noted, we all realize that Slithergadee caught the speaker, but there is really no clear evidence that he did. Indeed, as Fodor might put it, we must deny that the speaker's stopping in midword could be a sufficient condition for our *believing* that the speaker was caught — even though we are sure that he was. Many other accidents or activities might have intervened to cause the speaker to stop in midsentence. Similarly, although many other interpretations could be made for a particular pictorial representation, we recognize the ones the painter is plugging for because we're not stupid. But the intelligence that we exhibit is not based on absolute knowledge or deductive logic. According to Walton (1990) "For purposes of divining fictional truths there is no substitute for a good nose: a combination of imagination and common sense, leavened within limits by charity and informed by familiarity with the medium, genre, and representational tradition to which the work in question belongs as well as by knowledge of the outside world — all combined, of course, with sensitivity to the most subtle features of the work itself" (p. 184). The same seems true of recognizing resemblance. The problem with admonitions to avoid using figuration and something's looking like something else as *criterial* to understanding what they represent (though these admonitions may be accurate) is that they thwart our looking at how resemblance figures in understanding *some* visual representations, and in the development of a capacity for representation in Walton's sense (see also Bretherton 1984, p. 36).

REPRESENTATION AS IMAGE MAKING

In particular, many artistic representations have realistic presentation as (part of) their aim. For the "Really Really Real" exhibit at the Indianapolis Museum of Art in the 1980s, one artist had created lifelike three-dimensional sculptures of human beings that visitors, for some if not all the time they were at the exhibit, inattentively took to be real people. Once they discovered these statues were fictional, viewers were amazed at the similarity of the (now obviously) sculpted figure to potentially real people. A sculpture-woman who sat silently inspecting a manual did not refer to any actual woman, but she was clearly intended to be momentarily mistaken for an actual woman. In realistic self-portraits no attempt is made to fool the viewer, but the artist usually intends that the viewer believe that the portrait looks like (at least in part) and refers to the artist. In pictures of still life, the artist wants us to imagine the scene presented, using our knowledge of what the objects in the still life represent.

One characteristic of *systems* of representation usually taken as standards of realism is that they are inductively learnable or more easily so than other systems. . . . It seems plausible that another characteristic of realistic systems of pictorial representation is that they make better use of habits and processes of perception that we have developed for dealing with ordinary objects. . . . [This is not to] claim that there are no differences between the processes involved in perceiving objects and those needed to interpret pictures. All that is required is that certain two-dimensional [or even 3-D] systems of cues and ways of rendering space, shape, color, size, and light take better advantage of our ordinary perceptual skills than other systems. If this is so, then given the processes by which we do see objects in the world, systems that can tap these existing skills and habits will be considered relatively realistic. (Schwartz 1974, pp. 508–509)

Indeed, "realistic" representations in pictorial and theatrical arts can be mistakenly viewed as nonrepresentational (Gibson, cited in Menzel, Premack, and Woodruff 1978, p. 242; Gleitman 1990, pp. 131–132), as can nonrealistic representations (Kaplan 1979, p. 224). And realistic pictures can be used to represent something other than what they resemble (Goodman 1968), as when an image of two lumberjacks working together to support or move a felled tree is shown to be analogous to a pair of opposing muscles moving a bone (Knowlton 1966, pp. 176–177). Although recognizing a realistic image as an image of something probably utilizes "ordinary perceptual skills," possession of such skills is not enough to explain how humans came to *make* images, in the sense of both the technology of image making and the "discovery of the *representational capacity* of lines, marks, or blots of color which *need* not and often *do* not have a representational status" (Davis 1986, p. 194; see also Davis 1989b, p. 181). It is the latter process that I focus upon now. (Note that I assume the notion of similarity to be an undefined primitive — see Quine 1969.)

Davis articulated a theory of how representation — in the sense of image making — is possible. For Davis (1986, p. 196), "image making might be regarded as a *further* stage in the *continuing* evolution of the modern visual system."

For many hundreds of millennia before the emergence of image making, hominids had been making marks and recognizing

degrees of similarity . . . between them and examining things in
the object-world, including manufactured things, and recognizing degrees of similarity between them as well. . . . [Once such
marking activities are present, the development of an] image can
be derived from simple identities and similarities between marks,
things, and the chance perception of marks as things. . . . [Assume
that] because of the unpredictable possibility that information in
the structure of the light array presented to the eye is ambiguous,
a mark is taken for a thing: [for example,] a blot is taken for a
hole, a line for an edge, and so forth. . . . although the "illusion"
vanishes, and ambiguity is resolved, one has already learned that
the nearby surface mark can stand for the distant thing or quality.
The ambiguity can be investigated and the "illusion" confirmed
as such by repeating the mark and the viewing conditions as far
as possible — and by varying the mark and the conditions. . . .
The more you mark, the more likely you will be to produce at
some point a mark which potentially will be seen, by you or by
someone else, as a thing. . . . [Thus,] the emergence of representation is the predictable logical and perceptual consequence
of *the increasing elaboration of the man-made visual world.*
(Davis 1986, pp. 199–201)

Davis's theory predicts that hominids should exhibit markings suggestive of a desire to reexperience marks-as-things, and indeed

for some tracings, we *can* observe a continual *replication* of an
object-resembling mark. At Gargas, drawn vertically, a double
S-curve can be seen as the curving horns, forehead and muzzle,
and neck of a bison. Drawn horizontally, the same curve can be
seen as the horns, humped back, and rising tail or descending
hindquarters. . . . As *marks*, these elaborations are indiscernible
from what had already appeared in (hypothetically) non-depic-
tional or even non-semantic tracing. We can infer that they were,
in fact, being made *for* seeing-as, i.e., as images, because in
replication the object-resembling properties were preserved in
the transition from surface to surface or the translation from one
tool and medium to another. (Davis 1989b, p. 182)

For the image maker, "mark becomes remark" (Davis 1989b, p. 182),
a description suggestive of the idea that a mark can comment upon
a previous similar mark (see the next section).

Once the experience of seeing a mark *as* a thing (that is, of *seeing-as*) is replicated, the image maker has developed representations in Walton's (1990) sense of props for games of make-believe:

> Starting from one set of perceptual identities and similarities, internal variations lead one to the possibility of other sets — which are now experienced not *in the world* as such but only in and by the image. In such processes, a series of marks may become similar to but not identical with things in the world. . . . One may seek out [that is, create] satisfactorily identical or similar marks for these things, which then become *merely represented* things (they have *no* perceptual identity in the world). (Davis 1986, p. 200)

Thus, we might anticipate that along with image making came and developed fictive acts of *perceiving*.

If we are to understand an image semantically, we are required to know its history — how it came to be seen as an image of something.

> Because of the possibility of ambiguity, fortuitous resemblance, and the variability of seeing-as, no amount of study of the visual properties of a mark can definitively tell us whether it is an image, and precisely what it is an image of. . . . *an interpreter must know the whole chain of replication and variation of all the properties of the mark right back to the initial moment of seeing-as in his or in someone else's "psychological" experience.* . . . To understand what an image signifies, an interpreter must know exactly in what way the properties of a visual display were produced for seeing-as [otherwise he or she cannot know what it is an image *of*]. (Davis 1989b, pp. 183, 186)

Indeed, our knowledge of the referent of *any* image created by an organism is dependent upon the history of that image in the organism's ontogeny (Mitchell 1986, 1987, 1990). But note that the use of *image* here (standing for the recognition of similiarity amid difference — that is, resemblance) in no way constrains us to talk only about pictorial and sculpted (that is, visual) representations.

SIMULATIONS, SIMILARITY, AND MENTAL REPRESENTATION

Davis's account describes how one could come to experience a simulation as an image of something general. What is involved in such seeing-as? To see something A *as* something else B means, in part, that one experiences A as saliently similar to B *while* experiencing A as saliently different from B, where A and B are perceived as different types of things (Mitchell 1991a). A is not confused with B because B is a different *type* of thing from A. This typological difference could be because A is caused by something different from what causes B or because A is in a different medium from B, or because A is known to refer to B. One recognizes the image of the bison as similar to yet different from a bison; a forger knows that a forged signature is not the same as a real signature. This recognition of similarity is usually unidirectional: the image is related to the "object" of the image rather than vice versa. Recognition of the unidirectionality of a similarity may result because one thing is taken for granted as being present independent of someone's actions (for example, a bison), and the other is recognized as having been produced (for example, an image of a bison). Bidirectional recognition of simulation can create problems, as in the following: "He was outside the Waffle, staring at the three rows of Dolly Madison ice-cream cones slowly revolving in the window before a background of prisms. . . . It was a pretty unnerving display, actually; the ice cream was so obviously pink-tinted cotton and the cones themselves made of the plywood used in orange crates that you instinctively shuddered at the oral damage they could inflict" (Perelman 1960/1985, p. 453). Unidirectionality is implicitly more likely than bidirectionality in simulations (although see Bunge 1973, pp. 118–119) because the mental representation of the *object* of the image is more basic than the image — perhaps because the object is experienced earlier or because the object is somehow taken implicitly as more fundamental or primary or because the context mandates it. Thus, one bison is not an image of another, but a photograph or painting of a bison or the buffalo on a nickel is an image of a bison. But when the object and image are both human-made, reference can ricochet between them, setting up a difficulty in determining reference: a painting of a nickel with a bison on it may be an image of a nickel or an image of a bison or both. Such is the fertile ground for the ambiguity of puns (Hammond and Hughes 1978, Redfern 1984).

The simulation need not be a visual icon (nor need the object be visual): "the logical properties of depiction are quite independent of its physical medium" (Davis 1989a, p. 141). Thus, gestural simulations of visual objects or their movements, intentional mimickings of my voice, and musical renditions of emotional experiences are as much images as are cave drawings of animals (when these are recognized or experienced as similar). Indeed, nonvisual image making is myriad in humans: auditory image making is present in puns, onomatopoeia and vocal imitation, satire and irony, and verbal (spoken) deception; kinesthetic image making in self-imitation, dance, and some pretense and behavioral deception; and cross-modal image making in expressive music. If any form of imitation or re-creation (usually of something other than one's own actions, although there are exceptions) is an instance of image making, then image making is also extremely prevalent in human adults as well as infants (or *Homo imitans*, as Meltzoff [1988] has redesignated them), somewhat common in apes and dolphins and less so in monkeys (Mitchell 1986, 1987, 1990; Visalberghi and Fragaszy 1990; Whiten and Ham 1992).

Yet the sorts of simulations used in various image-making activities are clearly different, although psychologically interconnected. To understand the possible varieties of simulations, one might look to an analysis of pictures (Knowlton 1966). Any visual simulation can be described as having three parts — the elements, the spatial arrangement or pattern of the elements, and the order of connection of the elements — and these three parts can be either perceptually similar to what they represent, iconically similar, or arbitrary. For example, a map might represent cities (elements) arbitrarily as stars or circles while retaining an analogical pattern for the cities and analogical connections among the cities. A photograph, in contrast, would retain realistic elements and their pattern, but in shrinking the image and making it two-dimensional would use analogical connections among the elements. This scheme delineates three types of picture: realistic pictures, which have elements perceptually similar to what they represent; analogical pictures, which have elements iconically related to what they represent; and logical pictures, which have elements arbitrarily related to what they represent. Each of these types of picture has nine subtypes, depending upon the nature (perceptually similar, iconically similar, or arbitrary) of the pattern and connections of the elements.

This analysis of pictures can be used to understand different types of visually based mental representations. In the development of concepts, "perceptual similarity based on [mental] representations of . . . more accessible surface properties provides an initial conceptual structure that will be integrated with and differentiated into the deeper conceptual knowledge that is acquired later" (Medin and Ortony 1989, p. 182). Perceptual similarity serves as "a good heuristic for where to look for deeper properties" (Medin and Ortony 1989, p. 182). If our conceptual development *begins* with perceptual similarity as a means of clumping and organizing phenomena, such similarity acts "as a constraint on the predicates that compose our mental representations" (Medin and Ortony 1989, p. 182), such that we are likely to encode a dog as "animal" rather than as "not a turnip." Recognition of similarity, then, appears to be a relatively basic human capacity, suggesting that children's frequent overextension and metaphorical utterances (Marschark 1985, Seitz and Beilin 1987) should not be the least surprising — we are biologically prepared to notice perceptual similarity (Quine 1969). Internal simulation of the external world is a primate adaptation upon which the human mind is based (see Langer 1967; Bretherton 1984, p. 36).

Some human children are able to internalize copies of external visual and auditory stimuli: in eidetic imagery, the child "sees" and can report on an image no longer externally present (Leask, Haber, and Haber 1969); and in the "auditory loop," the child "hears" words that are no longer being said, even before the child has inner speech in the form of talking to itself (Hitch et al. 1991). Thus, internal imagery used in planning and pretense may derive from more primitive imaginal processes (Bickerton 1981, pp. 224–225; Holland 1990). Later internal speech, in which the child controls its auditory rehearsals, presumably utilizes and transforms the auditory loop (Hitch et al. 1991), and dreams apparently utilize and transform internal visual and auditory imagery.

Because visual mental representations are interconnected with auditory, tactile, and kinesthetic mental representations, the number of types of representations and interconnections is greatly expanded. Very young human infants recognize similarity between modalities such as hearing and vision or touch and vision (Marks, Hammeal, and Bornstein 1987), chimpanzees and rhesus and cebus monkeys also recognize similarity in matching from tactile to visual stimuli (Elliot 1977, Jarvis and Ettlinger 1977), and great apes and humans match between kinesthetic and visual stimuli (Meltzoff 1988, 1990; Mitchell 1987, 1992, 1993c). Humans and other primates can, then, map bits of information from the same or different modalities quite easily when the similarity between the bits is salient. The complexity of mental representations, like logical or analogical pictures, could be based on transformations of images initially derived from perceptions of the external world. Thought and intelligence can be viewed as based on simulations of the external world: transformations of perceptions, transformations of and mapping relations between the resulting representations, and mapping relations between these representations and the world (Piaget 1947/1972; Piaget and Inhelder 1963; Millikan 1984, pp. 240–244, 1986, 1989b). Indeed, manipulation of something like "logical pictures" from several modalities seems evident in scientific theorizing and artistic creation (Hadamard 1945/1954, Langer 1967) as well as in everyday life (Wittgenstein 1921/1961, pp. 15, 19; Millikan 1984).

REPRESENTATION AS SIMULATION

Once knowledge of similarity is present in mental representation, it can be a precursor to another form of representation — that involved in intentional communication. If organisms use their knowledge of similarity to recreate behaviors that they know from past experience influence other organisms, they would be on their way to an intentionally communicative system based on simulation. The use of metacommunication would allow for just such a development of communication through simulation (Bateson 1955/1972, 1956; see also Mitchell 1991a).

Metacommunication is communication that is "about" or refers to other communication. In Bateson's view most mammalian communication consists of involuntary mood-signs. If organisms voluntarily simulated these mood-signs to influence another organism, the voluntary simulation of these mood-signs would make reference (for the simulator at least) to these mood-signs. For example, human babies cry involuntarily when distressed. Once the babies can cry voluntarily, however, they can simulate the involuntary cry to get attention even though they are not distressed in ways that previously caused them to cry involuntarily. In voluntarily crying they emit a message (to themselves and perhaps to another) that indicates (perhaps nonvoluntarily) that this cry is different from an involuntary cry. This message, which indicates that the cry is a simulation, is metacommunicative — it informs the hearer that the new cry is not the same as the old cry, and therefore makes reference to it. Bateson, then, presented the origins of reference through simulation — reference began as a recursive form of symbolizing, by self-reference to "natural" signs.

Bateson's most well-known and controversial example of metacommunication is playfighting by monkeys. Playfighting monkeys simulate fighting and, in Bateson's view, this simulation is recognized by the playing monkey itself, by its partner, or by both. Simply stated, monkeys imitate fighting in play and either recognize the imitation themselves or expect their partner to recognize it as an imitation and to infer from this recognition that a fight is not intended. What makes playfighting fun for monkeys is the pretense that they are really fighting. (Similarly, for some ironic statements — "Let's go swimming" said during an electrical storm — what is funny is the pretense that one really believes what one says, even though one does not.)

That playfighting monkeys or other animals *usually* metacommunicate is unlikely (Symons 1978) in that they probably do not recognize the relation between fighting and playfighting (Mitchell 1991a). However, they may eventually recognize the similarity between fighting and playfighting and use this recognition for their own ends (Breuggeman 1978, Mitchell 1991a). Whether right or wrong in this particular instance, Bateson's ideas indicate that simulation can be an initial form of reference: in metacommunicating one makes reference to (the meaning of) the simulated communication because the simulation is an image of the simulated, though slightly discrepant with it. Once organisms recognize metacommunication, they recognize that "the other individual's and its own signals are only signals, which can be trusted, distrusted, falsified, denied, amplified, corrected, and so forth" (Bateson 1955/1972, p. 178). Hence, evidence of metacommunication indicates that "the gestures themselves are . . . recognized by their producers or recipients as the means whereby meanings are communicated" (Noble and Davidson 1991a, p. 226), a recognition that Noble and Davidson claimed is only present in, and criterial for, language.

Note that in Bateson's analysis there is no need for a simulator to communicate with anyone other than the simulator itself — for the playfighting monkey or the voluntarily crying baby, the connection between simulation and simulated may be "communicated" to itself alone. Similarly, for the human image maker alone, perhaps, "mark becomes remark" (Davis 1989b, p. 182); there need be no intention to communicate simulation to another (Mitchell 1991a). "Someone can make and see a depiction that no one else can interpret as such, that is never communicated to anybody; the logical properties of a picture are independent of their communicability" (Davis 1989a, p. 141).

The use of simulation to communicate is also prominent in Grice's (1982) account of the evolution of nonnatural meaning, which Grice (1957, 1968) believed to be an essential part of normal human communication. Nonnatural meaning occurs when an utterer U, by uttering something y, intends for his audience A to do or believe x, intends for A to recognize that U intends for A to do or believe x, and intends for A to do or believe x based upon U's intention for A to recognize U's intention that A do or believe x (Grice 1957, 1968). Grice (1982, pp. 232–235) suggested that such a convoluted mutual understanding could come about when an organism voluntarily produces an involuntary sign, such as behavior indicating pain. (Involuntary signs have *natural* meaning in that, as dark clouds are a natural sign of rain, no intentions are involved in recognizing the correlation between the sign and what the sign indicates.) If another organism A were observing the organism P that is simulating pain, A would recognize that P is voluntarily exhibiting pain-behavior yet wonder why P is "announcing" that it is *simulating* pain. A might at first suspect that P is offering a game of make-believe with A but might, upon further reflection, believe that P is trying to inform A that P is in pain. In Bateson's terminology P is metacommunicating. Once organisms are metacommunicating, they might begin to use signs that are not natural expressions of internal states such as pain but that "would be suggestive, in some recognizable way, of the state of affairs without being a natural response . . . to that state of affairs" (Grice 1982, p. 235). In other words, the vehicles for communication would become more and more arbitrary in what aspects of the simulated they emphasized. (See Parker 1985, pp. 622–623; Davidson and Noble 1989; Donald 1991; Noble and Davidson 1991; and Kendon 1991 for similar analyses of the development of referential communication via simulation.)

Why would organisms initially resort to anything as convoluted as communication of nonnatural meaning via simulation? The answer may be that animals had to "communicate" with two audiences at once: for one audience information was hidden, for the other information was manifested. For example, adult rhesus monkeys and humans appear to use playfighting as a threat to others when a direct threat would be problematic: they playfight in such a way as to intimate real fighting to their "play partner" while appearing to be merely playfighting to an ally of that partner (for example, the partner's mother or a teacher) so as to avoid the ally's intervention (Breuggeman 1978; Sluckin 1981, p. 78). A similar maneuver is used when a simulation simultaneously hides information and presents misinformation to the same individual, an apparently common activity for chimpanzees (de Waal 1986). When enacting these deceptions, the organism had to retain one interpretation for him- or herself and saliently present a different interpretation for another — *the same action came under two descriptions for the same organism* (Mitchell 1993b). With such dual description, the organism is able "to recognize that the other individual's and its own signals are only signals, which can be trusted, distrusted, falsified, denied, amplified, corrected, and so forth" (Bateson 1955/1972, p. 178).

Bateson and Grice suggested that the very simulativeness of simulation, once recognized, is used to communicate. How is the simulator to make the simulation *as such* salient to the other? The most obvious way is to *exaggerate* the activity so that its real-world interpretation is accessed by the other but also recognized as inappropriate, such that the viewer is compelled to an alternative interpretation (Bateson 1955/1972, 1956; Grice 1982). As with a pun that a listener (or observer) fails to understand, a double meaning can be communicated by *emphasizing* or *exaggerating* the simulation and thereby indicating that, and to what, the hearer must attend to understand fully. And, indeed, conventions arise from such exaggeration, as when a punster "often either leers, or enunciates slowly, idiot-fashion, to underline the subsidiary meaning" (Redfern 1984, p. 29). Exaggeration is a relationship between what is currently exhibited and what is normally or standardly exhibited and is therefore metacommunication derived iconically. Similarly, ironic or satiric statements are metacommunicative: what is understood derives from the incongruity between the expectations about what is presented and the impossibility of these expectations' being taken at face value. Of course metacommunication may also be a marker of some sort that *signals* a simulation (see Bekoff 1975). Such a marker may develop because one type of behavior is consistently observed with another type of behavior.

If a communicative act is simulated, the act is presented in two forms: one marked (metacommunicatively by *being* a simulation) and the other unmarked. The metacommunicative marking is *part* of the simulation of the communicative act (the part that informs of the fact of simulation) and indicates that the simulation is different from the communicative act. With the recognition that a communicative act can be marked comes the recognition that the "same" act can have dual interpretation depending upon whether the mark is present or absent. The metacommunication — the mark — can be viewed as a comment upon the communicative act, to be used as a guide to its interpretation (compare Peters 1974; see also Parker 1985, pp. 622–623, on combining known words to create new meanings). In Bateson's (1955/1972) probably false interpretation of monkey play-fighting, the playful quality that accompanied the aggressive acts in playfighting was the metacommunicative mark that distinguished the play from real fighting. This marking would allow monkeys to distinguish what was indicated by the aggressive acts — a fight — from what these aggressive acts tell them to do — act submissive or run away. As a result, the presence of the mark would make monkeys distinguish between indicative and imperative moods of signs, a distinction so far described only for human communication (de Laguna 1927, Millikan 1984). Skepticism toward alarm calls that are sometimes used deceptively apparently leads birds already protected from danger to look to see if the danger indicated in fact exists (Munn 1986), suggesting that even without marks indicating simulation, nonhumans may already distinguish indicative and imperative senses of significant utterances.

To avoid the problem of dual interpretations for simulations based upon communicative acts, let us assume, as a beginning, that simulative communications simulate noncommunicative phenomena (and thus would not be *meta*communicative but would be referential). Indeed, some modern scenarios for the evolution of language assume that simulation of *noncommunicative* things is essential to the origins of language, and they describe an evolutionary path to symbolization that mimics the developmental path described by others (for example, compare the evolutionary account of language described by Parker and Gibson 1979, Davidson and Noble 1989, and Noble and Davidson 1991 with the developmental account described by Werner and Kaplan 1963). Simulation of noncommunicative phenomena would seem an obvious place to start for communicative reference because it is the essence of most pretend play. Pretense in humans builds upon the child's understanding of real-world events (Bretherton 1984, Leslie 1988). If one knows an individual well, it is often clear what the individual is pretending because one knows the real-world events with which the individual is familiar. Thus, it would seem that simulative pretense could quite easily be used to communicate, provided the simulation was of some commonly known action (see discussion below on communicative simulations).

Without common knowledge, simulative communications are not always easy to interpret. For example, a 12-month-old boy J got his father from another room to retrieve a block that had landed behind the bookcase when the block flew off the top of the boy's jack-in-the-box. Because his father did not understand what the boy wanted (not having seen the original launching), J recreated the block's trajectory behind the bookcase: "J . . . takes his father's hand in his own, places them both on top of the jack-in-the-box, makes a kind of explosive noise, and moves his and his father's hand in an arc toward the bookcase. J then reaches his own hand down behind the bookcase, making somewhat conventionalized effort sounds to signal reaching. Still looking at his father, he says something like, 'Block.' " (Rubin and Wolf 1980, p. 18). J intended for his father to get something that had been shot behind the bookcase, and he also intended for his father to recognize this intention, and he attempted to realize these intentions by simulating for his father the events that led to the block's current unavailability. Although J used some conventionalized means to indicate a story of what happened (he used an explosive noise that was never made in the onset of the block's flight to indicate a "launching"; he said something like "block"), he also used simulation of the block's trajectory to communicate its movement to his father.

It is obvious that J used a simulation of the block's movement (with additional touches) in an attempt to communicate to his father that the block was behind a bookcase. How did J simulate these actions? That is, how did he recognize the similarity between the block's movements and his own movements of his father's hand, given that his own hand never experienced the movements he portrayed? To effect his simulation, J must have used kinesthetic-visual matching abilities: he spontaneously recognized and produced the similarity between the visually perceived movement of the block and the kinesthetically perceived movement of his own hand. He must have recognized a similarity (and perhaps at some level an identity) between his actions and the previous movement of the block. J must, in Wittgenstein's terminology, have had a picture of reality that related the movement of the block and his own (mimetic) hand movements. The "picture" that allowed J's hand movements to stand for the block's movements is usually called a schema and need not be a consciously present mental image.

J's hand movement represents or symbolizes for J the block's movement in the absence of the block and its movement. Although such decontextualized simulations can be used to communicate symbolically, the problem for the communicator is to get the perceiver to recognize that which is being simulated, and this problem is compounded when symbols are decontextualized. J's father faced the problem of *interpreting* J's intended but nonconventional and decontextualized communication through simulation, in that J's hand movements could indicate a vast array of possible events with the same shape. J's problem, and his father's, was that the father did not see the launching of the block and so could not recognize the resemblance between that and J's hand movements. If J's father is a good example, recognition that the *intent* of a communicative simulation is to communicate does not pose any difficulty for humans, whether or not the simulation as such or its meaning is recognized. Interpretation of J's "picture" shares the same problem as asyntactic utterances: "*Speech which is limited to the sentence-word can mediate coöperative action only with reference to a situation which is perceptually present to both parties concerned*" (de Laguna 1927, p. 94) or to a situation that is highly circumscribed (Bates 1979c; Tomasello et al. 1985).

Of course simulative communication would become extremely cumbersome if it were the only means of information transfer, and nonnatural meanings and simulative communications clearly must, and do, give way to conventionalized utterances if there is to be a continuing and extensive communication system (Werner and Kaplan 1963; Knowlton 1966, p. 169; Efron 1972/1978; Frishberg 1975; Grice 1982; Millikan 1984; Tomasello et al. 1985; Noble and

Davidson 1991a, p. 245). Any simulative vehicle of communication naturally becomes more and more arbitrary in its resemblance to its referent, so that the initial simulative element may be lost. Still, simulation is a place to begin to look for evidence of nonnatural meaning and metacommunication, and once the capacity for simulation is present, it is unlikely to disappear: a communication system that used iconicity *in addition to* conventionalized utterances would be more productive and useful (and entertaining) than one that used only conventionalized utterances. And, apparently, human verbal communication is one such system (Efron 1972/1978, Ekman 1977/1980, Van Meel 1982, McNeill 1985).

The same movement from simulation to nonsimulation is present in pretend play. Whereas early pretending derives from simulations in which the child acts as though *x* were the case or as though she or he were doing *x* (where *x* is some real-world phenomenon), once the understanding of pretense is in place, there is no requirement that it be simulative — that is, dependent upon the real world for primary representations. (Similarly, in Davis's view pictorial images may become independent of their real-world connection.) Indeed, much pretense makes no reference to the real world (Walton 1990). One can pretend even when there is no real situation the pretense simulates. For example, Timmy's father pretends that he is a monster, and Timmy (if he is older than 3 years) reacts toward his father in many respects as he would if a "real" monster appeared, but both Timmy and his father know that no such real monster is going to appear because no such real monster exists. Make-believe can, and usually does, require more than an attitude toward primary representations of real-world phenomena (Walton 1990, pp. 144–150). However, such forms of pretense derive from earlier cognitive structures that are themselves derived from simulation.

REPRESENTATION AS SYMBOLIZATION

" 'I am not merely saying this, I mean something by it.' — When we consider what is going on in us when we *mean* (and don't merely say) words, it seems to us as if there were something coupled to these words, which otherwise would run idle. — As if they, so to speak, connected with something in us" (Wittgenstein 1953, p. 139). In the experience of seeing-as, or recognizing simulation as such, the viewer recognizes the similarity between the representation (or vehicle) and what it represents (the referent). According to Werner and Kaplan (1963), the same recognition is present in any symbolization, including naming and syntax, and is rooted in an internal match between vehicle and referent:

> the *establishment of semantic correspondence* between the vehicular pattern and the referential object — the formation of the symbol — comes about through the operation of schematizing, form-building activity which shapes the pattern on one hand and the referent on the other. Correspondence is achieved when both pattern and object are rooted in similar or identical organismic states, the formation of these states being directed and regulated by the underlying activity of schematization. (Werner and Kaplan 1963, pp. 23–24)

Naming as well as any other forms of symbolizing derive from this inner dynamic schematization. This schematization is based upon "natural symbols" such as "deictic, onomatopoetic, and physiognomic" signs, which "most tellingly impress upon the child the inner relationship between name and thing" (p. 110). "If one accepts . . . that symbols are not simply substitute signs, pointers, labels, or coins, but rather represent or connote the referent, symbolization must always, and basically, include an element of depiction — however dissimilar the 'picture' and the referred object may appear to be" (p. 109). The internal "picture" is continuously transformed, in that *"available means are exploited for new ends until function-specific instrumentalities are forged for the novel function"* (p. 175). The development of speech shows a "distancing between vehicle and referent with respect to space and time [which] reflects the increasing use of the linguistic medium as a means for transcending the immediately given" (p. 113).

The ideas of Werner and Kaplan contrast sharply with the idea that names are merely labels for things, learned by contiguity. To depict the inadequacies of any contiguity-learning theory of symbolization, Werner and Kaplan (1963, pp. 110–111) present Helen Keller's famous account of her discovery of arbitrary naming. Helen's teacher, Annie Sullivan, had attempted vainly for a month to teach Helen through contiguity that, for example, the finger-spelled word "m-u-g" stood for mug and "w-a-t-e-r" for water, but Helen learned these spellings laboriously. Her learning was extremely rapid, however, after the following incident, during which she discovered arbitrary naming.

> Someone was drawing water and my teacher placed my hand under the spout. As the cool stream gushed over one hand she spelled into the other the word *water,* firstly slowly, then rapidly. I stood still, my whole attention fixed upon the motions of her fingers. Suddenly I felt a misty consciousness as of something forgotten — a thrill of returning thought; and somehow the mystery of language was revealed to me. I knew then that "w-a-t-e-r" meant the wonderful cool something that was flowing over my hand. (Keller 1902/1965, p. 21)

The transformation in Helen's understanding of words was apparent to Annie Sullivan, who had taught Helen thirty words in the preceding month; after the water incident, Helen learned thirty words in a few hours. "All the way back to the house [after leaving the well] she ... learned the name of every object she touched"; the next day Helen "flitted from object to object, asking the name of everything" (Sullivan, in Keller 1902/1965, pp. 187–188). In the use of contiguity to teach naming,

> there was an entirely external connection between a static object and certain tactual signs. In the "water" situation, [however,] the object was in itself an *event:* the flowing of cool water over the hand. And as Miss Sullivan "poured" the tactual signs over one of Helen's hands, and the cool liquid over the other, there perhaps occurred two tactual experiences that may have been internally linked in the child's mind — a linkage based on common dynamic features. (Werner and Kaplan 1963, p. 112)

In the case of Helen Keller, the similarity between the water and the fingered letters, in "pouring" over her hands simultaneously, effectively made the connection between them.

Once an inner correspondence between vehicle and referent is understood, an underlying experience of similarity is created between vehicles and referents. This underlying similarity manifests itself in nonlinguistic gestures during speech that support the meaning of the speech (Ekman 1977/1980, Van Meel 1982, McNeill 1985). The idea that words and sentences (as well as other symbolic vehicles) have an internal similarity with their referents is supported by the young child's ability to map the vehicle to aberrant referents that bear some resemblance to the "normal" referent. For example, in early pretense, a spoon can be called "ax" if chopping motions are made with it because the manipulation of the object maps to the inner schematization; yet the spoon will still be recognized as a spoon. Without the manipulation of the object — that is, without some form of similarity being created between vehicle and referent — such nonliteral naming is difficult for a young child (Werner and Kaplan 1963, p. 120; compare Killen and Uzgiris 1981). The inner dynamic schematization upon which nonliteral interpretations map seems likely to be some form of similarity between the object or its movements and the child's movements or between some other aspect of the object and of the child or context. The external situations in pretend and literal uses of objects are clearly different and are differently understood, but just as clearly they map to the same internal representation.

This similarity in the internal dynamic schematization is what allowed Leslie (1988) to claim that the *it is raining* of pretend and the *it is raining* of normal discourse are the same proposition, even though they are markedly different in their contextual support and referents. Pretense requires two *internal representations*: a primary representation (which has as its object the real) and a metarepresentation (which incorporates an attitude toward the primary representation with the representation itself). For the young child, "there is a non-arbitrary relationship between *pretending it is raining* and *believing it is raining*. What they have in common is the proposition *it is raining*" (Leslie 1988, p. 206), which is a primary representation. Any primary representation *x* can be combined with an attitude toward *x* to create a metarepresentation such as *pretending x, believing x, not believing x, expecting x, hoping x,* or *recognizing x.* Pretending, then, is a metarepresentation that "presupposes primary representation" (Leslie 1988, p. 207) and utilizes "inferences . . . drawn from the set of real world inferences . . . [but which can also] apply to representations of the pretence and not to representations of what is really happening" during pretense (Leslie 1988, p. 203). One can bake in a real oven, and one can bake in a pretend oven; both activities access the inner dynamic schematization of *baking.* If one can translate between Leslie's view and that of Werner and Kaplan, the latter would likely view the primary representation as the inner similarity between vehicle and referent and would also view metarepresentations as merely different potentialities of the primary representation, rather than viewing the manipulation of the primary representation as an operation separate from the primary representation. In Werner and Kaplan's view, the primary representation is a *dynamic* schematization, which means that its functions are variable and multifaceted.

In the instances of Helen Keller and the child J, the similarities between vehicle and referent are present externally, and the inner correspondences — the "common dynamic features" — can be inferred. However, such similarities are often not apparent: "linguistic forms . . . typically manifest no trace of similarity to the object they signify" (Werner and Kaplan 1963, p. 26; although see Foster 1990). That linguistic forms appear arbitrary is often used as an argument against representational or depictional theories of symbols and leads many thinkers to define symbols as "things that stand, by convention, for other things" and "have the property of being arbitrarily related to their referents, though conventionally bound to them" (Noble and Davidson 1991a, p. 226; see also Saussure 1916/1959; Holloway 1969; Lamendella 1980, p. 165; Savage-Rumbaugh 1986; compare Peirce 1895–1910/1955, p. 113; Deuchar 1990). (Note that in Goodman's [1968, p. 38] view, even relatively exact simulations such as pictures are conventionally interpreted as well as arbitrary: "almost any picture may represent almost anything. . . . [Interpretation of a picture] depends at any time entirely upon what frame or mode is then standard." Thus, simulations would be just as arbitrary and conventional as nonsimulations!)

Werner and Kaplan (1963, p. 60) as well as others (Peirce 1895–1910/1955; Langer 1942, 1967; Jakobson 1966/1990; Wescott 1971; Jakobson and Waugh 1979; Foster 1980; Foster and Brandes 1980) have argued that the relationship between vehicle and referent is not, in fact, arbitrary, and that "any symbolic system, including language[, is] basically 'analogical,' " but this analogy is *from the symbol user's point of view.*

> Insofar as words are taken as products or impersonal signs, that is insofar as language is abstracted from concrete, active, living individuals, this argument is undoubtedly valid: with the exception perhaps of onomatopoetic word-forms, one can see no *external similarity* between the properties of objects to which reference is made and the properties of the forms used to refer to

these objects. However, ... one may entertain the possibility that an *inner similarity* between vehicle and referent may occur without this similarity being apparent to an observer who regards solely the external, geometric-technical properties of word-form and object. (Werner and Kaplan 1963, p. 26)

As the child develops, the "similarity" relation may become more and more abstract (Werner and Kaplan 1963), such that the way we "picture facts to ourselves" (Wittgenstein 1921/1961, p. 15) is less and less direct (see Knowlton 1966).

Even outside child language development, evidence suggests that words and even syntax are not entirely arbitrary in relation to the world (see Peirce 1895–1910/1955, Werner and Kaplan 1963, Jakobson 1966/1990, Wescott 1971, Jakobson and Waugh 1979, Foster 1980). For example, English speakers were likely to guess which antonyms in a foreign language (Chinese, Czech, Hindi, Japanese, Hungarian, Polish) corresponded to English equivalents they were given (Brown, Black, and Horowitz 1955). Such findings suggest that "there may be some features of phonetic symbolism which have a universal validity. It is possible that speech originated in symbols imitative or somehow suggestive of their meanings and that traces of these 'appropriate' linkages survive in all languages today" (Brown, Black, and Horowitz 1955, p. 388). Infants' words for mother (such as *mama* or *nana*) as well as father (such as *papa* or *tata*), which clearly become symbolic for infants and others, are also cross-culturally extremely prominent (Murdock 1959), suggesting some nonarbitrary basis for these terms. The explanation for the cross-cultural consistency in symbolization for mother, for example, supports the idea that symbols *become* conventionalized and differentiated from their simulative origins. Originally,

the sucking activities of a child are accompanied by a slight nasal murmur, the only phonation which can be produced when the lips are pressed to mother's breast or to the feeding bottle and the mouth is full. Later, this phonatory reaction to nursing is reproduced as an anticipatory signal at the mere sight of food and finally as a manifestation of a desire to eat, or more generally, as an expression of discontent and impatient longing for missing food or absent nurser, and any ungranted wish. (Jakobson 1960/1990, p. 309)

Thus, the sign *mama* is not arbitrary in relation to its meaning but derives from simulations of naturally occurring sounds. However, although these sounds "come into existence only because . . . young children pronounce them easily and therefore frequently and easily, . . . they become kinship terms only because they are appropriated as such by adult relatives" (Freire-Marreco 1915, p. 202). Apparently speaking of his own child, Lewis (1936/1951, p. 130) wrote, "The child in discomfort makes urgent nasal sounds, he cries *mama*; to his mother this means that he is 'calling her', the more so when she finds, of course, that her attention to him results in the cessation of the cry." The process of repeated use may *create* the conventionality of symbols rather than that "conventionality" creates symbols. Thus, arbitrariness is not essential as evidence of symbols, especially when simulative signs used repetitively in a communication system naturally become more arbitrary in form.

Conventionality and arbitrariness criteria are often used to deny symbolization to nonhumans (see, for example, Davidson and Noble 1989, Noble and Davidson 1991a). Yet in vervets, each distinct call is arbitrarily related to its referent by convention, in that young monkeys have to learn to use the sound correctly (Seyfarth, Cheney, and Marler 1980). And nothing could be more arbitrary than the differences among alarm calls by vervet monkeys that differentiate predators—leopards, eagles, and snakes (Cheney and Seyfarth 1990) — in that the calls bear no resemblance to their referents. How innately constrained these vocalizations are is unknown, but because Japanese macaques raised by rhesus macaques adopted rhesus food calls (Masataka and Fujita 1989), other primates' vocalizations may be similarly learned by experience. (Note that alarm calls of other organisms, such as birds and ground squirrels, also differentiate type of predator [Munn 1986; Slobodchikoff, Fischer, and Shapiro 1986] and even individual predators of the same species [Slobodchikoff, Kiriazin, Fischer, and Creef 1991].)

Rather than define symbolization in terms of conventionality and arbitrariness, others have suggested that symbolization depends upon having "cognitive stand-ins" (McDowell 1980, p. 135) for their referents that can function independently of their referents: "symbols are representational acts that can be carried out without perceptual/contextual input from their referents" (Bates et al. 1983, p. 87; see Piaget 1945/1962; Werner and Kaplan 1963). In this view, symbolization can be simulative and provides support for linguistic

development: " *'decontextualized'* *gestures without perceptual sup-*
port from the object are the ones most closely linked with language
development at 13 months" (Bates et al. 1983, p. 88). Such decontextualized gestures are so common in children's pretense that "symbolic representation in play is essentially a particular form of speech
at an earlier stage" (Vygotsky 1935/1978, p. 111; see also Shotwell
et al. 1980). (Decontextualized representation also occurs, though
involuntarily, in dreams [Freud 1900/1963].)

Pretense shares specific structural correspondences with language as well. According to Millikan (1984, 1989a), "the cognitive
structures with which man is endowed were originally nature's solution to some very simple demands made by man's evolutionary niche.
. . . That solution involved the introduction of representations, inner
and/or outer, having a subject/predicate structure, and subject to a
negation transformation" (Millikan 1989a, p. 294). Millikan is describing the cognitive structures underlying language — a representational system that affords both external and internal
representations. (Internal representations differ from external representations in that the devices that produce and interpret the representations are in the same organism rather than being in different
organisms.) When an organism's representations have subject-predi-
cate structure, the organism has internal replacement rules such that
any subject can be replaced by any other subject, and any predicate
can be replaced by any other predicate. Such internal replacement
rules are strikingly reminiscent of Leslie's account of pretense, which
assumes the flexibility of metarepresentations in accepting different
primary representations. The negation transformation may also derive from qualities present in pretense: "In iconic communication,
there is no tense, no simple negation, no modal marker. The absence
of simple negatives is of especial interest because it often forces
organisms *into saying the opposite of what they mean in order to get*
across the proposition that they mean the opposite of what they say"
(Bateson 1971/1972, pp. 140–141). Communicative pretend play is
one example of such oppositional meaning.

From this review, I do not wish to suggest that once pretense is
present, language, pictorial image making, and all other simulative
skills people use are soon to follow. The skills required to act *as if*
one is doing something or *as if* something were true are only part of
the skills required to simulate, in two-dimensional form, the profile
of the side view of an animal or to understand syntax. But if Walton

was correct that make-believe is, in humans, normally experienced in relation to (external as well as internal) representational props in numerous and diverse media, using any or several perceptual modalities at once; if Davis was correct that recognizing or creating an image in any medium can lead to "seeing as" or make-believe; and if Werner and Kaplan were correct that what connects a symbol to what it is a symbol of is an inner dynamic schematization derived from experienced and created similarities both within and between modalities, then the recognition or creation of similarity is the basis for symbolization and make-believe. If Bateson and Grice were correct that simulation can lead to nonnatural meaning in nonhumans, then evidence of communicative simulation and its pretend precursors should be observed in nonhumans. Thus, knowledge and use of similarity in any modality or modalities by nonhumans would suggest that the basis for seeing-as, representation, and fictive acts of perceiving is present prior to the evolutionary advent of humans.

THE EVOLUTION OF ICONIC REPRESENTATIONS

The significance of simulation in the development of pretense, symbolization, image making, language, and intentional communication should by now be apparent. Any evidence of knowledge of simulation by human and nonhuman organisms is relevant to the evolution of pretense and symbolization. Clearly, humans generate a great deal of simulations. What remains to be shown is that simulations have some of the same significances for nonhumans as for humans.

Nonhumans both recognize and create simulations of other activities (Mitchell 1986, 1987, 1990, 1991a, 1993a, 1993c) even though these simulative activities are not as extensive as those of humans. These simulations depend upon repetition of the organism's own actions or, in more developed instances, repetition of another's actions, and they can be encompassed under five interdependent rubrics: imitation, pretense, deception, mirror self-recognition, and communication of nonnatural meaning (see Mitchell 1986, 1987, 1988, 1990, 1991a, 1991b, 1992, 1993a, 1993b, 1993c). Examples of these activities may be classified under four understandings of simulation that are hierarchically dependent: (1) self-simulation to influence others and recognition of simulation of nonself objects and activities by others; (2) simulation of others' activities dependent upon kinesthetic-visual matching, use of props for oneself, and recognition of self dependent upon kinesthetic-visual matching; (3) communicative simulation and use of props for others; and (4) creation and subsequent use of props external to or on the body as representations for others as well as self (see Table 6.1).

Table 6.1   Steps in the Evolution of Simulation, with the Types of Evidence of Each Step and the Types of Organisms Exhibiting Such Evidence

Steps	Types of Evidence	Types of Organisms
1a. Self-simulation to influence others	deceptive self-pretense	mammals, birds
	self-pretense in problem solving	chimps, humans
b. Recognition of simulation of nonself objects and activities by others	skepticism	mammals, birds
	use of mirrors to locate objects	monkeys, great apes, humans
2a. Simulation of others' activities dependent upon kinesthetic-visual matching	imitation of others	monkeys (rare), great apes, dolphins, humans
	pretend play	great apes, dolphins, humans
b. Use of props for oneself	imitative play with object	monkeys (rare), great apes, humans
	planning	great apes, humans
	insightful tool use	great apes, humans
c. Recognition of self dependent upon kinesthetic-visual matching	mirror self-recognition	great apes, dolphins, humans
	recognition of other's imitation of self	humans
3a Communicative simulation	conventionalized signs	great apes, humans
	pointing, eye gaze	great apes, humans
b. Use of props for others	teaching with objects	great apes, humans
	requesting with objects	great apes, humans

Steps	Types of Evidence	Types of Organisms
4. Creation and sub-sequent use of props external to or on the body as repre-sentations for others as well as self	bodily disguise	humans
	artistic representations	humans
	bodily decoration	humans
	social norms	humans

1.a. Self-Simulation to Influence Others

In general mammals are remarkably good at reenacting their own previous actions that have consistently caused others to engage in particular actions. What primates are especially good at, among all mammals, is *simulating* their own actions. They are apparently able to produce actions outside their (presumed) "normal" context in order to make others respond to the actions as they would under "normal" circumstances (Piaget 1945/1962; Quiatt 1984; Mitchell and Thompson 1986, p. 149; Whiten and Byrne 1988b; Mitchell 1993a). (Note that I use the term *normal* in Millikan's [1989b, p. 300; also 1984, 1989a] sense, such that "normal conditions" are "conditions in which [these actions] have *historically* . . . actually aided survival.") Some of these deceptions are common among mammals: a 19-month-old boy teased his mother by offering a toy but withdrew it when she reached for it (Chevalier-Skolnikoff 1986), enacting a game of "object-keepaway" common among mammals (Mitchell and Thompson 1991); a chimpanzee imitated his own limping gait that earned him sympathy from others (de Waal 1982, 1986, pp. 237–238); and similar limpings are present in dogs (Romanes 1883/1906, p. 444; Morgan 1900/1970, p. 280; Fox 1962; Goodall 1986, p. 581; Rollin 1989, pp. 151–152). Already presented was rhesus monkeys' simulation of playfighting to bully younger monkeys without being reprimanded (Breuggeman 1978). In other instances, yellow baboons and chimpanzees moved calmly toward a conspecific to get close enough to attack (Whiten and Byrne 1988b, p. 239); a gorilla male acted as though foraging to get near an infant whose mother was very protective, and another male acted as if interested in the infant in order to tease her mother during play (Mitchell 1991b); and a person acted as though running away from

207

a dog he was playing with to induce the dog to chase the person and thereby bring an object the dog held close enough for the person to grab (Mitchell and Thompson 1991). Primates are also good at simulating only *part* of a behavioral sequence in order to influence others, as when a gorilla, observed by people to slam her hand into the window of her cage and run, periodically stopped her hand just before it hit the window, presumably to keep her audience's attention (Quiatt 1984, p. 32).

In some instances primates used behaviors specialized for communication: a chimpanzee imitated friendly facial expressions and gestures to lure another chimp near enough to be able to attack her easily, and another chimpanzee imitated playful gestures to distract others from aggressive acts toward himself (de Waal 1986); two gorillas acted friendly toward a person to obtain an object or tease her, and a female gorilla repeatedly solicited sexual interaction with a male gorilla by presenting her rump but ran away when he approached (Chevalier-Skolnikoff 1986, p. 213); sign-taught apes frequently use their signs to deceive (Miles 1986, Savage-Rumbaugh and McDonald 1988); and chimpanzees, guenons, and vervets called false alarms for varying purposes (de Waal 1986, Whiten and Byrne 1988b), a deceptive activity also found in birds (Munn 1986).

In all of these instances, the primates engaged in what could be called *deceptive self-pretense* (see Jolly 1988, p. 376; Mitchell 1993a): they acted as though doing one thing while actually doing something else, where this something else benefited them in some way (see Mitchell 1986, 1993a). The organism seemed to use its bodily actions or voice as a prop in a game of make-believe directed toward another organism; they engaged in instrumental kinesthetic and/or vocal self-imitation. Although pretend play is infrequent in great apes in nature and less extensive than in children when great apes are reared by humans (see below, and Premack and Premack 1983, p. 107; Mignault 1985; Miles 1990, p. 533, 1991a, p. 16; but compare Patterson and Linden 1981), apes' pretend abilities are elaborate in their deceptions (Mitchell 1993a). This constrained use of pretense in deception and the unconstrained use of pretense in innumerable activities by humans suggest that pretense evolved initially for manipulation and deception but became independent of this function in apes and more extensive in hominids (Mitchell 1993a), perhaps as our ontogeny became prolonged (Gould 1977; Parker 1987, p. 22). Pretend enactment of one's own actions also may assist in problem solving, as when the chimpanzees described by Köhler (1925/1959, pp. 64, 117) acted out what they would like to do to solve a problem (see also Reynolds 1981, pp. 161–164, 210–212; Kendon 1991). Such self-pretense should not be conceptually confused with pretend enactment of another's actions (which would usually involve kines-thetic-visual matching), even though the two behaviors can be diffi-cult to distinguish empirically (Bretherton 1984, pp. 9–10). But as discussed below, the ability for self-pretense, when combined with abilities for kinesthetic-visual matching, leads to abilities for planning based on imagined scenarios for the imaginer's activities.

1.b. Recognition of Simulation of Nonself Objects and Activities by Others

Surprisingly little mention is made of nonhuman primates' recognizing another's simulations. However, it is likely that many nonhuman primates can recognize simulations: skeptical attempts to verify are common in many species (Mitchell and Thompson 1986, p. 359), as when the sign-taught orangutan Chantek sometimes expressed skepticism and attempts at verification when his caregivers pretended (Miles, personal communication). Humans, apes, and dogs recognize when an organism could be simulating its own actions, and they respond with counterdeception and inhibition of responding (Menzel 1974, Saarni 1989, Mitchell 1991a, Mitchell and Thompson 1991a). Primates can presumably recognize some instances of others' self-simulation and learn (as Bateson suggests they do) to mistrust particular gestures made by deceptive individuals (see Quiatt 1984). Some monkeys and apes also recognize simulations of objects (in the case of monkeys, only objects other than themselves) in mirrors and videotapes and can use the image of these objects to locate these objects (Menzel, Savage-Rumbaugh, and Lawson 1985; Anderson 1986).

By exhibiting self-pretense and recognition of simulation, pri-mates indicate that they have a kinesthetic sense of themselves, a visual sense of similarity, and in some instances a recognition of the impact of their own vocalizations. They know that if they re-create a kinesthetic or vocal experience, they can produce a particular effect on another. Thus, these primates exhibit a kinesthetic self as agent. They *may* also be aware of how the other visually perceives their action known to themselves through kinesthesis; but the deceptions alone typically cannot differentiate this interpretation from the less complex interpretation that they simply know that if they do a particular action (known kinesthetically and/or auditorally), they can influence another to perform a desired action (see Mitchell 1986, 1987, 1990). In one sequence of deceptions by chimpanzees, however, the deceptive actions of a female suggest that she planned her deception and recognized how her actions were interpreted by a male (Menzel 1974): the female led the male in a direction away from where she knew food was hidden and returned to the food when the male was searching in the wrong area. One could interpret this deception as resulting purely from the female's recognizing the effect on the male of her action of leading, but in this instance that interpretation is cumbersome for several reasons: we know that the female knew that the food was not located where she was leading the male; we also know from past observation that the male knew when the female's actions of leading (usually) indicated hidden food, and that the female (usually) intended for others to know that she was leading them to hidden food; hence, her use of actions to lead away from food actions that she usually employed to lead to food suggests planning and a recognition of how others perceived her actions (for further discussion, see Mitchell 1986, forthcoming a). The next understanding of simulation allows us more readily to interpret an animal as planning and knowing how its actions look to another and to itself, because at this next understanding the organism can map or translate between its kinesthetic sense of self and its visuallike image of its own body. The next section details the creation of a maplike connection between the kinesthetic and visual modalities.

2.a. and b. Simulation of Others' Activities Dependent upon Kinesthetic-visual Matching and Using Props for Oneself

Instances of kinesthetic-visual imitation by nonhuman primates raised with conspecifics are infrequent. When a rhesus monkey with a distinctive and unique tic of touching his forehead with his head when frightened was introduced into a colony of thirty-two rhesus, within three months another monkey was making similar movements, and within 1.5 years three additional monkeys were also making the same movements (Rivers, Bartecku, Brown, and Ettlinger 1983, p. 8). Kinesthetic-visual imitation is much more common and elaborated in chimpanzees. Young captive chimps imitated the "crooked" gait of an older female and the limping gate of an injured male (de Waal 1982, pp. 80, 135). At Gombe an adolescent male who had a favorite older male he followed imitated three times in succession the older male's display of "vigorous leaping and jumping from branch to branch . . . , [after which he] jumped down to the ground and slapstamped down-slope" (Plooij 1978, pp. 128–129); and lone adolescent chimpanzees enacted aggressive displays on three occasions (Goodall 1986, p. 590). These activities, especially those by Gombe chimps, suggest that the animal is *pretending* to be another animal. Such pretense implies "intrinsic representation" of the other animal's actions (Hallowell 1960, p. 352). From Plooij's description of the juvenile chimp's emulation of the older male, it is clear that some chimpanzees do, contra Clementson-Mohr (1982, p. 70), "strategically *seek out* someone to watch in order to learn from . . . that model."

Combination of kinesthetic-visual imitation in pretense with the use of a prop by primates in "natural" groups is rarely reported. At Gombe, a male chimpanzee repeatedly imitated a fellow chimp's aggressive display with a kerosene can, and another chimpanzee imitated her mother's ant fishing with a stick, but to "an imaginary nest" of ants (Goodall 1986, pp. 590–591). An infant orang "mimicked almost to perfection the adult threat response" of its mother, which it had just "watched . . . with great interest": the infant "grabbed onto a tiny twig . . . and after waving this back and forth several times managed to break if off and let it drop" on the infant's human observer, watching it intently as it fell (Horr 1977, p. 308). On Cayo Santiago, one free-ranging female rhesus monkey holding half of a coconut shell followed behind a mother carrying her own infant; the following female held the coconut shell in the same position the mother held her infant, and when the mother moved the infant and held it in a different way, the female moved her coconut shell to conform to the position of the infant on the mother (Breuggeman 1973, p. 196). In these instances, the objects were apparently used as ad hoc props in games of make-believe. Note that the rhesus monkey described by Breuggeman mapped the coconut's location on her body to the infant's changing location on its mother's body, thereby providing a "chain of replication and variation" (Davis 1989b, p. 183) for the simulation — we know that the coconut represented the baby by the microgenesis of the action sequence. Given monkeys' apparent difficulties with imitation of novel tool use

(Visalberghi and Fragaszy 1990, Whiten and Ham 1992), the latter instance is remarkable and might make us question whether imitation of tool use is more difficult than imitation per se. In almost all of the above instances, the pretenses are imitations of functional activities — maternal behavior, feeding, aggressive displays — which may be understood easily by the animals (see McCabe and Uzgiris 1983, pp. 79–80). These imitations also seem based on identification (Valentine 1930, Mussen 1967) with the primate imitated, which may be more difficult in imitation of tool use in that the observer desires the result of the tool use or the tool itself. Identifications are themselves based upon pretending (perhaps unintentionally) to be another, a common role-playing activity in humans (Garvey 1977, Bretherton 1984).

Once they have interacted extensively with people, the imitative pretend play of apes can be elaborate and shows a surprising similarity with human pretense (Mitchell 1990, 1993a). Orangutans frequently imitate human behaviors and often use novel objects to effect the same ends (Russon and Galdikas 1993). The chimpanzees studied by Köhler (1925/1959) moved their arms in sympathy with the actions of other chimps solving problems, as if enacting the same actions (see also Kendon 1991). The home-reared chimp Viki repeatedly pretended to have a pulltoy (Hayes 1951, pp. 80–84), and sign-trained pygmy and common chimpanzees pretended to eat, hide, and/or attack nonexistent objects (Savage-Rumbaugh and McDonald 1988). As with imaginary figures in human children (Singer and Singer 1990, pp. 109–110), Viki's play with this imaginary object stopped when someone participated directly in the fantasy (Hayes 1951, p. 84). Sign-trained apes — chimpanzee, gorilla, and orangutan — all played with dolls and appeared to attribute psychological qualities to them (Gardner and Gardner 1969, Patterson and Linden 1981, Miles 1990). However, the capacity of these apes for pretense, like the apes in more "natural" circumstances, appears limited to activities that are repeatedly enacted in their lives and/or for which the functional relevance of imitation for the ape is apparent or interesting. Thus, it is not surprising that chimpanzees as old as 26 months failed to engage in even rudimentary pretend modeling of human actions (Mignault 1985) with which they were largely unfamiliar. Although human children commonly enact scenarios from their everyday lives, they also invent scenarios they have never encountered (Garvey 1977, Mignault 1985). By 12 months of age children are adept at pretending to use prototypical objects in pretense (e.g., using a spoon to "eat") based on imitation of others' actions (Fein and Apfel 1979). Imagination is socially encouraged in humans and sometimes elaborated through traditional techniques (Lewis-Williams 1981, 1983, 1986; Noll 1985).

The enacting of pretense ontogenetically leads to the ability to plan (Vygotsky 1933/1976), an ability present in the pretend play of 2-year-old children (McCune-Nicolich 1981, p. 790) and perhaps earlier in nonplay activities (Piaget 1936/1979, p. 339, 1937/1954, p. 297). By *planning* I specifically mean "formulating an extended sequence of steps in advance in order to reach a goal or solve a problem" (Wellman, Fabricius, and Sophian 1985, p. 124). A plan is an imaginal (pretend) scenario in which a series of events occurs in the mind of the imaginer, with the recognition that the scenario is imaginal but could be (or is going to be) enacted by the imaginer (Mitchell 1993a). Along with planning comes the ability to translate back and forth from the visuallike image of oneself into the kinesthetically based movements of one's own body (actions) needed to effect the plan. Indeed, extensive planning may be impossible without kinesthetic-visual translation, which is present in the imitation of things other than oneself. Among modern-day humans, the capacity to imagine oneself in various scenarios is extensive and elaborate. Actors, for example, learn to "make-believe situations, emotionally charged human relationships, and to respond freely within the imagined situation" (Fergusson 1949/1972, p. 238). The words of 12- to 18-month-old children are "concerned with PLANS, that is, courses of action that are intended to achieve aims" (Gopnik 1982). For example, these children use the word *there* to indicate that an action has been accomplished, and *oh dear* to indicate that an action has failed. The child "develops a more conceptual and reflective system" after 18 months: "the child moves from solving problems by trial and error to solving them by using insight [that is, making plans]. The most plausible explanation is that he reflects on his own behavior and experience and constructs a model of the world which includes his own actions. . . . In this way he can do experiments with his head rather than his hands" (Gopnik 1982, p. 314). This "construct[ing] of a model of the world which includes the child's own actions" would be impossible without kinesthetic-visual matching. Planning is "representational" in that it allows us to "anticipate the outcomes of actions" (Cocking and Copple 1987, p. 428). But it is also representational in Walton's sense of a prop in a game of make-believe: in a plan one *imagines* oneself engaging in courses of events.

Planning seems especially evident in some activities of chimpanzees and hominids. When chimpanzees discovered food in the presence of other chimpanzees, they often waited until the others were not around before they obtained the food. Although such evidence of control in the presence of food can be thought of as learned, in some instances the chimpanzees moved away from the area and only hours later returned to obtain the food (Goodall 1986, de Waal 1982). Although the chimp may have *learned* not to touch the food in the presence of others, it also *anticipated* returning for the food when others were not around; if others were around again, presumably the chimp would go away and return later. Thus, that "chimpanzees do indeed make plans for the immediate future is obvious" (Goodall 1986, p. 588), and sometimes these plans involve tools (Goodall 1965, pp. 433–444; McGrew 1974, p. 503; Boesch and Boesch 1984). Objects and tools can be considered props not only when an organism pretends with them but also when the organism *plans* to use them (Mitchell 1993a) and thus engages in an imaginal game of make-believe — seeing-as — about the objects or tools: "foraging with tools . . . requires a propensity to *see* the potential of objects such as rocks, branches, leaves, and shells *as* tools, wipers, and containers" (Parker and Gibson 1979, p. 372, emphasis added). A similar idea was expressed by Bates (1979a, p. 326) when she suggested that "tool use is a type of symbolic activity" and by Köhler (1925/1969, for example, p. 106) when he described chimpanzees as "*see[ing]* . . . *as a stick*" objects such as a branch of a tree, a wire, a rock, and even a box. Of course, if objects are used as tools without pretense or planning but through trial and error, such use would not be evidence of seeing-as (see Parker and Gibson 1977).

More elaborate planning may have been present in early hominids. For example, "some degree" of planning seemed required in shelter construction by *Homo habilis* (Parker and Gibson 1979, p. 378), and complex forms of planning ("reconstruction of a scene displaced in space") and communication seemed necessary for cooperative hunting and "big-game drives" by *Homo erectus* (see Parker 1985, p. 625). However, evidence for shelter construction is contested (Noble and Davidson 1989b), and it is now believed that "there is little indication that *Homo erectus* hunted large animals systematically[;] it can be argued that advanced hunting techniques were not developed until late Pleistocene times" (Tattersall et al. 1988, p. 258). Still, short-term plans for "one or two days" were probably present among middle Paleolithic *Homo sapiens*, although "the *ability to anticipate events and conditions not yet experienced* was not one of the strengths of our [middle Paleolithic] ancestors" (Binford 1982, p. 178), and different levels of planning may be present in earlier and later hominids (Parker and Milbrath 1993). Although some have claimed that "the planning that makes human hunting possible . . . can only be realized linguistically" (Noble and Davidson 1991a, p. 227), the similarity in hunting strategies between humans and wolves indicates otherwise (Hall 1978; McMahan 1978; Peters 1978; Sharp 1978, p. 76).

During the early and (perhaps) middle Paleolithic, archaic *Homo sapiens* (including Neanderthals) as well as modern *Homo sapiens* appear not to have created or used external props (Lindley and Clark 1990) other than tools. However, later hominids' imaginative plans involving tools were more elaborate than those of earlier ones: "later Stone Age peoples tended to concentrate more on finer-grained, higher-quality rock sources, often quite localized in distribution and transported some distance to the archaeological site, than did hominids in the earlier phases of the Paleolithic, who appear to have exploited available rock sources in a more opportunistic fashion" (Tattersall et al. 1988, p. 543).

In addition to understanding pretense and planning, chimpanzees also understand that highly realistic representations of reality can be used to locate objects depicted in the representation. When shown a video image of their caregiver who disappeared from sight in a familiar area, chimpanzees were almost as good at finding the caregiver as when they had watched the caregiver disappear from view while directly watching the familiar area (Menzel et al. 1978). This ability to use highly realistic representations is probably based upon a similar ability to use mental representations to plan. Chimpanzees have more difficulty with three-dimensional models (Premack and Premack 1983, pp. 101–104), and some chimps recognize objects depicted in photographs (Temerlin 1975; Premack and Premack 1983, p. 101) whereas others do not (Winner and Ettlinger 1979). Human children at 2½ years of age can recognize spatial models as representations of something else through their perception of similarity between the model and what it represents, although difficulties can intrude upon their recognition (Blades and Spencer 1987, DeLoache 1990). Unlike apes, human children create external representations in pretend play starting at 1 year of age (Shotwell, Wolf, and Gardner 1980). Humans also map between internal simulations (internal props) in analogy and metaphor when they elaborate the connections between one domain and another (Gentner 1989).

2.c. Recognition of Self Dependent upon Kinesthetic-Visual Matching

The kinesthetic-visual matching abilities that humans, chimpanzees, orangutans, and gorillas exhibit when they imitate allows these organisms to recognize how their actions look visually — from the kinesthetic feel of their movements they can "conjure up" a global visuallike image of their actions. In all of these primate species, at least *some* individuals show mirror self-recognition (Suarez and Gallup 1981, Gallup 1985, Patterson 1990, Miles 1991b, Swartz and Evans 1991), and this global visuallike image is presumably what allows them to recognize their body image in the mirror (Mitchell 1992, 1993c). Thus, these organisms appear to recognize a simulation of their own body as such (Mitchell 1993c). Young humans are proficient imitators by 14 months of age, at which point they can not only imitate innumerable actions through kinesthetic-visual matching but also use this matching ability to recognize that they are being imitated when another is mimicking their actions (Meltzoff 1990). By 15 months some human children recognize themselves in mirrors, and by 18 months most do (Lewis and Brooks-Gunn 1979). Once the organism recognizes itself in the mirror, it visually objectifies its whole body (Mitchell 1993c).

The ability to enact or to pretend to enact *another's* actions as well as to recognize an imitation of oneself by another indicates a kinesthetic-visual translation module that allows for a visuallike mental representation of the organism itself that can be used to imagine itself (Mitchell 1993c). Those primates that used purely kinesthetic means to recreate their bodily expressions to manipulate others could now recognize that their kinesthetically derived replications are *visually* identical to their "normal" versions. Because of kinesthetic-visual matching, some primates have a mental representation of their body's actions (and therefore a self-representation), which they can use as an iconic prop in thinking about their activities

(Mitchell 1993c). Not surprisingly, deficiencies in symbolic play and mirror self-recognition in autistic children (Spiker and Ricks 1984, Baron-Cohen 1987) are tied to deficiencies in kinesthetic-visual matching in imitation (see Mitchell 1993c).

In simulating others' activities, primates exhibit their abilities for image making in Davis's (1986, p. 193) sense: their image "refers to its object iconically or nonarbitrarily"; it "stands for or denotes that which it necessarily also resembles" (Davis 1988, p. 224). Thus, if image making is independent of its medium, as Davis (1989a, p. 141) argues, then image making is present in nonhuman primates (Mitchell 1991a). In pretend play as well as in mirror self-recognition, non-language-using organisms clearly recognize or "see a resemblance" between a representation and what it represents, which indicates that "seeing a resemblance" can, contra Davidson and Noble (1989, p. 150), occur without language. As a result, the capacity for image making per se in hominids does not need to be explained. However, the capacity for image making on external objects does need explanation, and indeed what is largely absent in nonhuman primates is the creation of communicative props external to or on the body (see section 4 below).

It is likely that pretend activities survived evolutionarily as a result of their utility in "insightful tool use" (Parker and Gibson 1979, p. 372) and planning. Pretense of others derives ontogenetically from self-pretense (Piaget 1945/1962; Bretherton 1984, p. 9) and is, as I have argued, initially based on the ability to match kinesthetic and visual modalities (Mitchell 1990, p. 215), although other modalities can also be employed, as in verbal pretense. One hypothesis is that pretense of others was selected "as an adaptation for practicing extractive foraging" or as an "adaptation for learning tool use for extractive foraging" (Parker and Gibson 1979, pp. 367, 375).

> Human children begin to display symbolic play during the sixth stage of the sensorimotor period when they become capable of deferred imitation of novel schemes. . . . Symbolic play (imitative make-believe play) . . . is important for the rehearsal of subsistence roles. Specifically, imitative tool use is important in practicing extractive foraging. . . . Symbolic (imitative) play was selected because it facilitated social learning of tool technology and language. (Parker and Gibson 1979, pp. 374, 381)

217

It is clear that pretend play is useful in learning from others' activities how to use tools. But the learning of tool use via imitation of others' actions *presupposes* an imaginal ability (seeing-as) based on kines-thetic-visual matching and thus must follow *after* the evolution of pretense; and, as we shall see in the next section, the creation of verbal communication presupposes symbolic play or pretense. (Note, how-ever, that tool use per se is, across species, an adaptation for extractive foraging [Parker and Gibson 1979], and not all tool use implies imagination.) Pretense of others was more likely selected for its functional utility in communication, deception, insightful tool use, and planning before it had any *practice* value at all. Indeed, by two years of age and earlier children today actively use their pretend abilities to plan and solve problems (McCune-Nicolich 1981), com-municate and deceive (Piaget 1937/1954, Chevalier-Skolnikoff 1986), and use tools and objects (Garvey 1977), suggesting that little practice is needed to exhibit these capacities.

3a. and b. Communicative Simulation, and Use of Props for Others

None of the simulations presented so far is intentionally com-municative; that is, they are not enacted to inform others about anything. Intentional communication involves "the symbolic projec-tion of experience in socially meaningful terms, that is, in a mode that is intelligible inter-individually" (Hallowell 1960, p. 352). According to Hallowell, intentional communication allows for "the repre-sentation and articulation of a sense of self-awareness" (p. 352). However, an organism with kinesthetic-visual translation already has self-awareness in the form of a self-representation: it has some idea of how its actions look, even though it cannot see them. Contrary to Hallowell, this skill in kinesthetic-visual translation renders interin-dividual communication *possible* (rather than vice versa) because the translation between a kinesthetic and a visual representation makes the representation intelligible for both the creator of the repre-sentation and the observer, as the representation is "the same" for both.

As would be expected from this idea, some nonhuman primates with extensive skill in kinesthetic-visual translation use simulation to communicate. They also sometimes use props along with these simu-lations. Two chimpanzee mothers very slowly enacted tool-using behaviors for their infants to teach them how to crack nuts, watching the infant's gaze during the simulation (Boesch 1991; see also Parker and Gibson 1979, p. 374). At Gombe chimpanzees acted as though grooming a leaf to gain the attention of others (Plooij 1978, p. 122). One chimpanzee solicited another's help in pulling an object by touching her on the shoulder and then illustrated what she needed: she "took her rope, braced herself ready to pull, and turned back to look at [the other chimp] as if expecting her to come up and help"; the activity of touching on the shoulder soon came to be a conventional sign of a need for assistance (Crawford 1937, p. 57). To get another chimp to play chase — a game in which leg grabbing was integral — one older juvenile chimp "offered his leg in an exaggerated way to his partner . . . and then feigned effortful attempts to run away" (Tomasello et al. 1985). A young pygmy chimpanzee simulated activities to get his caregiver to enact the activity: "he made twisting motions toward containers when he needed help in opening twist on lids, or made hitting motions toward nuts he wanted others to crack for him" (Savage-Rumbaugh 1986, p. 386). In order to get his caregiver to give him milk, the orangutan Chantek eschewed using the sign for milk and instead recreated part of his normal milk-getting situation: he "gave his caregiver two objects needed to prepare his milk formula and stared at the location of the remaining ingredient" (Miles 1990). Chantek also invented signs based upon various types of resemblance: his sign for a man who was missing a finger required Chantek to bend that finger on his own hand; the sign for contact lens solution used the DRINK sign applied to his eye; and the sign to indicate that he would not bite was a closing of his lips with a finger, a copy of the action caregivers used to tell him to stop biting (Miles 1993). The combination of these examples with the recognition that all the signing apes (at least) learned to name themselves argues against the idea that "no . . . chimpanzee or any other non-hominid has developed a traditional means whereby it is possible for an individual to represent himself and other objects and events to himself as well as to others" (Hallowell 1960, p. 353). Evidently apes *do* invent symbols, contrary to conventional wisdom (for example, White 1960, p. 250; Donald 1991, p. 217; Noble and Davidson 1991a, p. 244), and their imitative communications are indicative of a "protolanguage" (Parker and Gibson 1979).

In a few cases apes apparently used their actions alone, without any subsidiary props, to communicate the "existence" of a nonexistent prop, an activity that in human children usually follows the use of atypical objects in pretend play (about 2 years of age) but may occur early in the first year soon after pretend play first begins (Piaget 1945/1962, Bretherton 1984). At times the chimpanzee Viki asked her human mother for assistance disentangling her imaginary pulltoy, and she seemed to know that her human caregiver recognized the pretense (Hayes 1952). In order to effect a reconciliation with a former adversary, chimpanzees have acted as if they have found something very interesting, even though nothing was there (de Waal 1986, p. 230); after the former adversary and others came near, only the adversary and the original "discoverer" stayed behind looking intently at the nothing, remaining "interested" only until they began to touch and reconcile. And the orangutan Chantek enjoyed pretending that cats and dogs were outside his home, signing *cat* and *dog* to his caregivers and acting frightened even when no cats or dogs were nearby (Miles 1986).

Pointing and eye gaze may develop from being simulations to influence others (as in the first understanding of simulation) to becoming simulations to communicate with others. Pointing in humans apparently derives from reaching to touch something interesting (Werner and Kaplan 1963, pp. 78–80; Leung and Rheingold 1981): when infant humans learn that reaching for something usually leads to someone's getting it for them or attending to it, they may begin to act as though reaching to create the same effect. (Note, however, that pointing may have extremely divergent interpretations in different cultures — "in certain South African tribes the object pointed at is thus damned" [Jakobson 1966/1990, p. 411].) Great apes point (even without interaction with humans), and some use pointing to deceive (Woodruff and Premack 1979; Patterson and Linden 1981; Savage-Rumbaugh 1984; Tomasello et al. 1985, p. 182; Miles 1986; Povinelli, Nelson and Boysen 1992). Pointing for objects can be learned by rhesus monkeys (Blaschke and Ettlinger 1987; Povinelli, Parks and Novak 1992) and cebus monkeys (Mitchell and Anderson forthcoming) as well. Apes also point protodeclaratively, that is, to indicate interest in something to another (Savage-Rumbaugh 1984, Miles 1990). Given that protodeclarative pointing implies awareness of the other's perception, it is not surprising that impaired protodeclarative pointing is associated with impaired perspective-taking in autistic children (Baron-Cohen 1989). Great apes and humans use eye gaze to communicate their intentions: for example, an ape may look back and forth from another's eyes to an object the ape wishes to obtain (Tomasello et al. 1985, Gomez 1991). In order for eye gaze to be influential, the observer of the eye gaze must recognize that the animal is not merely looking at the observer as the animal would another object, but that the organism is using its normal means of looking at the world to influence the observer; in effect, the organism is simulating its own action of looking at something.

Intentionally communicative simulations are, of course, extensive in humans. When people respond sympathetically with motor mimicry to another's pain, for example, they communicate their sympathy to the other by "portraying" their similar response, "sending the message, 'It is *as if* I feel as you do' " (Bavelas, Black, Lemery, and Mullett 1987, p. 329). Even young children recognize that the repetition of others' speech or mannerisms, often with exaggeration, is a way to insult them; and more exact repetition can be an attempt at humiliation. In 2-year-olds, imitation "may function in interpersonal situations to communicate mutuality and sharing of understanding": "to do the same thing as the other is to confirm a similarity in capability, interest, or feeling with the other" (Uzgiris 1981, p. 7). Humans replicate their actions when teaching others how to do something, so much so that in "many languages the word for 'teach' is the same as the word for 'show,' and the synonymity is literal" (Reichard 1938, p. 471). Deaf children raised by parents who elected not to teach them sign created their own sign system based in large part upon simulation. In one instance, the child David "points toward a plate of food, puts his fist to his mouth and makes biting motions, and then points at one of us" (Feldman, Goldin-Meadow, and Gleitman 1978, p. 360). Before learning to finger-spell, the deaf and blind Helen Keller apparently used pantomimic signs, such as "making a sign of turning a key" after finding a keyhole in a locked suitcase (Sullivan in Keller 1902/1965, p. 178). Even deaf children taught sign language create simulative gestures to communicate, as did one 4-year-old who used "an empty drinking gesture, without an object, to convey to the hearing adult his need for a glass of water" (Bates et al. 1983, p. 76). Nondeaf children also develop simulative communications (Van Meel 1982, Acredolo and Goodwyn 1985), as the previous description of the child J's re-creation of the trajectory of his block to communicate its location to his father indicates. And adult humans use iconic gestures in their normal communications to support or add meaning to their speech (Efron 1972/1978, Ekman 1977/1980, McNeill 1985).

Speech is also used simulatively to play, understand, or humiliate via irony, satire, metaphor, parody, and punning, and to demonstrate in quotation (André 1882–1883, Macdonald 1960, Allen and Stephens 1962, Redfern 1984, Clark and Gerrig 1990). Some simulative speech is observed quite early in humans; for example, a young deaf child created a pun when she "shortened the sign LONG instead of signing SHORT to tease her father when he asked for a long kiss" (Maxwell 1983, p. 290). Puns and simulative irony are amusing because the simulation hides yet shows off what is simulated and thus offers a double meaning (André 1882–1883; Shipley 1972, p. 59; Capelli, Nakagawam, and Madden 1990, p. 1837). Double meaning is also present in visual puns and rebuses, though it is not usually as well hidden (André 1882–1883, Hammond and Hughes 1978). Metaphors sometimes offer double meanings, but one meaning is hidden from an outgroup (Crider and Cirillo 1991, pp. 184–185). In irony, punning, and metaphor, one discovers (or uses) an attitude taken toward the simulation that differs from the attitude that one takes toward what is simulated; thus, these forms of simulation are playing with perspectives (see Crider and Cirillo 1991).

Communicative simulations are, many have argued, the basis for the evolution of language in humans (Parker and Gibson 1979, p. 374; Davidson and Noble 1989; Kendon 1991; Mitchell 1991; Noble and Davidson 1991a). What factors led to the selection of organisms that could communicate simulatively are, however, unclear. The evidence of communicative simulation in nonhuman primates indicates that apes are over the "Gricean hump" separating natural from nonnatural meaning (Tennant 1984, p. 85; Davis 1989a, p. 140). Thus, theories that assume that only humans create communicative simulations (for example, Davidson and Noble 1989; Kendon 1991, p. 214) are inaccurate. Apes and humans, then, know or expect that the other perceives the world in much the same way that they themselves do and can use this knowledge to inform the other. Thus, these organisms have knowledge of other minds: they interpret the other as having perceptual states and can share knowledge. Such knowledge of other minds, coupled with knowledge of how one looks visually, derived from kinesthetic-visual matching, allows for greater intentional communication as well as intentional deception, because now the organism recognizes how it appears *from the other's point of view* (see Van Meel 1982; Mawby and Mitchell 1986; Mitchell 1993a). Given that social knowledge is useful and was probably selected for in primate evolution (Whiten and Byrne 1988b; Steele 1989), such a means of increasing social knowledge is significant. Also, mutual knowledge engenders the development of conventions, which require that individuals in a group recognize that each individual is using a sign in the same way that every other individual is using the sign (Lewis 1969). This ability to "communicate mutually" is necessary for personhood and a sense of moral responsibility (Ishiguro 1980, p. 74; see also Dennett 1978; Mitchell 1993b).

The other's point of view is elaborated via pretense, in what Gordon (1986) called "practical simulation" of another's activities. One pretends that the other has or does not have particular knowledge and desires and, by imagining how the other would react in particular scenarios, thinks through what the other will do given this knowledge and these desires. Much as the organism with kinesthetic-visual matching can imagine a plan for itself, the organism with perspective-taking can imagine its own activities *and* those of the other in a given scenario. Such practical simulation is a model of the other, and "building" such a model depends upon imaginative pretense (Perner 1988). In Gordon's (1986, p. 169) view, "to attribute a belief to another person is to make an assertion . . . *within the context of practical simulation.*" Practical simulation of the other does not (contra Gallup 1985) employ analogy with one's own mind or projection from one's experience: "To make decisions in such a pretend-world is not to decide what *I myself* would do much less to reliably know what *I myself* would do 'in that situation.' . . . in such a case I cannot be making an implicit *comparison to myself*" (Gordon 1986, p. 165). The practical simulations of human children and adults are quite extensive: they can model another person's perspective, another person's perspective of another person's perspective, and imaginal persons' perspectives (Merleau-Ponty 1960/1982; Landry and Lyons-Ruth 1980; Sacks 1980; Reynolds 1981, p. 210; Watkins 1986; Perner 1988; Fridlund, Sabini, Hedlund, Schaut, Shenker, and Knauer 1990). Once an organism has practical simulation, "the simulated perception of how others will regard [the organism's] action" can be used instrumentally to manipulate these others (Reynolds 1981, p. 210).

One salient difference between human and nonhuman primates is that the latter rarely use props to communicate simulatively. When they do, the props are ad hoc (as in chimpanzee leaf-grooming) or are identical to the objects that were (1) used by the recipient of the communication (as in the chimpanzee mothers' tool use) or (2) used in a normal situation (as in Chantek's milk-getting communication). Remember that some props used noncommunicatively by nonhuman primates in pretense are also ad hoc — for example, the coconut shell the rhesus monkey used to represent the baby. Thus, nonhumans' props are either direct or arbitrary in their reference, whereas for humans communicative props are largely iconic, though they can be direct or arbitrary. What is interesting about the integration of props into plans or games of make-believe is that the varying possible uses of the props are greatly magnified by an imaginative component: although the props themselves may suggest particular uses by their shape and feel (Walton 1990), imaginative uses of props are not constrained by the props themselves (see de Laguna 1927, p. 314). As a result, objects used as props for imagination may come to have more diverse and potentially beneficial uses than objects not so used.

4. Creation and Subsequent Use of Props External to or on the Body as Representations for Others as Well as Self

However much they might use *available* objects as props for pretense or communication, nonhuman organisms apparently rarely *create* props for communication and thus rarely create extrabody representations for others and themselves (Hallowell 1960, p. 353). Although apes given crayons or other markers create markings on diverse media, these markings are not representational in Walton's sense; given enough time, the ape will cover the entire medium with markings (Köhler 1925/1959, Whiten 1976). Apes apparently do not use "extrinsic symbolization" (Hallowell 1960, p. 356) or "impos[e] arbitrary form upon the environment" (Holloway 1969); "apes can iconically manipulate their bodies, whereas humans can iconically manipulate both their bodies and external constructions" (Reynolds 1981, p. 164). Current-day *Homo sapiens* clearly create external representations for communication in artistic representation, ritual, and narrative (Biesele 1983; Liska, Chapter 7 of this volume). For example, !Kung rock art uses elaborate metaphors in its creation and understanding (Lewis-Williams 1981, 1983, 1986; Huffman 1983). Current-day humans also use external representations to disguise, such that "there is abundant ethnographic evidence of depictive mimicry in connection with hunting and fishing, often involving the use of animal skins, feathers, horns, etc., for disguising hunters or decoying game" (Hewes 1989, p. 145). For example, although they now simply use stealth to capture prey, !Kung in the nineteenth century disguised themselves as ostriches and antelopes while hunting these creatures, and represented these disguises depictively on walls of caves (Thackeray 1983). (!Kung believed that their imitations of prey were effective because the animals were under the control of shamans [Thackeray 1983], which suggests that knowledge of simulation was unimportant in the disguise. However, it seems likely that people developed their belief in shamanistic control to explain the variability of their success in deceiving the animals via simulation and the apparently mesmerized look of the animals watching them, a look that derives from the animals' curiosity [Thackeray 1983] about the familiar yet unfamiliar creature moving toward them.)

The representations of animals in upper Paleolithic hominid cave art and the life-size and miniaturized sculptures of animals and humans indicate the creation of external props, which were probably used, at least sometimes, for communication or group activities, or for associations between groups — that is, for visual consumers (see, for example, Marshack 1976; Stoliar 1977; Conkey 1983, 1984, 1985; White 1985). The artworks are "formalized, repetitive systems of visual imagery" that suggest "ritual communication" (Conkey 1985, pp. 307, 317). The two-dimensional upper Paleolithic representations are often extremely realistic depictions (simulations) of animals (Marshack 1991), in some instances even representing movement (Proudhommeau 1990). Calendars containing images of the waxing and waning of the moon were also made (Marshack 1976, 1991; Tattersall et al. 1988, p. 420). Animal representations are not all simulative, however: "imaginary or fantastic creatures" are also found (Tattersall et al. 1988, p. 423), suggesting imaginative processes unconstrained by perception (Hallowell 1960, p. 355). (The monsters of current-day children apparently have a long lineage.) Indeed, humans created "graphic representation of . . . animals . . . which could not have been present in the perceptual field of the artist when the drawings were made" (Hallowell 1960, p. 354), although it is impossible to know if the cave drawings were reworkings of drawings made while observing animals. Note that "the bulk — if not all — of the Paleolithic art as we know it is a behavioral phenomenon that clearly postdates the establishment of modern *H.*

sapiens, and that — because of its relatively limited distribution — is most likely the particular product of one or more regional socioecological contexts" (Conkey 1983, p. 222; see also Conkey 1984, 1985). External or extrinsic symbolization, then, may not be a defining characteristic of modern *Homo sapiens*.

At the same time that some humans created external representations in the upper Paleolithic, some also created " 'items of personal display' . . . that can be assumed to be for body adornment and 'decoration,' in a very broad sense" (Conkey 1983, p. 213; see also Mellars 1973, p. 259; White 1982, p. 170). Pendant beads used as "personal decoration" may also have been "made and worn by Neanderthals" during the middle Paleolithic (Tattersall et al. 1988, p. 428; but see Mellars 1973, Lindley and Clark 1990). Self-decoration suggests not only that one recognizes how others view oneself but also that one can take that point of view toward oneself: "self-decoration presupposes self-awareness and the capacity for objectification, which are both necessary and sufficient conditions for art" (Conkey 1983, p. 215).

The decoration of one human's body may have resulted from another person's decorating it, however, rather than from *self*-decoration (see Polhemus 1978, Fallon 1990). In the past and even today, many practices to transform the body image are mildly or even horrifically painful; individuals are forced by others to endure pain and torture, or willingly torture themselves, to satisfy cultural standards of beauty, masculinity, femininity, or whatever (Polhemus 1978, Fallon 1990, Herdt 1990). Many human children today have their clothes, jewelry, makeup, and hairstyles selected for them by their parents. Thus, in ontogeny and probably in history, transformation by others as well as creation of props in extrahuman media precede self-transformation. The existence of self-decoration and other-decoration in humans is intriguing because it defines, in our current state of knowledge, a distinct difference between human and nonhuman organisms: humans alone create images of themselves (and others) that are aesthetically or culturally satisfying to themselves and/or others (Reynolds 1981, p. 210; Mitchell 1993c).

Human adornment of the self has an aspect of "self-objectification" (Hallowell 1960, p. 344) — the subject must view his or her body as an object to be transformed. Such an objective view toward one's body likely develops from others' perception of one's body as a transformable object and from their actions upon one's body, and

226

this view results in recognition of the other's perspective on the self (Duval and Wicklund 1972; Reynolds 1981, p. 210). Among current-day humans, adornment and changes made to the body seem directed toward creating, evaluating, and posturing an ideal image of oneself either for others or for oneself. Combing, cutting, or otherwise styling one's hair; putting on makeup; having tattoos; and deleting or adding to one's body in other ways (or having someone else do it for one) suggest that one is trying to make one's image conform to a socially or personally acceptable ideal. When humans look at themselves in mirrors, they are usually not, like apes, merely recognizing their own image (Mitchell 1993c); rather, what "moves us to pride or shame is not the mere mechanical reflection of ourselves [in the mirror], but an imputed sentiment, the imagined effect of this reflection upon another's mind" (Cooley 1902/1983, p. 184). As a result of our attention and conformity to others' perceptions, fads in personal style develop; one need only observe the changes in fashion and hairstyles in human history to recognize our consistent desire to please others (Wax 1957, Meyersohn and Katz 1957, Fallon 1990). Initial transformations of the body may have resulted in sexual selection that favored those who were more desirably adorned (Hutchinson 1959, Low 1979), and such changes may account for the diversity of images humans consider erotic (see, for example, Krafft-Ebing 1893/1978, Stekel 1923/1952).

The human concern for visual conformity in the body's "look" and decoration exemplifies our desire to live up to (or defy) others' desires concerning our appearance and standards of good conduct (Freud 1930/1962; Goffman 1959; Hallowell 1960, p. 346; Nagel 1969/1979, pp. 44–47). Humans also experience self-evaluative emotions, such as shame and mature embarrassment, that incorporate self-awareness and cultural standards (Bennett 1989; Lewis, Sullivan, Stanger, and Weiss 1989). For humans, the experience of self incorporates the other's perspective into one's self-image (see Cooley 1902/1983; Mead 1934/1974; Wallon 1954/1984, p. 128; Goffman 1959). Humans thus evidence a "reflective self-awareness" (Katz 1937/1953, p. 178; Mitchell 1993c) that implies "a psychological system of standards of correctness that is possessed by each person" (Duval and Wicklund 1972, p. 3). The term *reflective* is used to imply both that one's self-image is experienced through others' perspectives (that is, the self-image is "reflected back" from the other) and that one is capable of self-examination (that is, one can "reflect on" and evaluate one's bodily image as well as one's thoughts). Mature empathy presumably develops from people's reflecting on the other's plight via practical simulation, along with "imagining themselves in the other's place" (Hoffman 1977, p. 183). Evidence of empathy in compassion toward impaired individuals and ritual burying of the dead in Neanderthals and other humans during the middle Paleolithic is ambiguous (Gargett 1989, Dettwyler 1991).

Given that great apes and humans alike have kinesthetic-visual translation skills and communicative simulations, they have a representation of their actions that eventually becomes recognizable as the same as the representation others have of them. Yet apes generally fail to show any evidence of adornment or reflective self-awareness (Reynolds 1981, p. 210; Mitchell 1993c), of "consciously relating or appraising [their] own conduct with reference to socially acquired values" (Hallowell 1960, p. 351). Their representation of self via kinesthetic-visual translation is not an ideal but simply a cognitive stand-in for their own body and its actions. Although apes recognize that others have perceptions and are influenced by these perceptions to present, hide, or provide false information, they are (as far as is known) not influenced by these perceptions to alter their bodies or their self-images (for example, experience shame). Apes may "adorn" themselves with fabrics and their bodies with colors, but they do not appear to do so to make themselves more attractive to others; rather, the adornment has a playful quality or enjoyable cutaneous feel to it (for example, Köhler 1925/1959, p. 93; Patterson and Linden 1981). Apes must discern whether their own actions conform to or deviate from their plans, and they have ideal interactional norms about relationships between particular individuals, such that if these norms are violated, efforts are made to reconcile (de Waal and van Roosmalen 1979); but they do not seem to have actions that conform to or deviate from an ideal self-image based on societal norms (de Waal 1991), specifically because they have no way to internalize an ideal self-image.

Of course humans have societal norms and ideal self-images, and they have ways to internalize an ideal self-image that allows for norms — external imagery and speech. By providing external images, image makers can represent their ideal image that observers can attempt to reproduce. With speech, others' expectations concerning the listener or observer can be made specific and applied to the self. Both imagery and speech, therefore, can be used to evaluate the self against a standard (see Cash and Pruzinsky 1990). Speech is also internalized directly in inner speech, which can function as a means by which "the self can acquire information about itself and form a coherent picture of what it is" (Morin and Everett 1990, p. 338) independent of others' expectations. Social norms are parasitic upon practical simulation, as when recognition of guilt and shame "puts the child in a position of *anticipating* how acting in a certain way might make the child or another feel, thereby giving the child the option of enacting or inhibiting behavior" (Ferguson, Stegge, and Damhuis 1991, p. 837; see also Harris 1989).

Because humans use speech and imagery to internalize and evaluate their bodies and selves, it is not surprising that apes given the basics of speech can use signs to internalize their caregivers' evaluations. For example, the sign-using orangutan Chantek labeled his own and others' disapproved behavior as *bad* (Miles 1993). Although variant interpretations can be made of this sign-use, it suggests that once an ape has a means of marking and internalizing others' perspectives on itself, it may, like a human, come to internalize norms via an ideal self-image that it can compare to its actual behavior and appearance.

SUMMARY AND CONCLUSIONS

I have tried to show the consequences of the development of an understanding of simulation — in particular, that more developed understandings of simulations lead to more developed understandings of self and other. The sequence presented in this chapter details a series of concentric circles, each more exclusive than the next in the type of simulative pretense that is present. That primates within these circles are closely related species and share psychological characteristics suggests that these characteristics (or their substrates) are homologous and were thus shared by a common ancestor.

In the outermost circle are many mammals and birds, which can replicate their own actions to repeat desired consequences or feel out a problem; here the organism is pretending to enact a previous behavior. Thus, pretense starts from kinesthetic simulation of one's own activities, along with recognition of the instrumentality of this simulation. Such pretenders can also recognize another's self-simulation by comparing past and current activities. This ability to recognize similarity and hence simulation in the external world is perhaps dependent upon thalamocortical representation, which allows for iconic representation of information in visual, tactile, and auditory modalities (Altman 1978).

In the second circle are great apes and humans, who can imitate quite generally others' actions and recognize themselves in mirrors, both via kinesthetic-visual matching. Dolphins also fit in this circle and perhaps the next as well (see Tayler and Saayman 1973; Marten 1991; Mitchell 1987, 1990 1993c). If we take the two descriptions of imitation in rhesus monkeys (Breuggeman 1973, Rivers et al. 1983) described above, we see that some rhesus monkeys appear to be in this second circle, but the extent of their ability for kinesthetic-visual matching may be much more constrained. Recognition of simulation in the external world is connected to internal simulation when kinesthesis is mapped to vision (and probably tactual and auditory modalities as well). (Dolphins also apparently engage in auditory-visual matching in echolocation, but their matchings in this cross-modality are much more extensive than those in primates.) Kinesthetic-visual matching spatializes the self and makes information about one's external appearance "accessible" (Rozin 1976) to internal imagery (conscious or unconscious). The capacity for planning one's actions beyond simple plans such as "do X to get Y" is a direct result of kinesthetic-visual matching, in that the organism now has a means of imagining itself in various scenarios. The organism can now pretend to be another, know that another is acting like itself, and imagine itself engaging in particular activities.

In the third circle again are great apes and humans, who can simulate actions to communicate to others. When an organism has kinesthetic-visual matching, simulative responses of others to its actions inform the organism that the organism and the others see its actions as the same. Thus, communication via nonnatural meaning becomes possible. Note, however, that only humans regularly imitate another's actions for the other, that is, present the other's visual appearance to the other, a sharing of the other's visage common in mother-infant play (Uzgiris 1981) as well as in humiliation and teasing. Knowledge of another's psychology can be elaborated by incorporating the other into imagined scenarios and focusing on its actions within the scenarios (practical simulation). The organism can pretend for the other, understanding minimally that the other perceives what is presented and in some cases taking the perspective of the other via practical simulation.

In the innermost circle are humans (and perhaps sign-using apes to a small degree), who have reflective self-awareness. The organism not only can imagine itself from the other's imagined perspective but tries to conform bodily and personally to the other's imagined perspective. Such transformations of the self by the self in adornment and cultural (normative) behavior show an objectification of the self and are derived from and dependent upon practical simulation. Remember, though, that transformation of the self *by others*, sometimes by force, is fairly common among humans and can be viewed as consistent with the human creation and use of external props: the other's body is just another prop. Transformation of the self's appearance may follow transformation of the self by others historically, as it does ontogenetically.

By focusing on the "analogical" nature of ostensively arbitrary symbolic systems, the simulative nature of primate imitation, deception, pretense, communication and self-recognition, and the basis of art in make-believe and of make-believe in simulation, I have delineated an evolution that shows human psychology to develop from the same simulative processes as those of other primates yet that indicates how the distinctly human creation of external props (including adornment of others) and self-adornment might have evolved from earlier forms of pretense. The simulative programs of humans and other primates that are responsible for pretense and thus for self-awareness, planning, mutual knowledge, and reflective self-awareness also serve as a basis for cross-modal matching, symbol formation and language, artistic expression, mathematics, science, religion, and the creation and understanding of analogy and metaphor (Hadamard 1945/1954; Werner and Kaplan 1963; Knowlton 1966; Kaplan 1979; Parker and Gibson 1979; Rappaport 1979; Kline 1980, p. 350; Bretherton 1984; Millikan 1984, pp. 240–244; Gentner 1989; Walton 1990) — a broad spectrum indeed.

In writing that "we picture facts to ourselves" and that "a logical picture of facts is a thought," Wittgenstein (1921/1961, pp. 15, 19) suggested that simulation is the basis for human psychology. He also suggested the recursive nature of thought by stating that "a picture is a fact," which, combined with his other statements, implies that we not only picture facts to ourselves but also picture these pictures of facts to ourselves, and so on. Yet pictures do not have to be accurate or realistic, and correspondingly we may have erroneous "pictures" of the world. But it is likely that we and other primates are able to understand the world, each other, and ourselves even a little bit because our pretenses are based on "pictures" that are "attached to reality"; in fact, as Wittgenstein (1921/1961, p. 15) so evocatively stated, they "reach . . . right out to it."

$$7$$

The Foundation of Symbolic Communication

JO LISKA

There is no knowing ourselves individually until we know ourselves as a species.

— Paul Shepard

ABSTRACT

The focus of this chapter is on the evolutionary foundations of communication. Specifically, my thesis is that a process of ritualization/representation laid the foundation for the evolution of word symbols and syntactic symbols. This process altered neural structures and processes in such a fashion as to afford increasingly arbitrary ways for characterizing relationships independent of reality. I offer a system for comparing the semiotic repertoires and abilities of various species.[1]

INTRODUCTION

Of all extant species, only humans appear to have the peculiar proclivity for creating symbols and sequencing them according to a set of rules that specify the relationship among those symbols, that is, grammar or syntax. Other animals use signs to exchange information, but our present and admittedly limited level of knowledge indicates that humans are the only species known to have created signs that are arbitrary and thus symbolic.

The ability to create symbols undoubtedly affected the success humans have had in colonizing almost every ecological niche on this planet. If evolutionary "success" is, particularly for mammalian species, in part attributable to the ability to process, store, share, and act upon information (in the form of signs), then the ability to create arbitrary signs (symbols) would have exponentially increased the information-processing capacity of early hominids, thus contributing substantially to their "success." Among other advantages, symbols provided the means for engaging in abstract thought, for categorizing, organizing, and storing information with increasing ease and efficiency, and for generating new symbols that represented new concepts.

Further, the development of what I refer to as syntactic symbols made it possible to represent counterfactual relationships, which make imagination, planning, and speculation not only possible but communicable. As Sebeok (1987) has suggested, "Syntax makes it possible for hominids to not only represent immediate 'reality' . . . but also to, among animals, uniquely frame an indefinite number of possible worlds" (p. 14).

It seems unlikely that language suddenly appeared as a fully developed system. Thus, I am interested in what I call semiogeny, that is, the evolutionary stages or phases in what can be described as a continuum of the development of semiotic systems. I call this conceptual system — which has proved useful for making comparisons across species — the symbolicity continuum.

In this chapter I focus on the evolutionary foundations of nonlinguistic signs that through a process of ritualization/representation laid the foundation for the evolution of word symbols and syntactic symbols. Specifically, my thesis is that the process of ritualization/representation produced signs I refer to as ritual and iconic semblances, respectively, which freed early humans from a dependence on what might be called "hard-wired" signs and, in the process, dramatically altered the character of neural structures and processes, resulting in increasingly arbitrary ways for characterizing relationships independent of reality.

Further, the evolutionary approach I take rests on several assumptions. First, the survival of any given species is largely dependent upon its ability to acquire, process, store, and share information. Second, there is considerable evidence for a phylogenetic increase in semiogenetic abilities in mammals generally and in primates specifically. Third, there was an evolutionary development of perceptual and neurocognitive abilities that are prerequisite to or facilitate information-processing capabilities. Fourth, the sign systems that have evolved are "particularly important in evolutionary theory because of the possible role the systems play in the origin of new species" (Wilson 1972, p. 4).

Finally, it is important to note that I am not concerned with the evolution of *speech*. Specific and preferred sensory channels various species use for processing and transmitting information developed as a consequence of constraints imposed by the environment and anatomical history of the species (cf. Hockett 1959, 1960; Hewes 1973; Milo and Quiatt, Chapter 10 of this volume). Theoretically, language, as a means for constructing and representing a largely symbolic reality, can be transmitted via any channel, and the channel selected depends upon the particular sensory potentialities of any given species as they developed within the constraints of particular environments. Primates are generalists characterized by a high level of sensory integration; however, all primates are especially sensitive to information presented in the visual and auditory channels and use both for communicative purposes. While my illustrations concentrate primarily on visual signs, the argument applies to all sensory channels.

SYMBOLICITY CONTINUUM

I have argued elsewhere (Liska 1986; 1987) that signs can be arranged along a continuum of symbolicity or arbitrariness and that this pattern is hierarchical as well as continuous. Signs that are symptoms bear a natural, functional, or physiological relationship to their significates, and these signs define one end of the continuum. Syntactic symbols define the most arbitrary end of the continuum and are characterized as symbols that specify more than the relationship among symbols; that is, the elements of syntax are symbols in and of themselves. In between these two extremes are ritual and iconic semblances, proper symbols, and conceptual symbols, in that order of arbitrariness.

I have diagrammed these semiotic stages in Figure 7.1. The underlying dimension is the arbitrariness of the relationship between

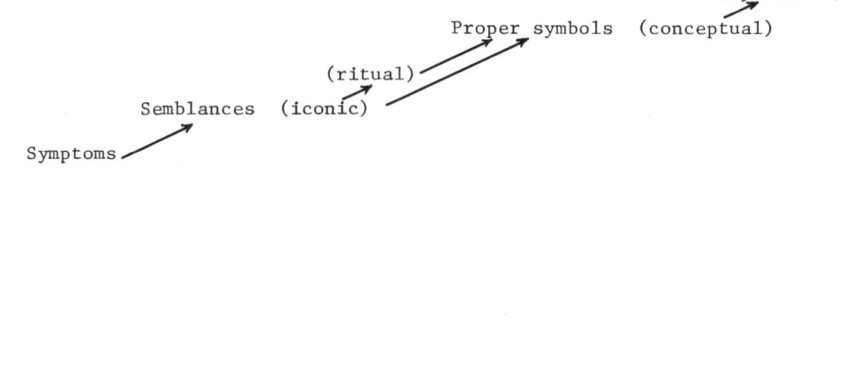

Figure 7.1 Stages in the symbolicity continuum.

sign and referent. In this case, arbitrariness refers to the extent to which the relationship is built on similarity or analogy. I have arranged the sign categories to reflect their order of relationship. I do not mean to suggest that these stages are separate and/or linear. They are highly interdependent, with each stage augmenting the preceding ones — that is, one stage did not replace another, but each expanded the existing communication potential. Thus, human interaction is an elaborate orchestration of the combined semiotic stages. I do suggest that the evolutionary pattern of semiotic stages follows a particular order: first symptoms, then semblances, and finally symbols.

It is much more difficult to determine the patterning of sub-stages. I assume that proper symbols preceded conceptual symbols just because naming seems to be more fundamental. Naming is fundamental, if not to category formation, certainly to communicating about categories and perhaps to thinking about them analytically, too; and so it seems likely that proper symbols would have preceded conceptual symbols in the symbolicity continuum. That vervet monkeys appear to "name" predators (Cheney and Seyfarth 1990) and dolphins adopt signature whistles apparently to name themselves and others (Tyack 1991) suggests that naming is fundamental to conceptual development. Of course, that dolphin and vervet monkeys may

236

use conceptual symbols does not obviate the possibility that conceptual symbols may be a part of their semiotic system. Additionally, human children appear to develop first a vocabulary of names for significant objects and people in their environment (*mama* and *dada* usually developing first) with the question "What's that?" appearing early in the process (Dale 1976, p. 11). Moreover, highly conceptual symbols such as function words (e.g., articles, conjunctions, affixes, and prepositions) are "universally missing from children's earliest utterances" (Cazden and Brown 1975, p. 299). However, fundamental rules of syntax such as word order probably predated rapid elaboration of conceptual symbols and antedated a basic vocabulary of names. I would suggest as well that once symbols first emerged, early humans would have quickly seen their advantage, so that names, concepts, and patterns for organizing them (syntax) would have coevolved closely and rapidly. The elaboration of a basic form of language such as pidgin into creole by the children of one generation (Bickerton 1990) is suggestive of the way language may have first evolved. The advantages of a symbolic language are clear, yet it appears that getting to the symbolic stage takes millennia and must have happened but rarely.

Signs that are symptoms are the foundation for all animal sign systems, regardless of the code in which those signs are encoded and decoded; that is, they may be auditory, visual, tactile, or olfactory. Such signs are the result of pure genetic transmission and are hardwired in the central nervous system; they require no learning whatsoever and are not at all arbitrary. "Symptomatic signs directly achieve their effects or, put another way, directly satisfy biological needs" (Liska 1986). Blushing, pupil dilation, eye blink rate, perspiration, and sex skin swelling are examples of hard-wired signs and are generally referred to as fixed-action patterns. Symptomatic signs may be governed by epigenetic rules, which are conceived as "genetically determined procedures that direct the assembly of mind, including the screening of stimuli by peripheral sensory filters, the internuncial cellular organizing processes, and the deeper processes of directed cognition" (Lumsden and Wilson 1981, p. 7).

Semblances are signs that either resemble the symptoms of the conditions on which they are based or bear a perceptual similarity or resemblance to their significates. Ritual semblances have their origins in symptoms but become exaggerated and transplanted into other contexts. This process of exaggeration and transformation is generally referred to as ritualization (Jolly 1985). For example, it has been argued that tongue showing begins as a means of ejecting the nipple and then becomes generalized to communicate rejection of unpleasant social stimuli as well (Smith, Chase, and Lieblich 1981). The play face, greetings such as hugging or kissing, and bluffs are semblances that are based on symptoms. Iconic semblances resemble their significates, as in the case of a line drawing of a person, a map of a territory, or a photograph of a gorilla. A semblance might also take the form of acting out, as in the case of showing someone how to make a stone tool or engaging in playfighting, an act of ritualized aggression. A semblance sign shares some quality or property with the entity for which it stands. A semblance, then, is neither totally symptomatic nor totally arbitrary.

Symbols are totally arbitrary signs. They may be proper symbols in that they refer to a specific stimulus (not an aggregation of stimuli, a stimulus complex, or an abstraction), or they may be conceptual symbols, which refer to highly abstract concepts and have no external referents. Examples of proper symbols are names of objects, a person's first or last name, and color terms, as least when applied to a specific colored light or color swatch. One crucial characteristic that distinguishes the acquisition and use of proper symbols from the acquisition and use of conceptual symbols is that conceptual symbols must be explained in terms of a concept, schema, or an "internal mediating response" (Osgood 1963). Proper symbols, in contrast, can be explained in terms of operant or classical conditioning or some combination of the two.

In addition to naming a specific stimulus, such as one's given name, proper symbols can name a class of stimuli, such as primates. Naming a class of stimuli based on some dimension(s) on which they are similar probably requires more cognitive complexity than does naming a specific object. Proper symbols that stand for classes of stimuli are a link between conceptual symbols, for which there are no external referents, and proper symbols that name specific objects or organisms.

There appears to be general agreement that "language" is fundamentally characterized as a set of symbols that can be arranged and rearranged according to a set of rules, typically referred to as syntax or grammar. Syntax is generally defined as a system of rules for expressing the relationships among symbols. I would go one step further and suggest that like word symbols, the elements of syntax

are clearly arbitrary and as such, conform to the definition of *symbol*
I and others have adopted. Thus, I have added syntax to the symbolicity continuum as a new category of symbol, henceforth referred
to as syntactic symbols. Syntax is clearly arbitrary in that it need not,
and frequently does not, reflect the order of events in reality. Further,
syntax *stands for* a relationship. A particular word order, for example, is a symbol. Syntactic symbols, unlike word symbols, encapsulate
who does what to whom. While the word symbol *hit* specifies a
relationship, it does not *stand for* the relationship among actors. The
relationship among actors is a function of word order, function
words, affixes, oral inflection, and written punctuation, in English at
least. Consequently, we can say "the boy was hit by the girl" or "the
girl hit the boy." Further, we can say that the "boy hit the girl" even
if that order does not conform to the actual sequence of events.

English speakers can recognize the difference between a declarative sentence and a question solely on the basis of the placement of
the verb. To form a question from a simple declarative sentence, one
places the first verb in the sentence at the beginning. Thus, to form a
question from the declarative sentence "Mary is here," one moves
the verb *is* to the beginning of the sentence and says "Is Mary here?"

Tense may be made explicit by the use of word symbols such as
today, *yesterday*, or *tomorrow*. However, word endings and verb
forms, which are best conceived as syntactic symbols, specify tense
without the use of the aforementioned word symbols. Thus, I can say
"I am going to a conference in Poland," and it will generally be
understood that I have not yet gone but will go sometime in the future.

Syntactic symbols also stand for concepts. That sentences may
consist of a subject, verb, and object is both a rule specifying a
relationship and a concept. Plural endings that indicate numbers of
entities or possession are created by syntactic symbols and represent
the concepts of number and ownership.

Function words, including conjunctions, prepositions, articles,
and pronouns, are other instances of syntactic symbols as they specify
relationships among word symbols and stand for concepts inherent
in those relationships. The function word *because* indicates a causal
relationship, whereas the function words *on*, *near*, and *above* indicate
spatial relationships. Function words develop rather late in the acquisition of languages and may well have been missing in the early
stages of linguistic development of early hominids. Fromkin and
Rodman (1983) wrote that "these first utterances of children that are

longer than two words have a special characteristic. Usually the small 'function' words such as *to, the, can, is,* and so on, are missing; only the words that carry the main message — the 'content' words — occur. Children often sound as if they were reading a Western Union message, which is why such utterances are called 'telegraphic speech' " (p. 330).

Mitchell (Chapter 6 of this volume) writes that "arbitrariness is not essential as evidence of symbols, especially when simulative signs used repetitively in a communication system naturally become more arbitrary in form." As I understand and use the term, *arbitrariness* is the degree to which a sign is independent of the reality to which it refers. Thus, the arbitrariness of any sign/referent relationship is continuous, not discrete. Mitchell seems to employ *arbitrariness* as an either-or proposition. He writes that "conventionality and arbitrariness criteria are often used to deny symbolization to nonhumans. . . . Nothing could be more arbitrary than the differences among alarm calls by vervet monkeys that differentiate predators." Using my system, vervet calls are proper symbols, which are certainly arbitrary. However, there are signs that are more arbitrary than proper symbols, and those are conceptual and syntactic symbols.

At some point signs become so arbitrary as not to reflect but to create reality. This point appears to be the juncture of proper and conceptual symbols. Proper symbols name reality; conceptual symbols create reality. Some symbols are a blend of proper and conceptual symbols in that they can serve as both. The word *king* stands for both a physical entity and a concept. And for most humans, the meanings of the entity and concept are highly interdependent. Possibly the same is true of language-using simians. But we cannot know unless we make the distinction. I used this distinction between proper and conceptual symbols in an analysis of data from ape sign language projects (Liska 1987) and found that most of the acquired signs named objects, people, and other apes (proper symbols) rather than expressing concepts and ideas. They did use signs to express internal states such as hunger, desire for attention, and so on, which appear to be more like conceptual than proper symbols.

I am not suggesting that arbitrariness is the only dimension on which signs differ. Other dimensions such as abstractness, the amount of pretense involved, and/or the degree to which the sign-referent relationship is based on similarity/simulation are probably highly correlated with arbitrariness. Grice's (1982) distinction between natural and nonnatural meaning as meaning based on analogy also appears to be highly related to arbitrariness as I have described it here. The less a sign resembles or is a simulation of the real object, the more arbitrary is the relationship. Of course, conceptual symbols do not have any external referents, as they stand for relationships that reside in the shared symbolic reality of the users. In Grice's terms, conceptual symbols have no physical analogs and are therefore nonnatural. They are defined in terms of other symbols. Symptoms are semblances that have natural meaning in Grice's terms because they have referents based on analogy.

Grice's distinction, though, suggests that nonnatural signs are artificial, that is, not real. Yet people act as if symbolic realities were real. Moreover, the ability to acquire, use, and create arbitrary signs has clearly become genetically wired into humans and is thus "natural." Using language is to humans what flying is to birds — we do it because we are programmed to. However, the integration of genes with culture obviously was a long time in the process and may well have passed through the semiotic stages I propose here.

RITUALIZATION/REPRESENTATION

I have elected to use the term *ritual semblances* to refer to signs that resemble the symptoms on which they are based. These relationships of resemblance have evolved in a process of ritualization Jolly (1985, p. 196) defined as "an item of behavior [that] has been subject to selection that has increased its communicative value, in particular, that has made it less ambiguous and more easily interpreted." The begging gesture, for example, is a ritual semblance based on reaching or grabbing for some object. It is a sign that has been modified, exaggerated, and transplanted into other contexts. Lorenz (1957) offered another example: "the numerous expressive motions which were originally mating intention movements in the females of Corvidae, cormorants, baboons, and others, no longer have anything to do with the quality of excitation specific for mating reactions. In their new form, they have gained a purely social 'meaning' " (p. 304).

Moreover, relationships of resemblance may take the form of echoic, iconic, olfactory (including taste) or tactile sensations, or images that resemble a referent found in external reality. Osolsobe (1986) wrote of "token:token modelling" in which one token stands as a "cognitive surrogate" of the original. That is, "we take some appropriate non-original and use it, cognitively, in place of the original" (p. 103). Put another way, "token:token models are by definition 'natural'; they are, rather, by-products and not artificial devices designed for the sole purpose of modeling" (p. 108). Thus, token:token models do not account for symbolic or arbitrary relationships but are designed to explain relationships of resemblance.

More familiar is Peirce's conception of relationships of resemblance, which he refers to as "iconic." An icon shares with its referent a quality or property found in the original. Peirce 1895 identified three types of icons: metaphors, diagrams, and images. The degree of similarity between an icon and its referent is relative. A painting may be so similar to the original as to be classed as "realism" or so dissimilar as to be classed as "impressionistic." Further, a painting may become so abstract that the original on which it is based may be undiscernible. Thus, an icon may become so arbitrary in its relationship to its referent that it is no longer considered an icon but instead a symbol. Iconicity, then, is a matter of the degree to which the representation and the original are similar to one another. The relationship is clearly continuous rather than discrete. I refer to these signs as iconic semblances.

Osolsobe (1986) wrote that "there is a gap between animal communication as a type of communication based on signalling and 'natural signs,' and human communication based on modelling. This is a gap that cannot be bridged. One can never arrive at language via natural signs and animal signals" (pp. 110–111). He went on to qualify this statement by noting one exception, play: "play is the direct antecedent of language. Nothing else is more deserving of this proud title. Language is nothing but animal play that has transformed itself in trying to cope with the expanding universe. And animal play is nothing but the antecedent of human language devoid of any temptation to denote anything beyond the domain of animal (competitive) interaction" (pp. 112–113).

Play is one example of ritual semblances formed as a result of the process of ritualization. Play is a process of make-believe in which individuals can test the limits of their own strength as well as experiment with the potential consequences of various behavioral strategies and tactics. Play is common in mammals, common in all species of primate, including (notably) human beings, and therefore presumably common in hominids now extinct. As Byers (1987) noted in his discussion of ungulate play, the physical and physiological consequences of play are of significant import to development: "The physiological and anatomical effects of such intense exercise are many and include influences upon the development of muscle mass, muscle biochemistry, muscle capillary density, bone size and shape, and heart size and pumping capacity. Play probably also influences the development of synapses in many areas of the brain" (p. 58).

Play undoubtedly facilitates the cognitive development prerequisite to symbolic ability. Play that involves manipulating objects provides important information about what those objects afford. Young chimpanzees learn, for example, that branches are good for throwing, for terrorizing others, and for reaching otherwise unattainable objects. Young chimpanzees learn that only certain long grass stems are good for collecting ants and termites and that chewed-up wads of leaves can be used as sponges to soak up drinking water. Goodall's chimps raided her camp and used empty kerosene cans as noisemakers. Captive chimpanzees, gorillas, and orangs are frequently provided with objects that appear to provide hours of entertainment. Their trainers suggest that object play appears more serious than social play and seems to proceed thus: first, the new object is approached cautiously; then it is maneuvered in such a fashion as to test its capabilities to see what it can do; finally, the individual takes more active control as if to determine "What can I do *with* it?"

Play has numerous social functions as well. It serves as an important aspect of bonding, and teaching how to get along with others, and it affords the opportunity to develop versatility and flexibility in dealing with others. Play also facilitates the development of confidence in engaging with others, certainly one of the conclusions to be gleaned from Harlow's famous experiments on social deprivation in rhesus monkeys. Play may also be important in the development and maintenance of dominance relationships, as has been suggested by Strum's work with olive baboons and Goodall's work with the Gombe chimpanzees. Friendships and alliances developed with one's playmates early in life generally come in handy later.

Especially important to my purpose is the notion that play is not real, but make-believe. Play is a pretense and, as such, is an instance of explicit framing and detachment. Playfighting is not real aggression, and its rough-and-tumble aspects are clearly ritualized. The extent to which the players are explicitly aware of the make-believe aspect of play is, of course, open to debate. However, it appears likely, based on the responses of the participants, that they understand the unreal character of play. One only has to observe an infant engaged in playfighting with an older, larger sibling and note the restraint with which the elder engages in play with the younger and weaker to conclude that both understand the detachment of the game from any real aggressive encounter.

Play is sometimes preceded by specific signs that appear to indicate that what is to follow is make-believe rather than real. At least these signs are not used in aggressive encounters. The most widespread is the play face, an open-mouthed, relaxed expression. Howler monkeys and some gibbons make a specific chirping noise before bouts of play, and rhesus macaques look at their partners upside-down through their legs (McFarland 1987). Goodall (1986) noted that "finger wrestling" frequently initiates play among adults. An adult male chimpanzee may initiate a session of play with an adult female by tickling her under her chin. The point is that specific ritual signs seem to have emerged for identifying a priori that the interaction that is to follow is make-believe.

Bateson (1973) concluded, in part based on observations of primates at a zoo, that the essence of play is its detachment from reality. Thus, for Bateson, the essence of play is pretense. Instances of pretense abound in the primate literature. Figan, one of the Gombe chimps, frequently retrieved his share of bananas by first leading the adults away and striding with apparent purpose into the woods. I described an encounter recorded by Strum (1987) as the pretense of helplessness (Liska 1988). In this instance, Olive was being harassed by Toby. Toby had been strutting his stuff, especially at Olive, who later in the day found herself in the vicinity of Toby and her older brother Sean. Strum reported that Olive "looked intensely at Toby, then at her older brother, then back again at Toby. Suddenly she screamed bloody murder. Simultaneously, the two males looked up. Her brother's actions indicated that only one conclusion was possible: Toby was the culprit and Olive was the victim, and Sean was not going to stand for it" (p. 134).

The difference of course is that in these examples the pretense is not detached from reality but located precisely in an area of reality that the pretender wants changed to his or her advantage. The examples serve to illustrate the lability of reality, but it is always with us; contra Bateson, it is not something we detach ourselves from even in play.

Play is an example of the use of signs to indicate reference contrary to fact and as such is an important step away from dependence on immediate reality. Play, as an instance of pretense, appears to be understood by the participants. Deception is also a form of pretense, but one in which at least one of the participants is largely unaware (see also Mitchell, Chapter 6 of this volume). In fact, if deception is to be successful, all but the deceiver(s) must be unaware. Deception is based on the modification of signs so as to misconstrue reality; it depends upon an individual's ability to represent a different reality, one contrary to fact. Menzel's experiments on deception among chimpanzees provide clear indication of the ability to deceive one's peers and control access to choice morsels. De Waal (1982, 1986), Krebs and Dawkins (1984), Strum (1987), Goodall (1986), Byrne and Whiten (1988b), Savage-Rumbaugh and McDonald (1988), and Cheney and Seyfarth (1991) have provided detailed accounts of deceptive attempts and successes in observations of nonhuman primates. Eco (1979) in fact defined semiotics as the study of anything that can be used to lie.

Animals, including humans, use ritual semblances to act out their interpretations of various aspects of their environment. Ritualized aggression is one form of acting out, as is inducing cooperation by showing another what needs to be done. Early hominids who fashioned stone tools presumably taught others by illustration. Indeed, this approach may be more effective than attempting linguistic descriptions of the process, especially for species so facile, as are all primates, in creating visual mental images. Young chimpanzees learn the art of termite fishing by imitating the actions of their elders. Koko, Patterson's gorilla, engaged in regular tea parties with her dolls, and small children act out the roles of their parents by playing house. Mimes tell entire stories in the form of ritualized semblances. A sequence of ritualized semblances in the form of pantomime serves as a useful form of communication for the traveler in a foreign land who does not speak the language. Further, American Sign Language, the language of the deaf, is based in part on pantomime. A great number of the signs used in ape language projects are ritual semblances, and those signs appear to be easily acquired by the apes involved in the projects (Liska 1987).

Using ritual semblances to act out stories probably provided the foundation for dances and ceremonies performed by early hominids, still evident in contemporary human societies. The norms and

245

traditions of the culture could be told and retold to successive generations and acted as a means for retaining and reinforcing collective memory and history. Every year at the end of the dry season, the Wodaabe tribe of the southern Sahara Desert have a ceremony in which mates are selected for the eligible young women of the tribe. As part of the ceremony, the men paint their faces in such a way as to exaggerate characteristics considered especially beautiful (e.g., long thin faces, long slender noses, thin lips, and white teeth). The women judges select winners based in part on the outcome of the men's efforts at cosmetic enhancements. The Wodaabe take great pride in charm and physical beauty, and the ceremony acts as a means of preserving these cultural standards.

The Pueblo Indians of the southwestern United States perform winter ceremonies including hunting or animal dances.

> These are dramatizations of the supposed relationship between the Indians and the larger game animals which, during the centuries, furnished their principal wintertime food. Dancers dress to represent bison (buffalo), deer, antelope, elk, and sometimes mountain sheep. They wear headdresses and horns or antlers which make the likeness to the various animals even more realistic. They carry slender sticks in the hands and by leaning forward on these as they dance, increase the likeness (Dutton 1983, p. 58).

The ceremony culminates in the enactment of the hunt. The game is ritualistically killed and the bodies carried away.

The hunting dances of the Pueblo Indians combine ritual and iconic semblances to represent the animals, process, and desired outcome of the hunt. The incorporation of iconic semblances into the sign repertoire of hominids constitutes yet another step away from dependence on immediate reality. Our early human ancestors left iconic semblances on the walls and floors of caves throughout western Europe. Some appear to be pictorial representations of hunts and may have been drawn during ceremonial reenactments of the hunts. Some of the drawings are so realistic and precise that contemporary researchers have used them, in combination with fossil remains, as a means of identifying long-extinct species. Some of the scenes depict stories of events that may have happened, or those stories may represent anticipated or projected outcomes, that is, alternative realities.

Ancient peoples may have drawn maps of their territory used to guide them during migrations or in preparation for a hunting trip. These maps may have been etched into bark or bone for a more lasting record or sketched into the earth if the trip was to be short and the route could be easily retained in memory. Maps have been drawn prior to a hunt to coordinate the movements of hunters or to plan food-gathering expeditions. These maps, if they ever existed, were apparently not made permanent enough to withstand the passage of time. However, these early hominids' ability to plan and manipulate mental images was clearly sophisticated enough to allow for the production of maps (see the section that follows).

Unlike ritual semblances, iconic semblances such as those left on the floors and walls of caves are relatively permanent records of the activities, beliefs, expectations, and traditions of the people who drew them. Thus, humans living 20,000 to 40,000 years after the originators of the cave art of western Europe can gain some insights into the lives and experiences of the people who lived in those caves.

Ritual and iconic semblances permitted the preservation and accumulation of the otherwise transitory neurological information. This information could be readily modified to account for new experiences or changes in the environment, whereas symptomatic signs must wait to be modified as a result of the slow process of genetic selection. Semblamatic signs provided for the transmission of collective knowledge and may well have been the foundation for the development of culture. The creation of even the simplest tool, the pretense involved in play, the telling of stories via pantomime, and the pictorial depiction of a hunt or religious ceremony is indicative of the ability to analyze, manipulate events and objects to some extent independent of reality, plan, picture different outcomes, and picture events in different contexts. These skills constitute the foundation for the evolution of symbolic capabilities. Sachner (1975) told the story thus:

> I suggest that until about 100,000 years ago the hominids were in a neurophysiological situation very like those other species, in that comparatively complex traditions of hunting, migration, social organization, toolmaking, and environmental control

were being implemented by comparatively inefficient nonsymbolic neural processes for learning, memory storage, and innovation, presumably based on highly evolved capacities for mental imagery, imitation and dreaming. (p. 417)

ROUTES AND MAPS

Primates, including humans, are clearly visuospatial in orientation, and their neurological structures and organization underscore this perceptual preference. Our origin as tree-dwelling animals placed intense pressure on the evolution of highly sophisticated and sensitive visuospatial and tactile skills. For primates, space and time are largely integrated by vision and touch and facilitated by binocular, stereoscopic, color vision accompanied by facile and mobile digits with sensitive fingerpads. Primates identify food, predators, and other members of their own species largely on the basis of visual cues. Indeed, the recognition of self and others is largely visual, and all primates, but especially humans, have evolved intricate facial musculature capable of transmitting a wide variety of expressions. A region that spans both hemispheres on the underside of the occipital cortex appears to be largely responsible for our ability to identify others by their facial characteristics and configuration. Humans who suffer lesions to that area suffer an affliction called prosopagnosia, literally, "not knowing people." The symptoms are quite specific in that the patients, while unable to identify even people they know on the basis of facial characteristics, can name objects and perform competently on other visual tasks. It is their ability to associate faces with identities that is impaired.

Further, body position and orientation yield important information regarding emotional states, predispositions, and responses to the action of others. Hewes (1978) suggested that "the forest niches of our primate ancestors are a key to understanding why we remain primarily visual in our thinking, and why human language, although chiefly auditory and vocal, is still largely a system for encoding information presented to us in the visual channel" (p. 1).

In fact, Hewes argued that elaboration of the visual system as a means for communication preceded elaboration of the auditory system. Indeed, it does appear that hominids were on their way to developing symbols well before contemporary vocal capabilities appeared. Consider, for example, that burials, tools, and other artifacts left by Neanderthals are evidence of rudimentary symbolic processes, yet their vocal anatomy did not provide for the kind or flexibility of phonemic expression evident in contemporary *Homo sapiens* (cf. Lieberman 1984). The vocal capabilities and anatomy of early hominids (200,000 to 3 million years ago) is akin to those of extant chimpanzees (cf. Lieberman 1984). While Hewes's contention is not free of controversy, the evidence he marshaled to support his claim is compelling. As I have said previously, the process I am describing is not channel dependent; rather, the process depends on extensive sensory integration and coevolving neural elaboration. Note, for instance, that the pattern of acquisition of American Sign Language by the congenitally deaf parallels that of hearing children's learning spoken language (Poizner, Klima, and Bellugi 1987). Such evidence further supports the notion that the fundamental processes with which I am concerned are enacted similarly regardless of the sensory transmission code.

Ritual and iconic signs may be viewed as episodes or event structures that represent visuospatial relationships frequently integrated with tactile information. Distinguishing episodes of play from other symptomatic encounters requires visual recognition of the ritual signs that accompany play (e.g., the play face and finger wrestling). Following the story told by a mime requires understanding events presented in a spatiotemporal sequence. Tool-using and tool-making require integration of visual, spatial, and tactile information, as well as the ability to image and/or imagine the desired product. Thus, ritual and iconic semblances are outward manifestations of the cognitive ability to map objects and events in space.

O'Keefe and Nadel (1978) suggested that the right hippocampus is responsible for spatiotemporal mapping in human and nonhuman primates. They distinguished between "routes" and "maps." A route, on the one hand, specifies a beginning, goal, and particular direction of movement; it is a list of stimulus-response-stimulus commands. A route may be exemplified as a linear description of how to get from point X to point Y. Relatively inflexible a route is rendered useless if damaged; it cannot be manipulated, requires only simple knowledge, and is exceedingly difficult to follow in reverse. A route is analogous to a symptom in that a symptom is inflexible, cannot be manipulated, requires only simple knowledge, is in this case genetically endowed, and is irreversible.

A map, on the other hand, stands as a representation of a part of space and is easily manipulated, flexible, requires considerably more knowledge to use, contains more information than does a route, and provides for reversibility. O'Keefe and Nadel wrote that "the constituents of space are places, and thus an alternative definition of a map is the representation of a set of connected places which are systematically related to each other by a group of spatial transformation rules" (p. 86).

Spatio-temporal maps are analogous to ritual and iconic semblances in that they connect episodes or events via patterns of signs. Ritual semblances are, to a greater degree than symptoms, flexible and malleable, contain considerable information, and require considerable knowledge for interpretation, especially contextual information.

Iconic semblances appear to provide much more flexibility and information capacity, require substantial information for interpretation, and are easily manipulated. In fact, a physical map is one example of an iconic semblance. Thus, the evolution of iconic semblances was an exponential leap in primate capability for representing and manipulating environmental information.

Another giant leap forward occurred with the advent of symbolic signs. Symbols, especially conceptual symbols, may be thought of as the "places" in a syntactic map, while syntactic symbols specify the relationships among those symbols. Therefore, what I refer to as syntactic maps, apparently designed and stored in the left hippocampus of humans, provided for models of relationships that exist independent of immediate reality, and they allowed for the creation of thought independent of visualization.

CONCLUSION

Humans have a particular interest in themselves, their history, and distinguishing characteristics. We take particular pride in our symbolic abilities and, of late, have set ourselves apart from other members of the animal world on that basis. As a consequence, we consider ourselves to be the "thinking," "reflecting," "linguistic" species. Yet it is clear that complex cognitive processes are evident without symbols. Ritual and iconic semblances stand as manifestations of such advanced cognitive processes as planning, imaging and imagination, manipulation of objects and events to some extent independent of reality, and the potential to picture different outcomes or picture events in a different context.

Still, if it is the case that humans are the only species to have created such a specialized worldview as symbols afford, it behooves us to attempt to understand the evolutionary conditions and consequences of this particular ability. I have suggested that symbols afforded a different kind and level of thought, further advanced by the evolution of syntactic symbols, that is, symbols can be manipulated independent of reality by means of syntax so as to create conceptions of relationships in reality that do not exist. Consequently, one can create conceptions of other possible worlds (counterfactuals) and their possible characteristics and consequences. Without syntactic maps, propositional processing, as exemplified by contingency relationships using the if . . . then format, appears quite unthinkable, so to speak.

Further, I have argued that the evolution of sign capabilities is best viewed along a continuum of symbolicity or arbitrariness. Support for the system I have offered is found in the evolution of neurophysiological structures and organization, the sequence of development of semiosis in human children, sign capabilities evident in extant nonhuman primates, and the artifacts and fossil remains of our long-extinct hominid ancestors. For my part, as there are clearly various levels of representational ability, the debate over whether only humans use language seems unnecessarily restrictive. A more fruitful endeavor involves systematic comparisons of the sign behaviors of other related species, other cultures, as well as the material remains of extinct human cultures to provide greater understanding of the possible course of evolution of symbolic capabilities.

NOTE

1. A version of this chapter was presented at the international conference on the Biology of Language: Essentialist versus Evolutionist in the Nature of Language, Czerniejewo, Poland, December 1–4, 1988.

8

"Ape Language" Studies and the Study of Human Language Origins

H. LYN WHITE MILES and STEPHEN E. HARPER

ABSTRACT

Investigators have used a variety of vocal and primarily nonvo-cal means to establish two-way communication with all four great apes: gorillas, orangutans, and two species of chimpanzees. These studies have shown that living apes can acquire vocabularies of symbols; combine symbols into rule-following combinations; make requests, answer questions, and comment on the environment; name and make reference to things, including things that are not present; invent new symbols; engage in some perspective-taking, deception, imitation, and pretense; show self-recognition; as well as develop other abilities. The extent to which apes can acquire linguistic abilities as shown by these results can serve as a model to reconstruct hominid language abilities. They suggest that early hominids could have possessed: (1) a basic, flexible semantic communications system with moderate levels of reference and (2) low levels of perspective-taking, imitation, and sequential organization, including protosyntax. Gram-mar may have coincided with, or followed, the origins of hominid culture and may not have become elaborated until much later periods through selection pressures brought about by the development of complex culture. The importance of the role of culture in human evolution is underscored by the possibility that some of the linguistic and cognitive abilities of enculturated apes may only be matched by late hominids.[1]

INTRODUCTION

Since we first became aware of the great apes, there has been speculation about ape-human similarities and the possibility that apes might be taught to speak. However, it was not until this century that major attempts were made to establish two-way communication with these animals, first by using speech and later by nonvocal means, which met with greater success. Eventually, approximately twenty apes received some degree of language training. Critics raised issues about whether or not the skills displayed by apes possessed the features of adult human language (Brown 1973; Epstein, Lanza, and Skinner 1980; Limber 1977; Mounin 1976; Terrace, Petitto, Sanders, and Bever 1979) and asked questions about the extent to which the results were simply achieved with a Clever Hans phenomenon, the well-known problem of unintentional cuing from a trainer (Sebeok and Umiker-Sebeok 1980). Because there was no generally accepted definition of language, a linguistic game of "Yes they can do that, but . . ." ensued between researchers and those who felt that human uniqueness was threatened by ape language (Miles 1983). Although some criticisms have persisted, particularly on issues of grammar (Wallman 1992), there is now general agreement that apes can use rudimentary symbol systems in some very interesting ways (Cheney and Seyfarth 1990, Parker and Gibson 1990).

Because apes can use symbol systems (whether or not one calls this language), it should be possible to use the results of ape language studies to begin to reconstruct the origins of language in hominids. More specifically, ape language studies can help to identify the hominoid and early hominid preadaptations that eventually led to symbol use through several evolutionary trends: (a) progressive dependence on symbolically mediated adaptations, (b) the need to communicate and coordinate actions with each other in order to meet the challenges of changing habitats, (c) the development of elaborate social interdependence while simultaneously acquiring the cognitive capacity for autonomous mental representations, (d) the extension of the memory of individuals to reside as common "traditions" or culture, and (e) the development of institutions whose role was to objectify ritual action and stabilize the intersubjective significance of those actions (Shore 1991). Ape language studies join other lines of indirect evidence with which to understand human cultural origins, including fossil interpretations, the protocultural aspects of ape behavior in natural settings (Goodall 1986), the ontogenetic development of language and cognition in children (Bates, Thal, and Marchman 1991), anatomical studies of the brain and vocal tract anatomy (Lieberman 1984), and foraging, tool use, social relationships, imitation, and cultural meaning (Greenfield 1991, Lancaster 1968, Miles 1975, Parker and Gibson 1977, Shore 1991).

Before we relate the findings of ape language studies to the origin of language in hominids, we first take an overview of important trends in this research and then examine specific ape language abilities. Ape language research is presently being continued by Roger Fouts, Francine Patterson, E. Sue Savage-Rumbaugh, and H. Lyn Miles. There are several important trends in this research. First, ape abilities are being compared with children's language development and not adult human language, that is, the research has a developmental focus. By removing the burden of comparing the limited language abilities of apes to adult human language, researchers can focus on comparing the *processes* of language acquisition and the cognitive foundations that are involved for children and apes. Developmental models also suggest areas that may be homologous, particularly if the developmental sequence in children and apes is similar. Second, the role of culture is beginning to be acknowledged (Miles 1976). We are recognizing the dual role of culture as it both sets up a level playing field for comparing ape and human cognitive development and is fundamental to language because of the necessity to establish elaborate shared frames of reference upon which meaning is based within a community of users. Furthermore, cultural institutions and rituals stabilize and maintain the subjective significance of these shared frames of reference, thereby creating a continuity of intersubjective significance through time.

Third, the relationship between language and cognition and how these have coevolved is being considered. By identifying the cognitive and communicative processes that are the foundation of ape skills and by examining their developmental coemergence, we are gaining insights into origins of hominid culture. While it is certain that a degree of cognitive development is necessary for language acquisition, less obvious is the process by which language develops the mind, although evidence already indicates that language training may enhance cognitive performance in apes. And finally, there is the issue of apes themselves. The research is beginning to look at ways that apes use symbols that might be unique to them, particularly in environments that attempt to reconstruct their natural lifeways. These, too, may provide clues for carrying out hominid reconstructions.

APE LANGUAGE ABILITIES

Nonvocal Two-Way Communication

There have been two major approaches to allow apes both to comprehend and produce symbols that employ either artificial codes or a natural language. The artificial codes consisted of either plastic tokens or lexigrams, that is, abstract forms arranged on a display panel. David Premack (1972) used plastic tokens with the chimpanzee Sarah. Duane Rumbaugh (Rumbaugh, Gill, and von Glaserfeld 1973) and E. Sue Savage-Rumbaugh (1986) used lexigrams with two species of chimpanzees at Georgia State University's Language Research Center, as did T. Matsuzawa (1990) with the chimpanzee Ai in Japan.

The natural language approach utilized the American Sign Language for the deaf, which is understood by millions of people and has the full properties of human language. Sign language was used to communicate with all three of the great apes. In the first study, Washoe and other chimpanzees were taught to sign by Allen and Beatrix Gardner (Fouts 1973, Gardner and Gardner 1969) and are now raised by Roger Fouts (Fouts, Hirsch, and Fouts 1982) at Central Washington University. In separate studies, Herbert Terrace (Terrace, Petitto, Sanders, and Bever 1979) taught a chimpanzee, Nim, at Columbia University; two gorillas, Koko and Michael, were taught by Francine Patterson (1978) at the Gorilla Foundation in Woodside, California; Gary Shapiro (1975) briefly taught signs to an orangutan, Princess, in Borneo; and H. Lyn Miles (1980, 1990) carried out long-term sign research with the orangutan Chantek at the University of Tennessee.

All of the ongoing ape language experiments use developmental models and address the issue of enculturating the apes, which will contribute greatly to our understanding of language origins. The natural-language approach takes advantage of the richness of natural ape gestures and creates a more naturalistic and portable mode of communication, replicating opportunities for intonation, grammatical encoding, and modulations of meaning. However, both the artificial-code and natural-language approaches have provided important data for comparisons with language learning in human children.

Only one of the projects specifically sought to answer evolutionary questions from the outset. The desire to utilize ape language research to understand the evolution of human language was one of the explicit goals of Project Chantek, the only anthropologically oriented ape language research effort (Miles 1976, 1980, 1991a). It was hoped that the development of Chantek's abilities would form a partial script to reconstruct hominid evolutionary scenarios so that we might begin to tease out the various technological, communicative, and social factors that led to the development of modern *Homo sapiens*. It is for this reason that we pay special attention to the findings of Project Chantek in the discussion of ape language abilities that follows.

Enculturation

Central to the notion of true language acquisition is the idea of shared meaning in communication, which must be constructed from both an intersubjective experience and a communal one. Culture is essentially a meaning-creating activity (Shore 1991), and by enculturating apes we are allowing their processes to filter through our culturally conventionalized object-world. While it is always important to standardize data when making comparisons between apes and children, more significantly, enculturation of language-trained apes is crucial to establish frames of reference for the participants in conversations and to establish a context for meaning. Teaching apes language necessarily entails teaching them the human cultural conventions that can schematize their object-world in a way that is similar to ours (Miles 1976). Furthermore, by considering the utterances of the apes within a known cultural context, we can better explore the boundaries of their mental states.

The signing projects, pioneered by Alan and Beatrix Gardner, placed early emphasis on the necessity of an enriched human cultural environment to test a human skill. This enrichment, which anthropologists call enculturation, refers to the manner in which human children are immersed in, and guided through, their culture (Lock 1980). Because the operant conditions of the artificial lexigram system hampered the role of enculturation and produced fewer languagelike results, researchers (Savage-Rumbaugh, Pate, Lawson, Smith, and Rosenbaum 1983) who had once extensively challenged the enriched conditions of the signing research turned to more humanlike rearing conditions for their language project with pygmy chimpanzees and found greater success. Also, cultural transmission in ape-to-ape and cross-generational signing is now being investigated (Fouts, Fouts, and Schoenfeld 1984; Fouts, Hirsch, and Fouts 1982).

Vocabulary

Early critical opinion held that apes would not be able to acquire signs, let alone extensive sign vocabularies. That apes might comprehend natural language or artificial lexigrams was not sufficient; language involves production, not just comprehension. Various ape studies placed differing emphasis on the number of gestural signs or artificial codes to be acquired and had different definitions of acquired, reliable, spontaneous, or frequently used gestural signs, tokens, or lexigrams. Further, a small vocabulary might be extremely meaningful and salient, whereas a large vocabulary might have many signs or lexigrams but be used rarely or with less reliability or merely be evoked by drills or linguistic routine exchanges.

The language studies reported that apes readily learned signs in the first months of training and after several years were able to develop vocabularies of several hundred signs or lexigrams. Despite some varying criteria (Hoffmeister, Moores, and Ellenberger 1975), the rates of acquisition are roughly comparable. During the first twenty-two months of learning, Chantek acquired forty-four signs, the chimpanzee Washoe acquired thirty-four signs (Gardner and Gardner 1969), and the chimpanzee Nim learned thirty-nine signs (Terrace, Petitto, and Bever 1976). In terms of total vocabulary, Terrace (1979) reported 125 acquired signs for Nim, and Gardner, Gardner, and Van Cantfort (1989) showed similar levels for their chimpanzees: 133 signs for Washoe, 168 for Moja, 140 for Tatu, and 122 for Dar. Fouts, Fouts, and Van Cantfort (1989, p. 286) reported seventy signs for the chimpanzee Loulis, about fifty of which he learned from direct observation of his foster mother, Washoe, and other chimpanzees (Fouts, Hirsch, and Fouts 1982). Premack (1972) reported 130 tokens for the chimpanzee Sarah. Miles, Mitchell, and Harper (1992) reported 150 signs for the orangutan Chantek. Patterson (1978) has extensively emphasized vocabulary size in her study and reported 224 signs for her gorilla Koko and later reported total vocabularies of over 500 signs for Koko and Michael, but it is unclear how many signs are used regularly in daily communications (Patterson and Linden 1981). In comparison, one study of a deaf child inventoried 132 signs acquired by the age of 3 years (Olson 1972).

What is significant is that apes are able to acquire natural signs or artificial lexigrams and that no cognitive limit to vocabulary size has yet been reached in these studies. The constraints to vocabulary size appear to be due to training emphasis and the number of signs that are relevant to the fairly simple daily activities of these apes. Some signs are more difficult to acquire than others, particularly pronouns, nontouch signs, and signs for more abstract concepts, such as pronouns that children also find confusing (de Villiers and de Villiers 1978). A more meaningful measure might be the number of symbols these animals use daily to meet basic needs during relaxed interactions without vocabulary drills, games in which the animals name scores of plastic toy animals, or other techniques to elicit communications. In Project Chantek this daily usage was about fifty gestural signs (Miles 1990). It should be noted, however, that when modulations of meaning are considered, these vocabularies are actually more extensive than when signs are counted merely in citation form.

Comprehension Versus Production

Ape language researchers initially emphasized production of symbols because this had not been demonstrated in animals before. However, language-trained apes were also tested on comprehension. Gardner and Gardner (1989, p. 181) carried out double-blind vocabulary tests as well as responses to Wh-questions with their chimpanzees (Gardner, Gardner, and Van Cantfort 1989, p. 198). Comprehension tests of responses to signed commands with chimpanzees were conducted by Fouts, Chown, Kimball, and Couch (1976) and Miles (1978). Chantek's comprehension of signs was also tested with double-blind format (Miles 1990). Patterson (1978, p.189) administered standardized tests to the gorilla Koko (with both speech and sign), and although motivational factors affected some outcomes, Koko was able to show comprehension of the items at the level of a 4- to 5-year-old handicapped child.

Spontaneous Signing and Discourse Ability

Discourse, or the ability to converse, is an important marker of language ability. Studies of child language have shown that by the second year of life, children engage in two-way exchanges with others (Brown 1973; Caselli 1983; Ervin-Tripp 1970; Ferguson, Peizer, and Weeks 1973; Kantor 1982; Lock 1980; Maestas y Moores 1980; Sachs, Brown, and Salerno 1976; Snow 1972; Snow and Ferguson 1977). Mandel and Fouts (1976), Miles (1978), Patterson and Linden (1981), and Van Cantfort and Rimpau (1982) concluded that chimpanzees and gorillas can engage in meaningful sign exchanges with their caregivers and each other, much like human children. For example, Fouts and Fouts (1989) showed that the signing chimpanzee Loulis could participate in turntaking (Baker 1977) with Dar, Moja, and other chimpanzees in simple group signing exchanges. In marked contrast, Terrace, Petitto, Sanders, and Bever (1979) concluded that their chimpanzee, Nim, had not mastered even the rudimentary skills of conversation because Nim interrupted and imitated his "teachers" and did not spontaneously initiate signing. Since other apes sign spontaneously and make fewer interruptions or pure imitations of the caregiver's signing, Nim's behavior was due to his enculturation. Nim's communication changed when he returned to the University of Oklahoma and was under less stressful conditions (Miles 1983, O'Sullivan and Yeager 1989). The style of an ape's discourse seems to match the culture of communication presented during the enculturation process.

Comprehension of Speech

Attempts to get apes to produce speech largely met with failure (Furness 1916, Laidler 1978) because of ape vocal tract anatomy (Laitman, Heimbuch, and Crelin 1978; Lieberman, Crelin, and Klatt 1972). It might be possible to teach apes a vocal system with a subset of human speech sounds that apes could make, but apes might still have difficulty producing the vocalizations readily (Miles 1978). Early in ape language research it became clear that enculturated apes exposed to spoken language were comprehending some speech (Fouts, Fouts, and Van Cantfort 1989; Miles 1990). Fouts, Chown, and Goodin (1976) tested this ability in the signing chimpanzee Ally. The researchers found that Ally could comprehend vocal commands and respond by locating or naming objects correctly, as well as making the signs that corresponded to the spoken words. Fouts, Fouts, and Van Cantfort (1989, p. 286) reported that all of their chimpanzees, including Washoe, the first chimpanzee to learn gestural signs, could understand spoken English. This has also been reported for the gorillas Koko and Michael (Patterson 1978, Patterson and Linden 1981), the orangutan Chantek (Miles 1990), and other apes.

Semantics and Context

Although much ape communication is about food and play, many other meanings are expressed in ape signing and artificial code use. Names for objects and actions (MASK, SCREWDRIVER, MONEY, RIDE, SHARE) constituted almost half of Chantek's sign vocabulary (Miles 1990). Apes also have signs for personal names, animals, foods, colors, place names, pronouns, locatives, attributes (GOOD, HURT), and emphasizers (MORE, TIME). Apes can over- or underextend the meaning of these signs, much as human children do (Bloom and Lahey 1978, Braunwald 1978, Clark and Clark 1977, Nelson 1973). Several examples of such overextensions have been reported, and many of them constitute class "errors" (Gardner and Gardner 1978, Patterson and Linden 1981). For example, Koko signed STRAW for plastic tubing, cigarettes, and a car ratio antenna (Patterson 1978). When full contextual information is available, it is possible to see shifting semantic domains in ape use of gestural or artificial signs. For example, Chantek used the sign DIRTY appropriately, then overextended its meaning to refer to soiled or disliked things. Later, when he learned the sign BAD, he shifted these meanings onto the new sign (Miles 1990). This shows a degree of semantic flexibility.

Symbolic Reference

Representational skills require an understanding that the sign or word stands for or represents a person, object, action, or idea. To make symbolic reference, a primate must create or understand a culturally formed concept with a semantic domain; use the symbol intentionally; generalize the meaning; understand that the referent can be classified into categories according to its attributes; decontextualize the meaning; and use the symbol when the object or idea is not present. In human children this begins by giving and showing objects and, at 9 to 12 months of age, by the act of pointing (Bates 1976, Clark and Marshall 1978, Hoffmeister and Moores 1973, Lock 1980). Earlier, Savage-Rumbaugh, Pate, Lawson, Smith, and Rosenbaum (1983, p. 462) found that the chimpanzees in their studies did not go through referential developmental stages, but more recently they have reported different results for the pygmy chimpanzees, whose enculturation was radically different. Although somewhat developmentally delayed, Chantek passed through stages of reference similar to human children, including spontaneous pointing at 21 months of age (Miles 1990).

The ability to refer to things or events not present is also an important indicator of reference (Bates 1979, MacNamara 1972). All of the apes enculturated in humanlike settings show the ability to make displaced reference. Chantek's signed displacements doubled in frequency as he matured (Miles 1990). The role of enculturation and context was clear when his displaced reference increased following his exposure to desirable locations and objects not in view such as CADEK-HALL (a student dormitory) and his YARD.

Invented Symbols

Many of the apes produced iconic gestures, some of which are also used by apes in natural settings, before they are formerly introduced as signs. These included gestures to express desires such as "give-me," "up," "come," and "go," which were accepted and adapted into their sign vocabularies. Most of the language-trained apes have also invented signs of their own. Early in their research, the Gardners felt that following a strict protocol was important in establishing the acquisition of proper American Sign Language. They actually discouraged ape sign inventions by explaining, "We corrected the cross-fosterlings when they offered their own nonstandard coinages, just as parents correct human children in similar cases" (Gardner, Gardner, and Nichols 1989, p. 62). Yet it is interesting that the chimpanzees persisted in attempting to invent signs, and at least two were eventually recorded, BIB and PEEKABOO (Gardner, Gardner, and Nichols 1989, pp. 61-62). Later, when under the Fouts's care, Washoe was allowed to invent or modify signs and did so readily (Fouts, Fouts, and Van Cantfort 1989, p. 185). Chantek invented five signs, NO-TEETH, EYE-DRINK, DAVE-MISSING-FINGER, VIEWMASTER, and BALLOON (Miles 1990), and Koko invented the signs BITE, TICKLE, STETHOSCOPE, and DARN (Patterson 1978, p. 191). Thus, apes have a productive capacity to coin new meanings.

Deception

Since deception requires a deliberate and intentional misrepresentation of reality, it is an important indicator of language abilities (de Villiers and de Villiers 1978, p. 165). Mitchell (1986) and Miles (1986) identified stages of the development of deception, pointing out that mature deception requires the ability to understand a construction of reality, how this is perceived by others, and how to negate this perception from the perspective of the other. Terrace (1979, p. 151) reported that the chimpanzee Nim signed DIRTY and SLEEP when he intended to do neither and was misrepresenting the situation in a deceptive way. Fouts (1973) and Patterson and Linden (1981) both reported incidents in which they believed that apes might be using their signs to deceive.

Many examples of Chantek's deceptions that have been analyzed show coordination and association of multiple behaviors with the proper signs. For example, on one occasion Chantek stole food from a caregiver's pocket while he simultaneously pulled her hand away in the opposite direction (Miles 1986). On another occasion he stole the caregiver's pencil eraser, pretended to swallow it and "supported" his case by opening his mouth and signing FOOD-EAT (Miles 1990).

Perspective-Taking, Attribution, and Mental Representations

Examples of signed and behavioral deception by Chantek and other apes show evidence of planning, taking the perspective of the other, and displacement. These processes require that some form of an internal mental image about the outcome of events and the actions of others be created. In apes or children such role playing and perspective-taking is indicative of symbolic functioning (Meddin 1979, Vasek 1986). Chantek began some limited perspective-taking at the age of 2 years, but his perspective remained egocentric, and he did not take into full consideration the perceptual orientation of his caregiver. His perspective-taking became more objective at approximately 4½ years of age, when, for example, he would alter his caregiver's position to improve the success of his deceptions (Miles 1990). Another example of the formation of mental images is found in Chantek's ability to respond to his caregiver's request to SIGN BETTER and improve the articulation of a sign (Miles 1990). Chantek showed evidence of perspective-taking by responding to this request by signing in a slow, emphatic way, using one hand to put the other hand in the proper sign shape while looking at his caregiver. His behavior provided important evidence that Chantek understood that his sign had to approximate an ideal shape and was not simply a trained motor response, that is, he had a mental image of the sign. That he would sign slowly and in a step-by-step manner while maintaining eye gaze with the caregiver indicated that he could also take the perspective of the other as he gradually constructed the proper form of the sign.

Additional evidence of mental representations came from an unusual aspect of Chantek's sign production: his execution of signs with his feet (Miles 1990). His foot signing indicates that Chantek understood the gestalt of a sign configuration and that a sign could be made in a number of ways with different body parts. Chantek even began to use objects in relation to each other to form signs (Miles 1990). For example, he used the blades of scissors instead of his hands to make the sign for biting, applied to his body or to objects.

Rimpau, Gardner, and Gardner (1991) also reported perspective-taking for signing chimpanzees. For example, when Dar made the sign BANANA out of the view of his caregiver and his caregiver replied CAN'T SEE, Dar removed his hands, which were hidden under the table, and signed SEE BANANA over his head (Rimpau, Gardner, and Gardner 1991, p. 243). Signing apes also offer their hands to be molded when wanting to know the name of a sign, and when they are adamant, they will mold the hands of the addressee directly rather than making the sign themselves (Miles 1990). For example, when one of Washoe's caregivers denied her request to go OUT, Washoe took the caregiver's hands and made them sign OUT, and then repeated this with the UP sign, to have the caregiver pick her up to leave (Rimpau, Gardner, and Gardner 1991, pp. 243–244).

Imitation and Symbolic Play

There is not a great deal of evidence for apes exhibiting imitation and symbolic play. Mitchell (Chapter 6 in this volume) argues that the basis for symbolic play and other cognitive skills is simulation, the ability to recognize and represent an aspect of something else. Pretense begins in early childhood and is an example of one of the first ways in which children represent knowledge to themselves. Chantek first engaged in active pretense at 20 months of age (Miles 1990). He would run about his yard, looking over his shoulder as if he were being chased; act afraid; and "feed" his toy animals. Patterson and Linden (1981) also reported that Koko and Michael would hide or disguise themselves in pretend games. The apes play with dolls and stuffed animals and pretend to feed them, but extensive symbolic play that is not heavily caregiver supported has not yet been reported for apes.

Imitation is another skill that has been linked to human uniqueness. Imitation functions to replicate something in context, to be later absorbed and processed. Imitation is the foundation of human culture and allows enculturation to continue from generation to generation. All of the signing apes used extensive imitation to learn signs, following an initial period where learning is mostly through having their hands molded. For example, Chantek performed his first sign imitations by 15 months of age, only a few months after we had begun teaching him to sign, and then continued to learn signs primarily through imitation. Chantek could also be instructed to DO SAME and then imitate his caregiver's action (Miles 1990). Chantek was even able to combine deception, perspective-taking, and imitation to outwit his caregivers (Miles, Mitchell, and Harper 1992). On one occasion during play, Chantek reversed roles by instructing his caregiver to DO SAME while he slapped his hand on his jungle gym. Although ostensibly Chantek indicated that the caregiver should slap his hand on the jungle gym, Chantek used this occasion to grab the caregiver and pull him back into a wrestling match. Like children, apes are told many times to do something or are asked many Wh-questions, such as WHO? WHAT? and WHERE? Children, in turn, ask extensive questions and far outstrip apes in imitative ability. A persistent question has been whether apes ever imitate these questions themselves through signing rather than a questioning intonation. It is significant that at least one signing ape was able to imitate the caregiver's command, reverse roles in the game, and use the command for deceptive ends.

Self-Awareness and Self-Signing

All of the ape projects have reported self-recognition by their animals, demonstrated by naming themselves, mirror self-recognition, and self-signing. Chantek began to sign to himself at 15 months of age to name objects to himself or describe concurrent actions (Miles 1990). Patterson and Linden (1981) reported that Koko signed she was ANGRY, SAD, or HURT, as have other signing apes. Washoe self-signed during play, while looking through magazines, and to correct herself after looking more closely at a picture (Gardner and Gardner 1974, p. 20). The chimpanzees Moja, Pili, Dar, and Tatu self-signed as well as signed to each other (Gardner and Gardner 1989). Both Chantek and Washoe have self-signed before carrying out actions. Chantek signed IN MILK RAISIN before going inside for a glass of milk (Miles 1990), and Washoe signed QUIET to herself while sneaking to a forbidden part of her yard (Gardner and Gardner 1974, p. 20). Thus, apes appear to be able to use their signs for internal reflection and show a degree of self-awareness.

Symbol Combinations and Protosyntax

All of the enculturated apes began to combine their gestural signs into sequences as soon as they acquired signs. Nim tended to produce long strings of sign sequences (Terrace 1979); however, the other apes generally used combinations of two to four symbols, characteristic of the length of combinations made by 2- to 3-year-old human children (Brown 1973), with occasional longer sequences. All of the apes have produced novel sign combinations that had not been presented to them previously. For example, apes have signed LISTEN DRINK for Alka-Seltzer plopped into a glass, METAL HOT for a cigarette lighter, and FINGER BRACELET for a ring (Gardner and Gardner 1980, Patterson and Linden 1981). Signing apes are also able to modulate their signs to indicate intensity, direction, and grammatical markers, such as questioning and other meanings.

Some sign order regularities have been observed and described as protogrammar. Greenfield and Savage-Rumbaugh (1990, 1991) reported classes of lexigram regularities in the pygmy chimpanzee Kanzi, who now uses a combination of mostly single lexigrams with pointing gestures and some ASL signs. Fouts (1974) reported sign order regularities in Washoe, and Fouts, Chown, Kimball, and Couch

266

(1976) showed that the chimpanzee Ally could master a simple syntactic system involving the use of prepositions. Miles (1976, 1978) reported regularities and rule-following semantic relations (for example, priority of agent-action sign order) in the communications of Ally and another signing chimpanzee, Booee. More than three-quarters of the communications of Ally and Booee showed that they placed the demonstrative point sign THAT before the name of the object and used subject-verb-object sign order in 89 percent of their communications. Booee, and to a lesser extent Ally, even substituted natural gestures for signs in proper sequence, thus adapting their natural communication system to subject-verb-object sign order (Fouts 1978, Miles 1976).

Chantek showed similar rule-following behavior in his communications (Miles 1990). Chantek was more likely to name an object and then sign GIVE if the object were present, and GIVE followed by the object name if it were not, a variation that was not modeled from the signing of his caregivers. If we keep in mind that the inflected nature of sign language does not require the order regularities found in English (Bellugi, Bihrle, and Corina 1991; Klima and Bellugi 1979), we may see this rule following as the beginning of semantically based grammatical combinational devices, or protosyntax, since it involves contextually dependent modulation of meaning analogous to the means by which a speaking child learns to use a certain word order.

An analysis of the markers, inflections, and modulations of meaning that occur in natural language provides another source suggesting nascent grammar. Sign language relies more heavily on inflectional devices than word order to convey meaning. Changes in the citation form of the sign, such as the use of two hands to make a sign rather than one, altering the movement or orientation of the sign, or simply reiterating the sign produces semantic, pragmatic, and grammatical effects (Fant 1972; Klima and Bellugi 1979; Poizner, Klima, and Bellugi 1988; Stokoe 1972). Chantek engaged in these sign modifications by, for example, signing BRUSH on the place where he wanted his caregiver to brush him, by collapsing the signs NUT and BUTTER for peanut butter, and by sustaining a sign for a longer duration, such as HURT when he was in pain. Extensive analysis of these inflectual devices in the signing of chimpanzees was done by Rimpau (Rimpau, Gardner, and Gardner 1991). The signing chimpanzees made directional changes to verb signs to indicate a reference to a person. These were systematic and expressed distinctions in meaning as well as other place modifications in articulation. For example, the chimpanzees signed HURT on the location of the injury, TICKLE on the hand of the person they wanted to play with them, or they would emphasize or exaggerate signs to make them more emphatic. By using a prolonged gaze, all of the apes could mark their communications as questions.

These modifications of signs in order to reflect variable meaning suggest a degree of plasticity in ape signing that utilizes many of the rudimentary grammatical features of ASL signing by deaf humans. In many cases the apes themselves introduced these features. This, in conjunction with invented signs, provides evidence for productivity, one of Hockett's (1960a) important design features of human language. Further, the rate of these systematic modifications indicates that some of the signing ape vocabularies have been underestimated and should be increased by at least one-third (Rimpau, Gardner, and Gardner 1991).

Insults, Curses, Rhymes, and Jokes

Patterson and Linden (1981) claimed that Koko used insults such as DIRTY after being falsely accused of breaking a doll and frequently used expletives such as YOU DIRTY BAD TOILET. Linden (1981) reported that some of Washoe's communications contained DIRTY as a form of cursing; for example, Washoe signed DIRTY MONKEY after a macaque threatened her and while she was pressed to name different primates, and she signed DIRTY ROGER when he refused to let her out of her cage. Chantek also signed DIRTY to indicate "bad" things and signed BAD BIRD at a bird continuing to give a noisy alarm call (Miles 1990). Patterson (1978, p. 191) also glossed a motion Koko made when she was frustrated as DARN, a motion Koko also used to refer to bird distress calls. These examples show the animal's ability to express negation (Miles 1986, 1990).

Patterson and Linden (1981) reported that Koko sometimes joked in a contrary way, for example, by deliberately misnaming the color of a frog, and made "rhymes," that is, sign-sign or speech-sign correspondences. Chantek also made sign-sign or sign-speech correspondences. However, such rhymes can be more conservatively explained by a matching routine in which the animal matches signs with their spoken English equivalents or produces a similar sign altered by only one aspect of articulation that would be analogous to phonemes in speech. These sound and sign correspondences may in fact be what human children are doing in their earliest rhyming. Matching sounds or signs does not imply the metaphorical, rhetorical, and literary aspects of poetry and rhyming in adult human culture, and in the absence of further supporting evidence, a conservative interpretation is appropriate. However, some of these uses of signs indicate the range of communicative intentions (Bruner 1975, Dore 1974) that apes can express or imitate in their human caregivers.

HOMINID LANGUAGE ORIGINS

The Case for Ape Language Models

The evidence to date demonstrates that the linguistic ability of enculturated apes is within the range reported for 2-year-old children, while their cognitive ability is within the range of 3- to 4-year-old children (Brown 1973; Hofmeister, Moores, and Ellenberger 1975; Miles 1991a). This seems more plausible to us now than in the 1960s because we have recently found that apes in the wild have a protoculture that includes toolmaking, cracking nuts with stones, observational learning, cooperative hunting, deceiving, gesturing with dialect variations, and idiosyncratic inventing (Goodall 1986, McGrew 1992). Thus, if all of the great apes, including orangutans, can acquire rudimentary language ability, then these abilities can become a set of clues for understanding the origins of hominid culture.

Some linguists hold that language is unique to our species but still maintain a Darwinian perspective (Bickerton 1981). However, many scholars who stress the uniqueness of human language subscribe to the generative grammar theory of language, which argues that language is discontinuous with human communication and cognition, that is, it had separate origins (Chomsky 1965, 1976). These scholars believe that language is a unitary, innate skill in our species with little or no input from learning. If we use this definition of language, then language was not constructed, even in part, from cognitive and communicative processes that apes could exhibit in some rudimentary way, either naturally or through enculturation. This generative approach demands a sudden qualitative change in primate linguistic abilities that can be accounted for only through divine intervention or by several simultaneous and coordinated

269

mutations. Given the morphological, neural, and behavioral factors involved in that very complex phenomenon of human language, it is highly unlikely that such a combination of events happened suddenly and synchronically (Bates, Thal, and Marchman 1991; Lieberman 1984; Miles 1983). Moreover, the definition of human language as discontinuous with speech or gestural communications of ancestral hominids prohibits a satisfactory explanation of the fossil evidence that shows morphological changes in human brain and vocal tract anatomy (Lieberman 1984; see also Milo and Quiatt, Chapter 10 of this volume).

It should be noted, however, that one can argue both that enculturated apes can acquire some human language skills and that adult human language abilities are qualitatively different from the natural abilities of apes who communicate in natural settings. This view stresses the continuity of communication and cognition across species and is the position held by most researchers, including the authors. In this model hominids developed enough quantitative changes in the degree of their cognitive and linguistic abilities to reach the threshold of qualitative change, that is, a change into contemporary adult human language.

Malmi (1976) has raised the question of whether the underlying neural and cognitive systems of apes are similar to those of humans. Greenfield (1991) has addressed this question by asserting that there are indeed some similarities and providing evidence of this from the use of tools and protogrammar in apes. Thus, humans may possess innate linguistic features with associated neural or anatomical features, but language has its roots in other cognitive and communicative abilities that have evolved in our species and have precursors in nonhuman primates (Bates, Thal, and Marchman 1991; Greenfield 1991; Hewes 1973a; Lieberman 1984; Malmi 1976; Miles 1991a). Old cognitive and communicative structures can be preadapted for the new functions of language. Since these old structures may be found both in the ontogeny of humans and in the ontogeny of enculturated apes, they suggest homologies that can serve as indirect clues for the reconstruction of language origins (Bates 1979, Lieberman 1984, Miles 1991a).

How can these old structures eventually yield language? Natural selection can act in different directions on the same structures, causing small, gradual changes that result in a more abrupt functional change. Lieberman (1984, p. 10) suggested that these different directions represent "functional branch points" where the structures remain the same but the evolutionary problem to be solved is fundamentally different. Changes in foraging, tool use, and social organization could have caused new selection pressures in hominids on the communicative and cognitive inheritance from hominoids. Undergoing a number of different selection pressures, language could ultimately become a new phenomenon, a "new machine built out of old parts" (Bates 1979a; Bates, Thal, and Marchman 1991).

Species Differences

The issue of species differences in the language abilities of apes also needs to be addressed. Savage-Rumbaugh, Pate, Lawson, Smith, and Rosenbaum (1983) speculated that any differences among the chimpanzee projects were due to experimenter bias and were "more reflective of ways in which different experimenters have interpreted the behaviors of the chimpanzees than of true differences in the capacities of the apes themselves" (p. 459). For some time chimpanzees have been assumed to be the most intelligent of the apes and thus most likely to acquire language skills, despite evidence that all of the great apes are highly intelligent. Studies with orangutans and gorillas followed those with chimpanzees and showed that chimpanzees were not unique in their abilities.

Lately, species differences have been highlighted once again with the speculation that the pygmy chimpanzee is unique in possessing certain behavioral, linguistic, and cognitive abilities. For example, Savage-Rumbaugh (1991) reported that the pygmy chimpanzees were the first to learn symbols through observation, yet learning signs primarily through imitation and observation has also been reported for all of the signing apes. Because of the varying methods of teaching and evaluating the animals, the small number of animals involved, and the varying purposes of the different projects, the results are far from conclusive. Also, the suggestion that pygmy chimpanzees are the most closely related to us has recently been challenged (Gardner and Gardner 1991). Furthermore, much of the perception that pygmy chimpanzees are unique is due to omissions in reporting about other species of apes. For instance, the orangutan engages in similar behaviors, such as long bouts of face-to-face copulation (Maple 1980), insightful cognitive style (Maple 1980), comprehension of speech (Miles, Mitchell, and Harper 1992), tool manufacture (Wright 1972), and imitation (Russon and Galdikas 1992). Recently, with the Language Research Center's change to more naturalistic enculturation methods, Savage-Rumbaugh (1991) has acknowledged the importance of enculturation rather than species distinctions, minimizing the differences among the apes.

Thus, it is likely that when we thoroughly enculturate all the great apes, and then test them with similar criteria, we will conclude that what we have been mapping is the entire pongid capacity and not the unique traits of a single species. Otherwise, how could we explain that the orangutan — who, based on fossil and biochemical evidence, is thought to be least closely related to humans — has similar skills (including advanced abilities such as imitation, pretense, rule-following protosyntax, comprehension of speech, etc.) as the pygmy chimpanzee? Slight differences in performance of different species of apes are more likely due to temperament, individual variation, or differences in enculturation. Indeed, it is possible that all of the variation in the results from the different language projects can be explained by enculturation alone.

The Role of Enculturation

In trying to assess language abilities among apes, we should increasingly consider the role of enculturation in these experiments (Van Cantfort and Rimpau 1982, p. 24). Is there a learning-to-learn phenomenon, or, as we are beginning to suspect, is the role of human culture, including its technology and social context, so powerful that it is the enculturation itself that can explain much of the language-trained apes' behavior? We know from studies of human children that the role of this "culture of communication" in language acquisition is extremely important (Bruner 1975, Dore 1974, Greenfield and Smith 1976, Miles 1983). A set of rules for noticing, sending, and receiving messages must be established if two-way communication is going to develop. It is increasingly possible to explain variations in the results of ape experiments in terms of the cultural setting established by the experimenter. We have seen that the cultural context of utterances largely frames the contours of ape discourse (Miles 1976, 1978, 1983). For example, Nim's failure to be spontaneous in his signing (or perhaps to invent signs) reflected the pattern of his instruction, including the rate at which he was interrupted by his trainers, as well as other aspects of the social context in which communication occurred (Miles 1983, O'Sullivan and Yeager 1989). As O'Sullivan and Yeager (1989, p. 278) confirmed after contrasting the signing of the chimpanzee Nim in two different environments, "the social ambiance or task elicits different behaviors." Likewise, the Language Research Center's success with both large and pygmy chimpanzees improved dramatically with its change in enculturational approach.

So strong is the role of enculturation that without the important scaffolding of the linguistic enculturation process, even early hominids may not have exhibited the rudimentary linguistic ability of today's enculturated apes. Ape language studies are showing that culture and the scaffolding provided by enculturation, the means by which we transfer culture, may be the foundation for the development of early hominid language and the primary selection pressure for later evolution of neurological linguistic features. The task then turns to discovering the likely preadaptations for more complex linguistic and cognitive ability in hominids and determining the extent to which nonenculturated apes in natural settings exhibit a similar complexity in communicative and cognitive skills. Unless we conduct further studies, it will be impossible to uncover the full extent of the role of enculturation.

What Enculturated Apes Lack

Despite remarkable skills that indicate some of the rudiments of language, apes have not yet shown evidence of a number of socially related cognitive abilities. These include selfless altruism, ethics, elaborated perspective-taking, regular vocal self-inhibition, complex imitation, extensive grammar or symbolic play, and representational art, although there have been suggestions of the beginnings of some of these abilities in apes (Lieberman 1984; Mitchell, in Chapter 6 of this volume). Many of the apes have made drawings comparable to the pre-representational level of 3½-year-old children (Miles 1990), but as yet there are no drawings that are widely accepted as representational.

Apes also differ from children in the rate at which their abilities develop and in the extent to which the symbolic aspect of their communications is elaborated, particularly in play. Human children are eager symbol users in play. While apes engage in some symbolic play, it is simple or more dependent upon caregiver support (Miles 1991a). Most particularly, the discourse of apes is significantly less abstract and contains fewer fantasies and narratives of make-believe (Lieberman 1984), although, again, this may be a result of enculturation.

However, the gap between the abilities of apes and human children continues to close, although no one has proposed that it will be eliminated. For example, Chantek's deceptive role reversal, which also showed perspective-taking and imitation, reveals greater mental ability than has been previously assumed for any of the apes (Miles, Mitchell, and Harper 1992). When they have had occasion to observe human behaviors, the orangutans at the Tanjung Puting Reserve have also engaged in very interesting extended imitations of those behaviors (Russon and Galdikas 1992). Although an adult ethical system is absent, enculturated apes have demonstrated that they have internalized some childlike concepts of good and bad when they chastise loud birds making alarm calls and label themselves as BAD when they have misbehaved (Miles 1990). Further, the readiness with which apes integrate their natural gesture communication system (and at least one vocalization) with their symbols (Miles 1976) indicates some linguistic plasticity. However, as in any case of negative evidence, there must be caution, and whether our social-related cognitive differences remain ones of kind rather than degree remains to be seen.

Hominoids and Parallel Evolution

The abilities of living apes can help us to build models for the earliest hominid communication systems, assuming that the early hominids possessed at least the skills of contemporary pongids. But how far have pongid communicative and cognitive skills developed since the hominoid-hominid split? Apes have undergone evolution themselves, and parallel evolution in both African and Asian apes may have independently resulted in the substantial cognitive and communicative skills they exhibit today. It is clear that contemporary apes are not identical to fossil hominoids from which they evolved because all of the great apes have some derived characteristics. In addition to biological changes, living apes may have developed their protoculture through selection pressures from changes in behavior, such as tool-using and more complex social organization. Povinelli and Cant (1992), for example, have suggested that orangutan intelligence developed from selection pressures related to becoming a large-bodied arboreal clamberer.

However, since all four of the great apes (orangutan, gorilla, large chimpanzee, and pygmy chimpanzee) show similar intelligence and susceptibility to enculturation, it is most likely that ape cognitive and communicative similarities are due to their common ancestry. We note that there has been a consistency in ape econiches and brain size over time. This suggests that although ape communicative and cognitive abilities may have undergone some evolution, Miocene hominoid forms had communicative and cognitive abilities similar to the most basic ones exhibited by living apes in natural settings. It follows then, that the roots of ape skills, as well as those of human language, are found in the preadaptations of the common hominoid ancestor, from which both the pongids and hominids evolved.

Large-bodied hominoids originated in Africa during the Miocene, about 20 million years ago. They dispersed to other parts of the Old World beginning about 16 million years ago, when contact between the African and Eurasian continental plates occurred, and split around 12 million years ago, when the Asian and African forms began to differentiate. *Sivapithecus* in southwestern Asia is linked evolutionarily with the orangutan, and *Afropithecus* and other hominoids in Africa are linked to the African pongids and hominids (Leakey and Leakey 1986, Pilbeam 1988). Like living apes, these hominoids may have used some tools, lived in social groups, and relied primarily on gestural communication, although these may not have been as well developed as they are in living apes. This (primarily) gestural communication coupled with some basic cognitive skills formed a set of preadapted abilities that could have served as the blueprint for later developments in pongids and hominids.

Early Hominid Protolanguage

Sometime around 4 million years ago the first hominids evolved from this hominoid population. These hominids were adapted more fully to a ground-living niche, displayed differences in tooth and jaw size, used bipedal locomotion, and possessed a slightly larger and perhaps differently organized brain (Johanson and White 1979).

Our hypothesis is that the earliest hominid ancestor would have possessed cognitive and communicative abilities similar to present-day apes in natural settings. These early hominids represent a functional branch point in which the hominoid preadaptations were challenged by the new problems resulting from a more terrestrial niche, upright bipedal posture, and perhaps greater reliance on scavenging. As hominid culture developed, the preadapted gestural and vocal signals of the hominoid legacy became more numerous and complex. Because early hominids may have needed a larger set of symbols to operate within a more complex set of social relationships, ape language research indicates that these symbols could have been rich and varied. As their tools become more complex with a wider range of materials, including stone, wood, and bone, their ability to label the environment, beyond immediate dangers or social relationships, may have developed. They may have had a small vocabulary of associative labels (both vocal and gestural) that could have been utilized in conjunction with one another and with slight alternations to express emphasis. Although they had a more complex technology and began to act upon the environment in more direct ways, their perspective may have remained somewhat egocentric and subjective, with limited awareness of self or other, and little altruism or ethical behavior. Thus, their abilities would be similar to those of apes in natural settings and enculturated apes in the earliest stages of acquisition of rudimentary language skills.

Hominid Language

Later hominid communicative and cognitive skills were probably similar to those of enculturated apes and formed a rudimentary language. A number of cultural developments associated with stone tool manufacture, such as selection and location of raw materials, played a key role in the expansion of communicative and cognitive skills. Hominids increasingly objectified and acted upon the environment, modifying it to suit their needs.

The current evidence from enculturated apes suggests that at this point hominids could readily acquire large vocabularies of communicative symbols and could combine these into short two- to three-symbol sequences or longer strings of symbols that slightly expand information. Since apes can also ask questions, make simple requests, and engage in communicative turn-taking, hominids that were making stone tools probably had developed discursive skills at least equal to those of enculturated apes. These hominids may have been able to modulate the meaning of symbols but had not yet developed an elaborate syntax or grammar such as is found in adult human language. Just as apes can use signs in new and interesting ways, overextend their meaning, and invent new signs for additional meanings, hominids would have had particularly rich semantic variations of their lexicon. Since most signs invented by apes are usually iconic in some aspect of their form, additions to the hominid lexicon were likely iconic as well. Hominids would have used these vocabularies to communicate with each other to meet rudimentary needs, solve problems, and comment on their environment.

As evidenced by use of untrained footsigning, enculturated orangutans can refer to the environment by pointing and can understand that a sign represents something else, that is, a sign is symbolic and not merely associative. This ability would be significant for hominids because their signals would become truly referential for the first time, giving them a means by which to refer to people, places, or things even when these were not present. Their subjective representations would become increasingly objectified as hominids developed in their perspective-taking, self-awareness, and ability to attribute mental states to others. This would provide obvious advantages in the development of culture as an externalization of memory, permitting greater individual and cooperative planning through stable and intersubjective mental representations of the environment. In addition, symbolic representations would have allowed hominids to engage in linguistic and behavioral deception.

Transmission of culture must have been an important element in the evolution of hominids. Early on, they must have relied extensively on imitation, just as enculturated apes do today. As culture became more complex, observational learning would eventually give way to deliberate instruction. As representational ability became more abstract, enculturated ape models show that hominids may have begun to engage in limited symbolic play. Greater representational complexity, combined with taking the perspective of the other, would have allowed hominids to develop not only a sense of self and other but to develop a sense of group identity and to engage in sympathy, nonparental care of others, and altruism.

Last, in a new functional branch point, probably in the upper Paleolithic, hominids developed new ways to transmit their symbols through changes in the vocal tract anatomy and restructuring of the brain for their communications (Lieberman 1984). Enculturated ape models imply that hominids must have exceeded ape combinatorial levels at this point. Physical artifacts of culture that were produced for functional reasons would begin to have symbolic significance, and hominids would begin producing representations through various markings and representational art (see Mitchell, Chapter 6 of this volume).

CONCLUSION

The evidence obtained from the ape language experiments suggests that communicative and cognitive abilities of early hominids were probably preadapted both for nonvocal forms of communication and for very good memories that gave them the ability to label or categorize the environment. Vocabularies of several hundred symbols might have been possible, with rich semantic domains and relations. These vocabularies were probably unstable, and these early hominids would rely on iconic meanings of invented symbols. Hominids would also have been able to refer to the environment and to place aspects of the environment in simple relation to one another in multisymbol, nonsyntactic combinations. The first hominids with language would also have had aptitudes for linguistic reference, pointing, imitation, displacement, and idiosyncratic symbol production. There would be some identification with the other and, perhaps, some social awareness beyond what we see in apes today. Thus, the first language of our ancestors could be characterized as a communication system that was rich in meaning, referential, and productive but not highly structured or symbolically reflective.

NOTE

1. This research was supported by the National Institute of Child Health and Human Development grant NICHD 14918, National Science Foundation grant BNS 8022260 and grants from the UC Foundation. The loan of Chantek was provided by the Yerkes Regional Primate Research Center supported by National Institutes of Health grant RR 00165. We thank Philip Lieberman for his initial support of this research, and we thank Robert W. Mitchell and the members of Project Chantek for their assistance in the enculturation of Chantek and the analysis of data collected from this study.

9

Tool-Using, Toolmaking, and the Evolution of Language

TIM INGOLD

ABSTRACT

In this chapter I explore the relationship between technology and language in the context of human evolution. I discuss six different ways of approaching this relationship, presenting each in the form of a hypothesis: (1) that language is necessary for the invention of novel design forms; (2) that language is necessary for the intergenerational transmission of designs and hence for their incorporation into culture; (3) that language and toolmaking are concordant derivatives of a single, more fundamental cognitive mechanism; (4) that language and toolmaking evolved under the same pressures of selection; (5) that the original language was not vocal but gestural; and (6) that toolmaking and language are linked in phylogeny, as they are in ontogeny. However, my attempt to synthesize these different approaches is frustrated by unresolved definitional ambiguities surrounding the concepts of both language and technology.

INTRODUCTION

"Technology," wrote Margolis, is "the practical capacity of a creature that has mastered language" (1978, p. 27). Contemplating the technical virtuosity of contemporary humanity from a standpoint of evolutionary hindsight, we would think that a being capable of such complex feats of design and execution must inevitably also be endowed with the faculties of language and speech. Surely an intelligence so advanced in the practical sphere must also manifest itself in the field of communication. Yet the connection between language and technology, if not merely fortuitous, is anything but self-evident, especially if they are viewed in the context not of their modern conjunction but of their concurrent emergence in the evolutionary trajectory of the human species. In this chapter I review some of the arguments that have posited a link between language and technology in human evolution. Such a link has often enough been suggested but remains to be positively demonstrated. Over a decade ago, Ashley Montagu (1976, p. 270) prescribed a "scientific study of tools," which he christened hoplonology, as a way of furthering our understanding of the relationship among hand, speech, and mind in the creation of human works. What follows could perhaps be regarded as a prolegomenon to the science of hoplonology.

Some might consider the enterprise doomed from the start. Atran (1982), for example, dismissed hoplonology with the remark that "such a science is virtually guaranteed to remain little more than mere conjuring as long as it is tied to the groundless presupposition that tool-making is, or should be, a guide to the nature of language" (p. 38). The illusion evidently consists in presenting surface resemblances as proof of some common cognitive or neurophysiological substrate. Yet granted the speculative nature of such inferences, the idea that there can be no connection at all between language and toolmaking is surely just as implausible. It can only be sustained on the assumption that the human mind is assembled from a number of more or less autonomous, domain-specific modules (cf. Fodor 1983, p. 37), one of which generates speech and another tool behavior. I prefer to go along with Count in his assertion that "we speak with the entire brain" (1976, p. 460). If that is so, then it is with the entire brain, too, that we must make and use tools and do everything else besides that qualifies as purposive action. For real people, what ultimately links these practices — of speaking, tool-using, and so on — is that they are all embedded in, and take their meanings from, shared contexts of everyday activity. With regard to any particular community, they are part and parcel of what Wittgenstein (1953, paras. 19, 23) called its "form of life."

One way of dealing with technology, of course, is to treat it as a *science* whose object is the technical activity of human beings. In that case it stands to technique as linguistics stands to speech — or, in other words, it is to linguistics and not to language that technology should be compared (Sigaut 1985, p. 122). Opposed to this is the view that technology can be constituted as an object of study distinct from technique, a distinction somewhat analogous to that in linguistics between language and speech (McArthur 1991). But just as *language* can refer either to a universal faculty of the human mind or to the code in use in a particular community of speakers (distinguished in French as *langage* and *langue*, respectively; see Lyons 1981, p. 2), so *technology* may connote either a human faculty or the practical arts of a particular culture. The questions posed by the relation between language and technology as aspects of the human "capacity for culture" are clearly different from those that arise from an investigation of how, in the culture of this or that group of people, their technical arts are ordered and classified in the categories of language, or of how language itself — as in the practice of magic — can have, or at least be seen to have, some technical efficacy. My concern here is with questions of the first kind; nevertheless, the capacity of both language and technology to assume a plurality of culturally specific forms is an important aspect of the link between them.

I proceed by focusing on six ways of approaching this link, which I present in the form of potentially falsifiable hypotheses, to each of which there corresponds a counterhypothesis that would deny the posited connection. In adopting this mode of presentation, I do not, however, mean to imply that the alternatives are in every case mutually exclusive or to deny that each may contain a grain of truth.

1. (a) Language is necessary for the invention of novel design forms.
   (b) The invention of novel forms depends on practical, non-verbal intelligence.
2. (a) Language is necessary for the intergenerational transmission of designs and hence for their incorporation into culture.
   (b) The principal means of intergenerational transmission of traditional designs are observation and imitation rather than verbal instruction.
3. (a) Language and toolmaking are concordant derivatives of a single, more fundamental cognitive mechanism.
   (b) The cognitive bases of language and toolmaking have nothing in common.

4. (a) Language and toolmaking evolved under the same pressures of selection.
   (b) Tools are for dealing with the physical environment, but the context of selection for language was social.
5. (a) The original language was not vocal but gestural.
   (b) Human language has evolved from the vocal communication of other primates.
6. (a) Toolmaking and language are linked in phylogeny, as they are in ontogeny.
   (b) In the evolution of language and toolmaking, ontogeny does not recapitulate phylogeny.

1. Language Is Necessary for the Invention of Novel Design Forms

The view that language, quite apart from its communicative functions, is the fundamental instrument of inventive insight has had many advocates. The philosopher Grace de Laguna (1927) devoted a whole chapter of a treatise on the function and development of speech to the use and manufacture of tools, arguing that the essence of making is that it is behavior "controlled by a *conception*" (p. 237), namely, of the nexus of relations to be set up between the modified object and other objects in the environment in a future context of use. This, in turn, requires an ability to focus attention on the properties of objects and their interrelationships, *independently* of the immediate situation of the perceiving agent. And it is language, de Laguna claimed, that makes this possible, in that it allows the subject to frame an anticipatory verbal response, prefiguring the expected course of action and its objective outcome. "It is scarcely credible," she declared, "that the art of chipping stone implements could have been developed by men who had not yet learned to speak" (de Laguna 1927, p. 218).

A rather similar view was advanced in a major work on language origins by the psychologist Géza Révész, who drew a necessary connection between language and "the power of constructive imagination." To make a tool appropriate to some given end, Révész argued, one must be "capable of surveying or visualizing the process of work to be carried out," and this is mental activity of a kind that requires "linguistic fixation" (1956, p. 93). Kenneth Oakley, approaching the issue as a prehistorian of lithic technology rather than a psychologist of language, came to much the same conclusion.

Toolmaking entails a "conception of the result to be produced before its realization in the material," yet thinking, planning, and inventing are virtually impossible without the use of verbal symbols: "most of our constructive thinking is done in unsounded words" (1954, pp. 14, 18). Hence, language is itself a tool, the tool of invention.

Evidently, the problem of the role of language in invention is inseparable from the more general question of the relation between language and thought. How much of our thinking is actually done, as Oakley presumed, in "unsounded words"? One influential approach to the question has come from the psychology of L. S. Vygotsky. In ontogeny as in phylogeny, Vygotsky argued, "initially thought is nonverbal and speech nonintellectual" (1962, p. 49). Citing the classic experiments of Köhler and Yerkes, Vygotsky concluded that the inventive capacities of the great apes were nevertheless prelinguistic, and that even in adult humans a considerable area of thought remains independent of speech: "the thinking manifested in the use of tools belongs in this area, as does practical intellect in general" (1962, p. 47). But speech originally evolved as a vehicle of sociability, lacking any significant intellectual content. Such "thoughtless speech" is apparent both in the babbling of the human infant and in the vocal communication of nonhuman primates, which is likewise emotive in content and social in function. Yet in the course of development, the faculty of speech is progressively (though never more than partially) co-opted by the intellect, coming to serve an individual rather than a social purpose. Appearing initially in the pseudo-communicative guise of the child's "talking to himself" as he goes about his business, this egocentric speech eventually "goes underground," taking the form of the silent, "inner speech" of the adult, entirely divested of its original social functions (Vygotsky 1962, pp. 18–19). In this kind of speech there is no simple translation, as in external speech, from thoughts to words. They are, rather, linked in a continuous, reciprocal movement from thought to word and vice versa, thus "thought is not merely expressed in words, it comes into being through them. . . . In inner speech, words die as they bring forth thought" (Vygotsky 1962, pp. 125, 149).

If this view is correct, then thought and language have converged from separate origins on the phenomenon of inner speech. Language would have emerged in hominid populations in the social context of individuals' dealings with one another, whereas intelligent toolmaking would have been promoted within the environmental context of relations with external objects. Our ancestors may have been both speaking among themselves and making tools for any length of time without involving speech directly in the production of design. We might ask, however, what the effects of such involvement would be, and whether they would leave their mark in the record of human activities. Again, an observation of Vygotsky offers an important clue. He noted how, in the development of the child, "egocentric speech first marks the end result or a turning point in an activity, is gradually shifted towards the middle and finally to the beginning of an activity, taking on a directing, planning function and raising the child's acts to the level of purposive behavior" (1962, p. 17). Transposed from ontogeny to phylogeny, this observation suggests that at some point in human evolution, the verbal representation of technical procedures overtook and hence came to lead, rather than to follow, the operations represented (Ingold 1986, p. 30).

When this happens, we can speak not merely of the making but of the *shaping* of tools, to adopt a distinction originally proposed by Irving Hallowell. To shape a tool, Hallowell argued, an individual "must have some image in mind which necessitates intrinsic representative processes." In fully human culture as distinct from the "protocultural" level of contemporary nonhuman primates and ancestral hominids, the repertoire of design forms is integrated — together with a body of technique and knowledge about the properties of raw materials — into an encompassing tradition. "It is difficult to see," Hallowell remarks, "how these factors could be integrated or transmitted without speech" (1956, pp. 98–99). As regards invention, the implication of Hallowell's argument seems to be that language is not (pace de Laguna, Révész, and Oakley) an essential prerequisite for the generation of novel conceptions through imaginative, "spur-of-the-moment" reconfigurations of the relations among things. To that extent Vygotsky's supposition that practical intelligence is predominantly nonverbal holds good. But language *is* necessary on what might be called a metatechnical level, if conceptual novelties are to be integrated within an evolving system of knowledge.

We may refer to such a metatechnical system as a *technology*. Like human kinship, which consists of relationships between relationships-between-persons (Wilson 1983, p. 60), human technology consists of relationships between relationships-between-things. Tool-shaping, as opposed to toolmaking, is constructive activity guided by a conception that does not stand alone but is situated — and acquires meaning and salience — within a technology. Where novelties are arrived at through reconfiguring relations on this metatechnical level, we may justifiably speak of technological innovation. And as Bloch (1991, p. 193) has suggested, linguistic explicitness, while not normally a feature of technical improvisation, may be a precondition for innovation on this level. Our conclusion would then be that language (including here both natural languages and artificial languages such as mathematics) is essential for technological innovation but not for inventions that have no technological grounding. When the point at which speech intervenes in practical activity shifts, as Vygotsky described for the developing child, from the end to the middle to the beginning of each performance, technology gradually acquires priority over technique. One likely consequence is a greater diversity and standardization of design forms (Isaac 1976, Guilmet 1977). Another is a major change in the tempo and mode of intergenerational transmission. This is my next concern.

2. Language Is Necessary for the Intergenerational Transmission of Designs

Even if language is not a prerequisite for the design of novel forms, it might still be necessary for their establishment within a population as part of an enduring tradition. Gowlett, for example, argued that the operations of early hominid toolmaking were of such complexity that they could only have been transmitted by means of language (1984, p. 55). Many have disputed this, however, claiming that the intergenerational transmission of toolmaking technique depends almost entirely on learning through observation and repetitive imitation rather than on verbal instruction (e.g., Davidson and Noble 1989, p. 127, Goodenough 1990, p. 601). Wynn (1993) provided examples and ethnographic documentation. According to Marshack (1979), "tool making and tool use are not learned through language, and the skills of the hunt are learned through example and participation, not by linguistic description" (1979, p. 395). In the same vein, Steklis and Harnad (1976) noted that "contemporary artisans are celebrated as the principal exponents of the so-called 'nonverbal' functions" (p. 450). Much traditional design would seem to fall into the category of what Alexander (1964) called unself-conscious, where "form-making is learned informally, through imitation and correction [rather than] taught academically, according to explicit rules" (p. 36).

Not only is such observational learning independent of linguistic formulation, it is also common in certain nonhuman species. The transmission of techniques of both tool-using and toolmaking has been especially well documented among free-ranging chimpanzees (see, for example, Boesch and Boesch 1990, 1993; Goodall 1986; Kortlandt 1986; McGrew et al. 1979; Nishida and Hiraiwa 1982; Struhsaker and Hunkeler 1971; Sugiyama and Koman 1979), although the interpretation of field observations as attesting to processes of intergenerational transmission has recently been disputed (Tomasello 1990, Visalberghi and Fragaszy 1990). Hallowell, however, was at pains to stress the limitations imposed on any technical tradition that relies *exclusively* on observational learning for its transmission. Traditions of this kind remain confined to what he regarded as a protocultural behavioral plateau; they lack the cumulative character of human technologies whereby every generation is provided with the means not merely to replicate but to advance beyond the performance of its predecessors (Hallowell 1962, pp. 248–249). Peter Wilson (1983) has argued that the achievement of this uniquely human level of culture depends upon "the art of teaching" (p. 146). To substantiate this claim, and thereby to specify more precisely the role of language in the transmission of design, we have to attend to the distinction between teaching and learning.

Much observational learning takes place through unintentional modeling, defined by Guilmet (1977) as "that process whereby one individual demonstrates through the display of his/her nonverbal behavior how to complete a task to another without the former purposely trying to do so" (p. 41). This is how, in a much-publicized instance of alleged nonhuman primate cultural tradition, a population of Japanese macaques is supposed to have learned the art of potato washing (Itani 1958, Kawamura 1959). A minimal criterion for teaching is that behavior should be intentionally displayed by one individual *in order that* another should learn by example. Here it is not merely the novice that attends to the model but also the model that attends to the novice, to the extent of intervening in the latter's behavior so as to bring it into conformity with a standard. This is what Premack (1984, p. 18) called pedagogy. It is notoriously difficult, in studies of nonhuman animals, to distinguish true pedagogy from simple imitation or observational learning, and opinions differ as to whether the former is confined to human beings or common to certain nonhuman species as well. Premack considered pedagogy to be uniquely human; however, Boesch and Boesch (1993) claimed to have witnessed instances of pedagogy — albeit relatively rare — among free-ranging chimpanzees. When it comes to humans, of course, pedagogy is a matter of common experience (Carrithers 1990, pp. 197–198) and characteristic of practically every situation in which novices are instructed through a period of apprenticeship (see, for example, Goody 1978, on Ghanaian weavers, and Lave 1990, on Liberian tailors). But although this certainly qualifies as teaching of a sort, it still does not depend upon the use of words (Premack 1984, p. 30), and its purpose may be limited to securing the replication of an established routine (Wynn 1993).

It is quite otherwise when what is taught is not the technique itself but the metatechnical principles, the relations between the relations-between-things, that allow for the generation — and not merely the implementation — of particular forms. Innovation, then, is not a deviation from tradition, originating in some chance insight or an accident of transmission. It is, rather, an exploration of the generative potentials of a received system of knowledge. The propagation of such knowledge is of necessity an activity distinct from its application, taking place — as Marshack put it (1979a, p. 395) — "before, after, and in between." To teach navigation (to borrow Wilson's example, 1983, pp. 145–146), one does not navigate. The Caroline Islanders, who are used to crossing great expanses of ocean without the help of charts or instruments, teach navigation "through designs made in the sand and by telling highly structured stories. . . . The designs and stories are about navigation but are not *of* navigation" (Wilson 1983, p. 146). That is to say, they speak not of voyages made but of the principles and guidelines for making them.[1] More generally, in this mode of teaching, rules, recipes, and procedures for the generation of objects and events are transmitted without the novice's having any direct experience of those objects and events through participation or observation. It is hard to see how knowledge of this kind could be passed on, *outside* the context of its practical application in concrete situations, without the facility of propositional language.

In short, the unintentional or intentional modeling of nonverbal behavior may suffice for the transfer of *technique* through observation, imitation, or wordless pedagogy, but the transfer of *technology* requires deliberate verbal instruction. Of course the balance between these different types of intergenerational transmission is likely to vary from one society to another. The mere existence of a technique (say, of making flaked stone tools) does not imply an organized body of technological knowledge relating form, function, and material. White et al. (1977) documented a situation of this kind among Duna people of New Guinea. Duna do not think of themselves as *shapers* of stone, believing rather that their flaking serves only to reveal forms that are already present in the material. Moreover "copying, and trial and error, rather than explicit teaching, are certainly the methods by which young Duna men learn about flaked stone" (White et al. 1977, p. 381). At the other extreme, in our own society an extraordinary amount of time is devoted to verbal instruction in a classroom setting institutionally separated from the practical situations in which, it is supposed, this acquired knowledge may be put to use (Guilmet 1977, pp. 43–44). Whether knowledge so acquired is *actually* applied in people's practical dealings in the world outside the schoolroom is, of course, another matter (see Lave 1990, pp. 320–323, on school math).

It would be wrong to assume, however, that people brought up in our technologically "advanced" culture should be more proficient, in terms of technique, than their predecessors. To operate an advanced technology does not require a correspondingly advanced technique. The construction of flint implements by the so-called Levalloisian method, practiced by human toolmakers some 600,000 years ago, calls for at least as much skill as is required of the modern car mechanic, and it can take as long if not longer to acquire. Yet by any criterion, the automobile is a vastly more complex artifact than anything produced by Mousterian flint-knappers. Thus, it is in the evolution of technology mediated by language, and not in the evolution of technique, that the progressive aspect of human culture is manifested. But the contexts for the intergenerational transmission of technology lie largely outside the contexts of its technical application. Doubtless the skills of hunting are acquired through participation — however much of the "ethnoscientific" knowledge that hunters bring into their activity comes from the recitation of narrative and myth — and is part and parcel of a far more comprehensive understanding of the nature of things, or what anthropologists call a worldview. It is because humans not only act in the world but also *talk about* their activity, in the company of others, that there exist technological traditions with an intrinsic potential for growth and development.

3. *Language and Toolmaking are Concordant Derivatives of a Single, More Fundamental Cognitive Mechanism*

In a classic series of papers, Hockett (1959, 1960b, 1963) attempted to isolate those "design features" that are unique to human language as a system of communication. Chief among these are displacement, reflexiveness, and prevarication. The first is the ability to refer to objects and events that are remote in time or space from the situation of utterance. The second is the ability to speak about speaking — to make one's own discourse an object of verbal commentary. The third is to engage in deliberate deception, to convey information whose meaning is based on illusion. Advocates of the view that language and toolmaking rest on a similar cognitive basis have been quick to point to parallel features of human constructional capacities (Ingold 1986, pp. 69–70). Kitahara-Frisch, for example, discovered displacement in the activities of both speaking and stone tool manufacture: "human language [requires] the ability for the speaker to detach himself from the ongoing flux of experience, from the present time and place, just as the stone toolmaker must divert his attention from the ultimate purpose of his toolmaking activity, while nevertheless keeping it in mind as his guiding principle" (1978, pp. 105–106). And reflexiveness depends upon a capacity to attend to one's own performance that is evident both in speaking about speech and in secondary toolmaking — that is, making tools with which to make other tools (p. 106). Kitahara-Frisch did not consider the feature of prevarication; I have argued elsewhere (Ingold 1986, p. 69), however, that its technical equivalent is the production of toys, which are objects with which we *pretend* to change the world. But pretense is not quite the same as deception, and a more apt analogy may be with such devices as hunting blinds and other forms of camouflage or counterfeit objects (Kendon, personal communication).

In an attempt to find parallels to toolmaking, Holloway (1969) has likewise focused on a number of Hockett's design features for human language. These include traditional transmission, arbitrariness, productivity, and duality of patterning. That an element of either learning or teaching is involved in transmitting the conventions of

289

both language and toolmaking is plainly evident yet begs a number of questions. For example, there is some evidence that the acquisition of language is governed by partially innate (genetically transmitted) structures: to what extent is this also true for the acquisition of technical skills? Is there an "acquisition device" for technology as well as for language? Wynn (1993) thought not, for whereas children can generally be relied upon to gain full competence in the language of their community of origin, this is far from the case with regard to toolmaking skills, in which there may be considerable variation in proficiency within the community, from incompetence to expertise. Moreover, as observed in the previous section, a learned tradition of toolmaking does not necessarily imply the existence of a body of societywide conventions analogous to those of a linguistic code, nor does it require language for its transmission.

The notion of arbitrariness, with regard to toolmaking, is equally problematic. For Holloway (1969), the arbitrariness of human toolmaking lay in the *imposition* of form on raw material, where "there is no necessary relation between the form of the final product and the original material" (p. 401). This is compared with the arbitrariness, in human language, of the relation between words and the objects they signify — there is nothing saline about the word *salt* or canine about the word *dog*, hence any other words would do as well so long as an established convention about their meanings exists in the community of speakers (Hockett 1960, p. 90). In this respect, however, I believe Holloway's comparison to be ill founded. The relation between a word and its meaning is analogous, if anything, to that between a tool and its use.[2] But whereas one word is as good as another to refer to a hammer, certainly stone is better than wool for hammering, and the carpenter's hammer better still. Thus the arbitrary relation in the linguistic sign, between signifier and signified, does not obtain when we turn to the relation between a tool and its uses. Following Saussure (1959, pp. 114–115), we may say the "value" of a word, that is, the area of conceptual space that it covers, is limited by its opposition to other words within the language; the use-value of a tool, in contrast, is limited by the physical properties of the object itself. Stone is useful for hammering not because it is opposed to wool but because it is hard and durable. The crucial implication of the comparison is that the manufacture and use of tools does not presuppose their incorporation into a total system of relationships validated by social convention. One cannot speak without a language, but one can perfectly well make and use tools without a technology.

The design features of productivity and duality of patterning are closely related insofar as the latter, in a system of communication that operates with discrete (digital) rather than continuously graded (analog) signals, is a condition for the former. Productivity (or "openness") means the capacity to vary the semantic content of utterances indefinitely without loss of comprehensibility. This is achieved by combining a rather small number of meaningless elements (phonemes) in different ways to generate a very much larger number of meaningful words or wordparts (morphemes), which can be discriminated in terms of their elements. Duality of patterning refers to the division between lower-level structural elements and higher-level meaningful units. Applied analogously to the domain of toolmaking, the structural elements compose a "vocabulary" of basic motor operations (Holloway 1969, p. 402), combined in different ways to produce dissimilar tools. As Noble and Davidson have argued, however, the scope for such combination in the manufacture of flaked stone tools, far from being virtually unlimited, is tightly constrained by mechanical requirements — "the physics of forces and rocks" (Noble and Davidson 1991a, p. 239). The mechanical conditions that, in speech, make productivity possible, actually *limit* productivity in the case of stone toolmaking. These limits were removed only with the introduction, in the upper Paleolithic era, of new stoneworking techniques involving grinding rather than flaking, and of new raw materials with greater plasticity, such as bone, antler, and wood. In this transition, as White (1985) remarked, "we seem to be going from the medium defining the possibilities to language and thought structuring the possibilities which are then realized in plastic media" (p. 107). The consequently enhanced productivity, however, is due not to the relaxation of constraints on the range of combinations afforded by a principle of duality of patterning but to the replacement of discrete operations (flaking) with operations that permit continuous gradation (e.g., cutting, grinding, polishing). In other words, it is the productivity inherent in an analog rather than a digital system.

To return to language, besides the rules for combining phonemes into morphemes, there exists a further set of rules, of grammar or syntax, governing the combination of morphemes into structured sequences (sentences). To appreciate the functions of grammar, and the kinds of pressures that could have led to its evolutionaryemergence, it is helpful to think of language (following Liberman 1979) as an interface phenomenon mediating between the intellect on the one hand and the bodily structures of message transmission and reception on the other. In the case of speech these are the structures of the vocal tract and ear, organs that evolved to very nearly their present form long before the appearance of language and that were originally adapted for quite different purposes (Bradshaw and Nettleton 1982, p. 176). Hence, there is a limit to what they can be made to do — to the number of discrete sounds that the voice can produce and that the ear can recognize, and to the speed with which they can be relayed. By contrast, there have been extraordinary advances in the structures of cognition in the course of human evolution, associated with the overall process of encephalization. The result is a potential mismatch between the capabilities of the intellect to generate and comprehend messages and those of the bodily apparatus to transmit and receive them. In agrammatic communication, with a simple one-to-one correspondence between message and signal, the number of transmissible messages would be limited to the number of available signals. But with a grammar that permits the rule-governed combination of elementary signal units into strings, the number of transmissible messages becomes effectively infinite. "Thus, grammar serves to match a message generator, at the one extreme, to a transmitter and receiver at the other. In so doing, it makes human communication vastly more various and efficient than it would otherwise be" (Liberman 1979, p. 686).

Is there a similar "grammar" of toolmaking, as Montagu suggested (1976, p. 269), indicative of some common derivation with language? Holloway (1981) argued that this is indeed the case and hence that "any theoretical model that describes language *also* describes stone toolmaking. Both processes utilise a limited number of basic units that are combined in a finite number of ways (with reference to a specific language or specific tool) and there is an overlying set of rules, or syntax (grammar), about how units combine and concatenate" (p. 290). Another author to have drawn the parallel is Lieberman (1975, pp. 163–170). Comparing the kind of encoding and decoding that goes on in fully formed human language with the Levalloisian technique of making flaked stone tools, he showed that both entail the operation of a transformational syntax (1975, pp. 12–13, 169–170; see Chomsky 1957). Arguably, the syntax of toolmaking could function in much the same way as the grammar of language, to overcome the mismatch between the potential of the intellect to generate designs and the limitations of the manual and visual apparatus co-opted for their implementation. As with agrammatic communication, so with nonsyntactical object manipulation, a one-to-one correspondence between design and movement (analogous to message and signal, respectively) would limit the number of feasible designs to the number of possible movements. Any new design could only be incorporated into the repertoire as a "total act" rather than as a resequencing of the elementary motor operations making up already existing routines. But by starting off with a limited set of such operations, each ineffective on its own (as phonemes are meaningless), it should in principle be possible to implement a wide range of design forms.

Yet for much of prehistory, from the first appearance well over a million years ago of stone tools bearing the hallmarks of formal design (bifacial hand axes of the so-called Acheulean type) to the dawn of the upper Paleolithic era, the most remarkable feature of hominid toolmaking was its spatial and temporal uniformity. For hundreds of thousands of years, whether from Africa, Europe, or South Asia, bifaces remained virtually identical. Though some variation exists within the overall form, "a particular biface shape would not be out of place in any biface assemblage, of any age, from any part of the world where bifaces have been found" (Wynn forthcoming). To account for this uniformity, as Holloway did (1969, p. 401), by invoking the power of semantic and syntactic rules, backed by social consensus, seems to me far-fetched. There is nothing even approaching this degree of conservatism in the archeological or ethnographic record of modern *Homo sapiens*. At present we have no satisfactory explanation for the regularity of biface form, although Davidson and Noble (1993) have recently suggested that the form may be no more than an unintended by-product of flaking techniques, combined with mechanical constraints imposed by the raw material. It seems possible, then, that the lack of a transformational component in the syntax of premodern hominids might have imposed limits on form simply by restricting the range of operations that could be performed upon a given raw material. But even if this explanation is accepted apropos the Acheulean hand ax, it still leaves us with the problem of accounting for the substantial spatial and temporal overlap of Acheulean and Levalloisian industries.

The notion that there exists an isomorphism between language and toolmaking remains highly speculative and has its critics as well as its advocates. Wynn (1993) approached the problem by thinking of both tool behavior and speech as "layered systems" that can be analyzed on a number of levels. Thus, tool behavior can be considered in terms of the biomechanical constraints imposed by the executive organs of the body (analogous to those imposed on speech by the structures of the vocal tract and ear), the ways in which elementary movements are linked into sequences (analogous to the syntactical ordering of speech), and the organization of knowledge (analogous, perhaps, to the semantic organization of language). But with regard to the latter two levels, the comparison with language reveals fundamental differences. Wynn argued that action sequences in tool behavior are *not* generated by rules of syntax but are merely "strings-of-beads," produced by chaining one action to the next and reproduced by rote memorization. There is, in Wynn's view, no syntax of technology (Wynn 1993). As to the organization of knowledge, he showed that although tool users and toolmakers do bring together ideas about form, function, and material into larger conceptual "constellations," these differ from the semantic paradigms of language first in that they are tied to particular contexts of use and second in that they are idiosyncratic — specific to their individual constructors rather than common to a technical community (see Daugherty and Keller 1982).

There is one way in which both advocates and critics of the thesis that language and toolmaking have a common cognitive substrate may be right. Little, so far, has been said about development. It may be, however, that language and tool behavior share a common basis in the early stages of development but that they subsequently undergo differentiation. Alternatively, they may be distinct to begin with and undergo subsequent integration. The latter, as we have already seen, appears to have been Vygotsky's view. However, Greenfield (1991) has recently put forward powerful arguments in favor of the former position. These are backed up by neurological evidence that appears to show that a substrate initially identified with Broca's area undergoes differentiation by creating separate though parallel circuits with anterior regions of the prefrontal cortex responsible for the distinct combinatorial properties of mature speech and tool behavior. If it is indeed the case, then, that making or using tools and speaking are generated by separate cognitive modules, we may have to conclude that modularity *develops*. The evolutionary implications of this conclusion depend on the validity of inference from ontogeny to phylogeny, an issue to which I return in a later section.

4. Language and Toolmaking Evolved Under the Same Pressures of Selection

In their application of the theory of variation under natural selection to the problem of human evolution, both Darwin (1874, pp. 195–224) and Wallace (1870, pp. 302–331) were agreed that the pressures that favored the advance of human intelligence lay in our ancestors' dealings with components of the physical environment, such as predators and prey. The use and manufacture of tools played a central role in the argument. Individuals who were "more sagacious" (as Darwin put it), with bigger and better brains, could design more ingenious tools, thereby securing a reproductive advantage. Intelligence-enhancing variations would consequently tend to be preserved in future generations, leading to yet further advances in the technical sphere, and so on through mutual reinforcement. This view of the evolution of intelligence continues to enjoy considerable support (Washburn 1960). For example, Parker and Gibson (1979) asserted that feeding strategies, and above all "extractive foraging with tools," were the primary determinants of the kind of intelligence and protolinguistic abilities of the earliest hominids, and that the further development of both language and intelligence was an adaptive response to complex hunting involving aimed missile throwing, stone tool manufacture, animal butchery, food division and shelter construction.

This scenario has not, however, gone unchallenged. In an influential paper, Humphrey (1976, see also Jolly 1966) has observed that the practical problems nonhuman primates face in the procurement of subsistence are not particularly difficult and do not call for a remarkable degree of intelligence. Most will already have been solved, so that each generation need only learn the solutions from its predecessors. Learning, however, entails close and prolonged social interaction in groups with overlapping generations. The successful management of relations within such groups entails considerable foresight, ingenuity, and skill. Humphrey's argument is that selection for "social skills" rather than for technical inventiveness has been responsible for raising the intellectual capacities of nonhuman primates to a level beyond that which is required for most practical purposes. And if this is so for contemporary non-human primates, then the same should be true in the evolution of our own species. In short, contra Parker and Gibson (1979) and others, intelligence is primarily an adaptation not for subsistence but for social life.

Recently dubbed "the Machiavellian intelligence hypothesis" (Whiten and Byrne 1988a), this argument has attracted a good deal of support. It is, however, an argument about intelligence in general and not specifically about language or toolmaking. Thus, the field of argument is in fact more complex than at first appears. For quite apart from the question of whether the critical selective test for intellectual advancement lay in individuals' abilities to manipulate the objects of their physical surroundings or to manipulate other individuals in their social group, there is the problem of the extent to which the evolution of language can be regarded as a specific corollary of the evolution of generalized intelligence. We might recall, in this connection, Vygotsky's view that early speech, though it functioned as a vehicle of sociality, was in essence emotional or affective rather than intellectual, and that the co-option of speech by the intellect was a late development and never more than partial. It would follow that an account of the evolution of language cannot be simply subsumed under an account of the evolution of intelligence. Ranged against this are views that regard both toolmaking and language as instantiations — technical and communicative, respectively — of a unitary "intelligence" that is uniquely human; these views are themselves divided, as we have seen, according to the primary selective pressures adduced to account for the evolution of human intelligence. To further exemplify this division, I present the arguments of two other scholars, Holloway (1981) and Lancaster (1968), both of whom regard language and toolmaking as closely related functions of the human intellect.

Holloway attributed the emergence of the symbolic intelligence that he supposed to underwrite both language and toolmaking to the operation of selective pressures inherent in *social* processes. "It was not the tools themselves that were the key factors in successful evolutionary coping. Rather, the associated social, behavioral and cultural processes, directing such activities as toolmaking, hunting and gathering, were basic" (1981, p. 288). In this, Holloway followed a line originally pioneered by Engels in his speculative essay of 1876 on "the transition from ape to man," in which he stressed the role of socially organized labor in the development of both toolmaking and speech (Engels 1934; see Woolfson 1982 for an assessment of the contemporary relevance of Engels's theory with regard to tool behavior and language origins). For Lancaster, by contrast, the adaptive value of language, like that of tool use (which she regarded as another uniquely human capacity), lay in facilitating the exploitation of the physical environment rather than the conduct of social interaction (1968, p. 456). The original function of language, she argued, was one of environmental reference or *naming*, allowing human hunters and gatherers to communicate information about raw materials, or animal and plant resources, of strategic value in the conduct of cooperative foraging activities. As such, language is contrasted with nonverbal communication, which serves primarily as a vehicle for the expression of emotion — that is for conveying information about motivational states essential for the management of ongoing social relationships in small and intimate groups. Lancaster was aware that there already exists among nonhuman primates a sophisticated, multimodal system of communication involving voice, gesture, and facial expression, which fulfills precisely this function. This basic primate system, she believed, persists in modern *Homo sapiens* — manifested, for example, in laughing, smiling, and crying — and with much the same form and function. But it persists *alongside* the novel capacity of language, remaining quite separate from it.

A directly contrary view has been put forward by Nancy Tanner (1981, pp. 127–130, see also Tanner and Zihlman 1976). She argued that language cannot be set apart, in the way that Lancaster suggested, from the body movements, facial expressions, gestures, and vocal qualities that accompany every utterance and provide essential clues to understanding its meaning. In this respect, human communication is just as multimodal and context-specific as that of nonhuman primates, and there is no breach of continuity from one to the other. Moreover, among contemporary hunter-gatherers language is little used in the learning and application of subsistence techniques, whereas it is central to the management of personal identity in face-to-face social interactions within the camp or band. As this is likely to have been equally true in the past, Tanner believed that the social relations of the band, rather than subsistence-procurement activities, formed the leading context for the evolution of language and that it evolved "in a matrix of extensive nonverbal communication" (Tanner and Zihlman 1976, p. 474). Thus, the functions of affective signaling and environmental reference cannot be neatly partitioned between nonverbal and verbal communication. Rather, the total system made up by the interweaving of verbal and nonverbal constituents serves at one and the same time as a vehicle of emotional expression and for conveying information about the environment.

Here, then, is an argument for the evolution of language that situates the conditions of selection in the social rather than the physical environment yet does not treat language as the vehicle of a manipulative "intelligence" divorced from the context of individuals' intimate and affective involvement with one another.[3] The argument clearly foregrounds questions, first, about the continuity, or lack of it, between human and nonhuman primate communication and, second, about the involvement of language in the emotional-affective domain. These questions have been further highlighted by the extensive debate surrounding one particular theory of language origins, according to which the evolutionary antecedent of speech was a form of gestural rather than vocal signaling. The theory is not new, having a venerable history ably reviewed by its leading contemporary advocate, Gordon Hewes (1976a). It stands most immediately opposed to the theory that human speech has evolved directly from the vocal communication of ancestral primates. The arguments for and against these alternative theories are finely balanced, and I do not intend to take sides. However, I mean to show how they implicate two other important issues, the first concerning the reasons for the pronounced lateralization of the human brain, the second concerning the role of tool behavior in the evolution of language. I begin with a brief summary of the gestural theory as presented by Hewes (1973a, see also McNeill 1979).

5. The Original Language Was Not Vocal But Gestural

The curtain opens on protohominids on their way to acquiring bipedal locomotion, freeing the hands for gesture and regular tool use (as well as for carrying things), and moving into open country where selection pressures were intensifying in favor of both improved skills of object manipulation and the emergence of a propositional language for communicating information about tools and their uses, environmental resources, and terrain. As to which came first, tool-using or language, Hewes was noncommittal; his point was that both developed concurrently and that both involved identical visual, kinesthetic, and cognitive pathways. Hence, the language of instruction and report drew on the same repertoire of gestures as those entailed in the actions directed or described, so that the translation (for example, in learning situations) from one to the other did not require the surmounting of any neurological barrier to cross-modal transfer — as when vocal instructions call for a manual response. Like Holloway, Hewes (1973a) believed that toolmaking and language manifest a single, more fundamental syntactic capacity "to acquire and utilise complex patterned sequences" (p. 109). Furthermore, skilled tool use puts a premium on preferential handedness, allowing a specialization of functions between the dominant (usually right) hand that operates the tool through a precision grip while the subordinate (usually left) hand anchors the object worked upon with a power grip (Toth 1985 documented this functional division for stone knapping). As each hand is controlled by the contralateral hemisphere of the brain, the development of dextrality entails a corresponding dominance of the left hemisphere. The division between anchor and operator functions of the hands in tool use is mirrored, in language, in the division between grammatical subject and predicate (see Greenfield 1978, p. 419): one more indication, for Hewes, that both are underwritten by the same cognitive mechanisms. Another is that certain kinds of damage in the dominant hemisphere can affect motor skills just as they affect speech — the patient can produce or recognize isolated words, or motor acts, but is unable to chain them syntactically into coherent sentences.

To continue the evolutionary story, increasing reliance on language and tools intensified the pressures promoting further cerebral enlargement, complexification, and lateralization, until eventually the requisite neural reorganization was established to enable the transfer of language from the visual-gestural to the vocal-auditory channel. Just how and why that happened is left rather open, but a number of advantages of the vocal-auditory channel suggest themselves: speaking is probably quicker and more energy-efficient than gesture, one can speak when the hands are otherwise engaged in carrying or tool use, speech can be employed to communicate in the dark or with people who are not in the immediate line of sight, and it can serve to direct others' attention to the speaker. One of the most interesting suggestions is that speech eliminates the ties of iconicity in the relation between word and concept that are invariably present to some degree in gestural communication, allowing for the unlimited diversification of natural languages – or "Babelization." With an extremely limited lexicon of mimetic signs, the gestural language of early humans may — like the tools of the period (e.g., Acheulean bifaces) — have been virtually the same for the entire species. It is tempting to see a parallel between the sudden diversification of tool traditions evident from about 40,000 years B.P. and the switch from a single, species-specific gestural mode of communication to the manifold, culture-specific, arbitrarily encoded spoken languages of modern humans (Richards 1987, p. 256). Both language (or dialect) variation and differences in technical practices may have played a role in signaling individual or group identities, or even in the demarcation of territorial boundaries, as indeed they do in many contemporary human societies.

"Gesturalists," as we may call the advocates of this theory (Hewes 1976a, p. 489), can point to a number of factors in their favor. First, there is the proven ability of chimpanzees to operate a gestural sign system, which contrasts with their marked inability to operate a similar system using sounds (this contrast, however, is disputed by Armstrong 1985). The same may have been true of early hominids. Second, the visual-gestural channel is still available for use in situations where speech and hearing are for some reason impeded. This is most obviously the case among persons who are deaf, who can communicate spontaneously and effectively in the gestural mode. Thus the gestural channel seems to have been held in reserve rather than closed down with the transfer to speech. However, gesture remains an almost universal accompaniment to face-to-face vocal communication, leading further support to the gesturalists' case. Thus Adam Kendon (1975, 1980) has argued that gesture is no mere embellishment on what is being said; rather, "speech and movement appear together, as manifestations of the same process of utterance" (1980, p. 208). His observations, that gesture tends fractionally to anticipate the spoken utterance, that it can continue smoothly even when speech is interrupted, and that it may sometimes precede any vocal response, all indicate the primacy of the gestural mode (Kendon 1980, pp. 226–227). Other scholars, such as Parker and Gibson (1979, p. 373), have supported the gestural theory on the grounds that in the acquisition of "protolanguage" by human children aged between 9 and 13 months, a complex of gestures for pointing, and for the showing, giving, and requesting of objects, appears well before the first spoken words. The relevance of this observation for the evolution of language depends, of course, on the legitimacy of ex-trapolating from ontogeny to phylogeny, a question to which I turn in the next section.

Rather more controversial backing for the gestural theory comes from the field of neurology. Myers (1978) has argued that neurologi-cal evidence offers clear proof of a dichotomy between the *emotional* and *volitional* use of both the voice and facial expression in human subjects. On the emotional side, laughs, shouts, screams, and the like are invariably accompanied by similarly expressive and apparently involuntary facial movements. In contrast, the use of the voice in speech is clearly volitional, though there is no obvious link between speech and volitional facial movement. For nonhuman primates, however, it appears that face and voice are only poorly under voli-tional control. "Rather, in these species, the face and voice function primarily as signalling systems, but apparently only within the realm of social and emotional behavior. . . . The facial expressions and vocalizations of nonhuman primates are closely akin to man's emo-tional use of the face and voice" (Myers 1978, p. 72). Yet chimpanzees have excellent volitional control over their hands and body move-ments. Now the mechanisms of the cerebrum that govern the kinds of emotional and affective behavior common to humans and nonhu-man primates are thought to be located in a relatively primitive region of the mammalian brain known as the limbic system, which is far removed from those regions of the neocortex that control volitional movements such as those involved in chimpanzee gesturing and human speech (Robinson 1976). For this reason, primate vocaliza-tion seems an unlikely precursor for spoken language (Steklis and Harnad 1976, pp. 447–448). As Hewes (1976a) remarked, "if spo-ken language had evolved directly out of the primate call system, we would not expect it to have become localised in an entirely different portion of the cortex" (p. 490). To this one might add that had speech actually evolved from primate vocalization, the expected result — to be consistent with evolutionary theory — would not be the myriad diversity of existing natural languages but a single, species-specific code with a limited vocabulary and finite generative potential (Jerison 1976, pp. 377–378).

Myers cannot be called a gesturalist. Like Lancaster, whose ideas I have already reviewed, he believed that speech developed de novo in man rather than from either primate vocalization or an earlier gestural form. Nevertheless, his conclusions are perfectly consistent with the gestural theory, which likewise presupposes a rigid division between language and nonverbal communication, such that the latter takes all the emotional load, leaving the former with the purely propositional function of environmental reference. Not that early hominids would have conducted their lives in eerie silence, for their gestural communication would have been carried on to the accompaniment of an outpouring of emotionally charged vocal (as well as facial) expression, just as our own speech is modulated by paralinguistic vocal qualities, supposedly under limbic control, that serve to convey affect. However, a key property of propositional language, according to Hewes (1976a, p. 490), lies in the capacity to deliver detached reports on the most highly charged situations without the slightest trace of affect. This is the property of decontextualization, by which language is said to be sharply distinguished from the context-specific vocalizations of nonhuman primates.

It is here that the gestural theory touches its weakest point. As we have already seen, it is debatable whether language can be clearly disengaged from its nonverbal matrix, whether the meanings it conveys are independent of relational context, and whether its primary function is really to convey information about the external environment.[4] Language, as Richards (1987, p. 260) so clearly stated, may be "primarily to do with inter-personal relationships, and thus with emotional life as much as, and more fundamentally than, with cognitive life" (p. 260). If this is so, both the limbic system and neocortical regions of the brain may be implicated in the production of speech (Marshack 1979a, p. 395), gravely weakening the neurological argument against its continuity with primate vocalization. Moreover, the neurological evidence has turned out to be decidedly equivocal. Steklis and Raleigh (1979) have shown that (contra Myers) nonhuman primates *are* capable of the volitional control of vocalization and facial expression. Furthermore, the limbic system, besides governing emotional reactions, is involved in memory and cognition and has not remained immune from modification in the course of cerebral evolution. Thus, we should not, as Bradshaw and Nettleton (1982) warned, accept too uncritically the belief that in nonhuman primates

> invariant emotional displays [are] mediated by limbic mechanisms, while man employs voluntary cortical mechanisms to

transmit information largely disassociated from his emotional state. . . . In all species cortical, thalamic and limbic structures may play a part [in the production of voluntary and spontaneous vocalizations], and the differences between the species with respect to the mediation and the nature of facial and manual gestures and vocalization may be quantitative rather than qualitative, relative rather than absolute. (p. 182)

(For a more recent presentation of the neurological evidence, see Deacon 1989).

Let me now turn to the "vocalist" theory, that there *does* exist an evolutionary continuity between speech and primate vocalization. A leading advocate is Dean Falk (1980). Not only has she said that language is as old as, if not older than, the earliest stone tools, she has also held that the first language was spoken. Her argument centers on the question of the origins of cerebral lateralization. Recall that for Hewes, skilled tool use favors handedness, which leads in turn to cortical asymmetry. But as Falk pointed out, cortical asymmetry has also been found in Old World monkeys, which do *not* exhibit systematic handedness (despite some individual preferences). Moreover, "regions related to cortical asymmetry in monkeys . . . are apparently homologous with human cortical regions that have important language functions in left hemispheres" (1980, p. 74). This observation led her to construct an evolutionary scenario of the following kind: selective pressures would have tended to enhance the vocal communication of the very earliest, monkeylike, and forest-dwelling ancestors of human beings, and the neurological correlate of such enhancement would have been cortical asymmetry. With the move to an open-country habitat, new pressures came into operation that promoted *both* bipedalism *and* a more complex and finely graded communication system. Thus (contra Hewes), "vocal communication systems had become more complex under the selective pressures that *led to* bipedalism" (Falk 1980, p. 70). Once the hands and arms were freed by the attainment of bipedalism, they, too, became subject to the neurological asymmetry that already affected speech. Through what Falk called a "field effect," left-hemisphere dominance ascended from association areas of the cortex controlling speech to those controlling the hand. In consequence, ancestral hominids became righthanded.

If Falk's argument is correct, then skilled tool use, far from propelling the evolution of language, came about through the extension of a neurological substrate for language, involving cortical asymmetry, that *had already evolved*. This is the precise reverse of Hewes's position, which is that language came into being through an extension of the neurological correlate of tool use into the arena of gestures. However, the vocalist argument does contain a significant lacuna in that it fails to explain *why* vocal communication, rather than preferential handedness, should entail cerebral lateralization. A possible answer was suggested by Passingham (1981). He noted that whereas peripheral organs such as the hand are controlled by the contralateral hemisphere, central organs such as the tongue and jaw are bilaterally innervated by both hemispheres of the neocortex. The same has been found to apply to the vocal cords. If an organ that can be controlled by both hemispheres is to execute highly complex sequences of movement, as in the production of speech, it is likely to be most effective if "directed by one central programme, located in a single hemisphere, rather than by two separate programmes which must use the long commissural pathways to co-ordinate their instructions" (Passingham 1981, p. 172). Additional support for this idea comes from the discovery of left-hemisphere dominance for vocalization in songbirds such as canaries, in the absence of anything analogous to handedness. The song of the canary, like the speech of human beings, is extremely intricate; moreover, the bird learns a new song each season (Nottebohm 1977).

One final area of controversy remains to be considered. This concerns the evolution of the anatomical prerequisites for rapid, encoded speech. Lieberman (1975) has shown that the production of speech depends on the capacity of the supralaryngeal vocal tract to act as an acoustic filter. By varying its length and shape, the speaker is able to generate the range of discrete sounds corresponding to the minimal phonemic units of the language. However, the evolutionary modifications to the vocal tract that have made this possible, notably the lowering of the larynx, were not brought about without cost. They have made swallowing more difficult, increased the risk of choking, and led to a crowding of dentition. Thus, they would not have taken place except under strongly positive pressures of selection (Lieberman 1984, p. 283). On the basis of fossil evidence, Lieberman posited that they took shape relatively late in human evolution in the line leading to modern *Homo sapiens*, subsequent to its divergence from the Neanderthal lineage.[5] If speech is thus a late development, the pressures that gave rise to it could be understood as a result of the growing inadequacy of the visual-gestural channel for increasingly complex verbal communication and the co-option of the vocal-auditory channel in its stead, leading in turn to structural modifications of the vocal tract concordant with its new functions. Vocalists, however, dispute the interpretation of the fossil evidence and argue that an ape vocal tract, properly wired neurologically, would be quite capable of generating comprehensible speech (Armstrong 1985, p. 627). Wind (1976, p. 626) has gone so far as to suggest that were it possible to transplant a complete chimpanzee vocal tract into an otherwise normal human being, apart from sounding a little odd, his speech would be perfectly recognizable.

6. Toolmaking and Language Are Linked in Phylogeny, As They Are in Ontogeny

Studies of the intellectual development of human children, adopting a conceptual framework from the psychology of Piaget, have sought to demonstrate the existence of an invariant sequence of cognitive stages through which all children pass, given appropriate environmental stimulation, in their growth toward adulthood. The Piagetian schedule of cognitive development is divided into four main periods: sensorimotor (birth to 2 years), preoperational (2 to 6 years), concrete-operational (6 to 11 years), and formal-operational (11 years upward). The sensorimotor period is divided into six sequential stages and the preoperational period into two subperiods: the symbolic (2 to 4 years) and the intuitive (4 to 6 years).

Both language and tool use are implicated in this developmental schema (Bates et al. 1979, Case 1985). By the fourth sensorimotor stage (8 to 12 months), the infant combines duplicated-syllable babbling ("dada," "gaga," etc.) with its first attempts at single-handed object manipulation. The propensity to babble appears to be innate (Gould and Marler 1987) and serves to engage the child's language-learning potential. The apparently spontaneous tendency to experiment repetitively with object-object manipulation (using one object to displace another), which is characteristic of the fifth sensorimotor stage (at 1 year to 18 months), may be similarly innate and may likewise serve to canalize the acquisition of tool-using skills. At this stage the child has an understanding of the relations among objects in space and of means-ends relationships, as when one object is used to retrieve another. It can also produce one-word, context-specific utterances, in combination with a complex of referential gestures. Realizing that by means of such an utterance it can elicit a desired response from a caregiver, much as it can obtain one object by means of another, the child becomes what Bates and colleagues (1979) called a "social tool-user." By the sixth sensorimotor stage (18 months to 2 years), the child develops an ability to form interiorized mental representations that allow for planned and insightful tool use, as opposed to the trial-and-error groping of the previous stage. Moreover, the child begins to combine words into longer, two- or three-word sentences that serve a range of different functions and may refer to objects and events outside the immediate context of utterance. As Greenfield (1978) has shown, there is a formal parallel between the hierarchical organization of these simple sentences and that involved in manual constructional tasks such as using blocks. The concurrent development of technical and linguistic faculties continues into the preoperational and concrete-operational periods, with an increasing degree of hierarchical complexity in the construction of both physical structures and verbal sentences.

If language and tool behavior are linked in ontogeny, might they be similarly linked in phylogeny? In a series of papers, Sue Parker and Kathleen Gibson have attempted to apply the Piagetian framework in order to construct a model for the development of object manipulation, language, and intelligence in human evolution (Parker and Gibson 1979; Gibson 1983, 1990, forthcoming; Parker 1985; Parker and Milbrath 1992). Their method consisted of establishing parallels between successive stages of human cognitive development and the mature capabilities of nonhuman primates. Thus, prosimians manifest stages 1 and 2 of sensorimotor intelligence and Old World monkeys stages 3 to 5, whereas the great apes progress through all the sensorimotor stages to reach the symbolic subperiod of preoperational intelligence. By assimilating the capabilities of these contemporary species to those of their putative common ancestors with humans, Parker and Gibson (1979, p. 370) arrived at a phyletic series of stages in the evolution of intelligence: the prosimian stage, Old World monkey stage, great ape stage, and hominid stage. Thus they supposed that the common ancestor of the great apes and humans possessed fifth- and sixth-stage sensorimotor and symbolic subperiod abilities but "that the addition of all the subsequent stages of preoperational, concrete operational, and formal operational intelligence must have occurred subsequent to hominid differentiation" (p. 370).

Of course the path of intellectual progress was not laid out in advance for evolving hominids to follow. At every point in the passage from the common ancestors of great apes and hominids to modern humans, specific selective pressures have to be invoked to account for the emergence of abilities characteristic of each successive stage, from a base in those of the previous one. In the evolutionary scenario of Parker and Gibson, fifth- and sixth-stage sensorimotor intelligence was selected because it facilitated the use of tools in the extraction of vegetable foods embedded in the earth or encased in hard shells. Apart from an understanding of the means-ends relationships involved in tool use, this subsistence strategy requires an ability to form mental representations of hidden objects and to handle the topological relations of proximity and enclosure. It also entails a rather long period of juvenile apprenticeship and dependency, and it is hypothesized that the first "protolanguage," equivalent to that of human children of between 1 and 2 years of age, emerged in the context of food sharing between mothers and offspring, or the offspring of their close kin (Parker and Gibson 1979, pp. 382–383). The subsequent development of intuitive intelligence (corresponding to the second subperiod of the preoperational period), which surpasses anything found in the great apes, is interpreted as an adaptation to the subsistence mode of *Homo habilis*, with its much greater emphasis on hunting. Both the use of projectiles in the hunt and stone tool manufacture entail certain Euclidean preconcepts such as straight line, angle, and section. The meat of animal carcasses, butchered with stone tools, is more widely shared, and "cutting substances into equal parts is an ability that emerges in human children during the intuitive subperiod" (Parker and Gibson 1979, pp. 376–378). Finally, Parker and Gibson stressed the importance of shelter construction for hominids inhabiting open country, and this, too, depends on the cognitive abilities of the intuitive subperiod.

All this, of course, is highly speculative. However, apart from doubts about the validity or universality (across cultures, let alone across species) of Piaget's schedule of cognitive development, the most controversial aspect of Parker and Gibson's argument lies in their extrapolation from human ontogeny to hominid phylogeny. This presupposes that the ontogenies of ancestors are reproduced in their evolutionary descendants but that they are extended in the latter by "terminal additions." Hence, the mature condition of the ancestral species corresponds to a stage in the development of the descendant species. As regards the evolution of intelligence, it implies that each stage represents a terminal addition to the one before. Thus, the development of the common ancestor of the great apes and humans culminated in the symbolic subperiod of preoperational intelligence, whereas that of *Homo habilis* was extended by addition into the intuitive subperiod. The concrete-operational and formal-operational stages were likewise "added on" in the subsequent evolution of *Homo erectus* and *Homo sapiens*.

The objections to recapitulationism, the doctrine that ontogeny repeats the trajectory of phylogeny, are numerous and have been debated for generations (Gould 1977). There are, for one thing, a great many exceptions. Certain features of modern humans, including the morphology of the brain, are better accounted for by the opposite principle of neoteny, according to which the final stage in the ontogeny of ancestors is *left off* in the ontogeny of descendants. Thus, the mature descendant resembles the immature ancestor rather than vice versa. Even if evolution does proceed by terminal addition, it is unlikely that this would occur without substantial modification or reconstruction of earlier ontogenetic stages (Gould 1979, p. 386). Natural selection, of course, operates on the entire process of ontogeny, not merely on the final, mature form — thus the course of evolution is represented not by a series of ontogenetic additions but by a *"series of successive ontogenies"* (Borchert and Zihlman 1990, p. 15). In other words, development itself has evolved (Sinha 1985, p. 682; Hinde 1991, p. 585).

Applying these general objections to recapitulationism to the specific case of language, it may be observed that the "protolanguage" of human infants prepares them for the acquisition of the actual language spoken by surrounding adults; as such, its adaptive function is quite different from that of the "protolanguage" of the great apes or the supposed "protolanguage" of the common ancestor of apes and humans. Moreover, the infant only succeeds in acquiring language because already proficient caregivers in its immediate surroundings actively intervene in providing "generous interpretations" of its spontaneous utterances as purposive and meaningful, guiding the infant into the discovery that this is indeed what they are (Graves

1991, p. 305; see Lock 1980). The species-specific behaviors of human infants — including whining, fussing, babbling, and spontaneous object manipulation — function precisely as ways of ensuring that they make the discoveries on which the subsequent attainment of mature linguistic competence depends (Borchert and Zihlman 1990, p. 36). Thus, infancy has evolved as an adaptation for learning in a socially structured world and is not simply a survival into the present of the adaptive capacities of mature ancestors.

To be fair, I must acknowledge that Parker and Gibson were aware of most of these objections. They did not treat recapitulation as a doctrine, nor did they view it as applying to every aspect of brain maturation. The essence of their argument is summed up in the following passage:

> Given the fact that in human children the abilities of each stage of intellectual development . . . are logical and structural prerequisites for the emergence of the abilities of the succeeding stage . . . , and given the fact that the abilities of each ancestral species were logical and structural prerequisites for the evolution of new abilities in descendant species, we must conclude that intellectual abilities develop in the same sequence in which they evolved. In other words, in the case of human intelligence, ontogeny recapitulates phylogeny. (Parker and Gibson 1979, p. 380)

A closer look at this passage, however, reveals the argument to be perfectly circular. For the conclusion, that ontogeny recapitulates phylogeny, is built into the very premise from which they derived the "fact" that the abilities of descendants are tacked onto those of ancestors. This premise, which allows them to infer a phyletic series of stages in the evolution of intelligence from a comparison of the mature capacities of contemporary primates, is precisely that intelligence evolves by terminal addition. If it does not, if in fact the ontogenetic prerequisites for mature intelligence evolve alongside that intelligence itself, as part of a total developmental program, then the reconstruction is no longer valid, and the argument based on it falls to the ground.

CONCLUSION

Despite the many attempts to link language and technology, we are still far from achieving a coherent synthesis. This is not merely for want of data, nor is it entirely because of the obstinate fact that we will never have the privilege of hearing our ancestors speak or of watching them make and use tools. The problem is to an equal extent one of our own making, and it stems from basic disagreements about the meanings of the terms we are trying to connect. *Language*, it seems, can mean almost what you will. Anthropologists have faced similar problems in defining *culture*, but as Spuhler lamented (1977, p. 513), there is no compendium of language definitions to match Kroeber and Kluckhohn's (1952) catalog of the definitions of culture. A lot of sorting out remains to be done. Thus, one authority (Lieberman 1975, p. 6) can define *language* as "a communications system that is capable of transmitting new information," of which human language is just one of many varieties in the animal kingdom, whereas another (Chomsky 1980, p. 230) can claim not only that language is an organ specific to *Homo sapiens* but also that it is not purposely built for communication or anything else. One wonders whether they can possibly be talking about the same thing, or indeed whether such a thing as language exists at all save as a construct of those who profess to study it.

The situation is no clearer when we turn to technology. Some authors are happy to speak of *technology* wherever tools are being used or made. For others, the important factor is skilled activity, or technique, which may — but need not — involve the manipulation of detached objects as tools (Ingold 1990a, p. 7; see also Ingold 1988). For others again, technology refers to a planning capability or mental constructional device that has nothing directly to do with either technique or tool use. The kind of confusion that can arise may be illustrated by Wynn's (1993) comment that the relevance of object manipulation to tool use is far from established. The essence of tool use, for Wynn, lay in the routine deployment of acquired, practical skills — that is, it consisted in the doing, not in figuring out what to do. Yet object manipulation may involve no *practical* skill at all; rather, it represents a solution to the problem of how to reach a desired goal that is not immediately attainable. Everything hinges on being able to solve the puzzle, to arrive at a plan of action whose execution is a mere mechanical formality. Thus, when tool use is identified with object manipulation, it becomes a kind of puzzle solving rather than skilled artistry. And it is just the same when speaking is treated not as skilled verbal performance but as the programmed behavioral output of a mental facility for producing novel, hierarchically organized word sequences. It is obvious that the link between technology and language will be construed differently, depending on whether one's concern is with objects used (tools, words), acquired motor skills (manual technique, speech), or cognitive functions (designing, thinking).

It is important to be clear about these kinds of distinctions, if only to avoid the illusion that there is a direct line of continuity, say, from the Oldowan chopper to the automobile, or from the hoot of an ancestral hominid to this sentence. We can be fairly sure that the difference between the chopper and the finely flaked stone tools of the upper Paleolithic reflect real changes in the innate cognitive and manipulative capabilities of their makers and that these changes took place concurrently with the evolution of the language faculty of modern *Homo sapiens*. But the gap between upper Paleolithic tools and the car poses a problem in the evolution of design, an evolution that was underwritten by a language capacity and a symbolic intelligence *that had already evolved*. The same goes, of course, for the evolution of writing and literacy, or what has been called "the technologizing of the word" (Ong 1982). Hence, one can no longer draw conclusions from the complexity of material culture about the language and intelligence of the people who made it.

In his comparison of the tool kits of free-ranging (Tanzanian) chimpanzees and modern human hunter-gatherers (Tasmanian aborigines), McGrew found that although some features of the human tool kit were absent from the chimpanzee corpus, "the difference is far from wide, and the gap between [ancestral] hominid and pongid is narrow and bridgeable" (1987, p. 256). However, the gap between the modern human and the chimpanzee is not, for even though chimpanzee material culture could approximate to that of certain hunter-gatherers, the latter — had they been placed in different historical circumstances — would have been quite capable of designing and operating a complex industrial technology, an achievement way beyond the capacities of *any* group of chimpanzees. And just as the capacities of chimpanzees and humans are different, even though their technical equipment may under some circumstances look the same, so likewise the capacities of modern humans are the same, even though — if we were to compare Tasmanian aborigines and contemporary Westerners — their equipment would appear vastly different. As Lieberman warned (1985), "Who would think that we had essentially the same biological endowment as the human populations that lived 30,000 or 20,000 or 500 years ago if all that he had to go on were the preserved artifacts — stone tools versus the ruins of great cities, dams, interlocking highways, etc.?" (p. 628).

NOTES

The issues surrounding the relationships between speech and tool behavior in human evolution touch upon the concerns of many disciplines, including archeology and physical anthropology, primatology, psychology, evolutionary biology, neurology, and linguistics. Addressing the issues from their disparate perspectives, and with their own batteries of concepts and theories, scholars representing these diverse disciplines have generated a formidable literature. Couching their arguments in mutually impenetrable technical languages, however, they have often talked past one another in the process. This chapter originated in my attempt, as a relative outsider — a social anthropologist interested in the relation between language and technology in human societies — to make sense of this literature and to bring it together into a form that others, too, might find comprehensible. The essay was originally written for and presented at the international conference on Biology of Language: Essentialist versus Evolutionist in the Nature of Language, held at Czerniejewo, Poland, December 1–4, 1988. I am most grateful to the organizers of the conference for the invitation to participate, and to other participants for their comments in discussion. Since then, the essay has lain dormant for far too long, waiting for the moment when I would have time to spare to make some necessary revisions. This moment did not arrive until 1991–1992, thanks to grants from the Leverhulme Foundation and the University of Manchester's Research Support Fund, which have financed a year's research leave. In the intervening period, a vast quantity of new literature has appeared; it would require a separate piece to review it all, and I make no attempt to do so. Thus, my references to the latest publications are necessarily selective. Moreover, my own ideas have moved on, in certain respects, from those expressed here.

In revising the paper for publication, I have benefited from conversations and correspondence with many colleagues, and I would like to extend my thanks, in particular, to Iain Davidson, Kathleen Gibson, Paul Graves, Gordon Hewes, Adam Kendon, Jean Kitahara-Frisch, Andrew Lock, Bill McGrew, William Noble, Graham Richards, François Sigaut, Nick Toth, Elisabetta Visalberghi, and Tom Wynn. Any remaining inadequacies are, of course, my own.

1. The classic ethnographic source on Caroline Islanders' navigation is Gladwin (1970). From what Gladwin reported, two qualifications should be entered here. First, although "formal instruction begins on land," there is also a great deal that can only be learned through practical experience of sailing. Therefore,

"instruction beyond the first stage of intensive learning takes place alternately on land and sea" (1970, pp. 128, 133). Second, unlike Western navigators, Caroline Islanders are not taught the general principles for *working out* a course from any point to any other. They are, rather, expected to commit to memory every possible course "for every pair of islands between which a navigator might conceivably find himself sailing" (p. 130). Prior to embarking on a voyage, then, the navigator does not make a plan but selects it from the very large number of ready-made plans filed in memory. Thus, although the knowledge of navigation is clearly abstract and theoretical insofar as "it is explicitly taught and conceptualized as a set of principles governing relationships between phenomena," it is also "prepackaged or predetermined," neither requiring nor offering scope for innovation. The navigator is "never called upon to have new ideas, to relate things together in new ways" (p. 220). As this example shows, the link between theoretical thinking and practical innovation is one that needs to be demonstrated, not assumed.

2. The arbitrariness of the linguistic sign, according to the classic Saussurian paradigm, lies in the relation between word-image and concept. The linguistic analog of the relation between the tool and its raw material would then be that between the word-image and the medium of sound.

3. Humphrey, it may be noted, modeled the primate group on the "collegiate community" (1976, p. 310). While the Machiavellian image may be appropriate to the politics of academic conferences and senior common rooms, it is hardly apt to characterize the experience of everyday life in small and intimate groups. For a critique of Humphrey's argument, see Ingold (1993b).

4. Consider the example that Hewes himself used to illustrate the property of linguistic decontextualization. "The message 'the house is on fire' can," he suggested, "be conveyed with no more excitement than the information that Paris is a city in France" (1976a, p. 490). Yet even to deliver these words in a flat monotone is an expressive act, conveying the speaker's feelings of indifference rather than a sense of urgency or anguish. And this, in turn, is likely to evoke a quite different response from the audience, compared with the response had the same words been delivered in an excited shout. It seems, then, that the work of decontextualization is being carried out not by the speaker or audience but by the analyst, and that it is the very process whereby "language" is *constituted* as a discrete object of study, consisting of a class of propositional statements that — like words printed on paper — have an existence of their own, independent of the conditions and qualities of utterance. By the same token, utterances considered to be devoid of propositional content, including what Myers (1978) rather primly calls "curse words or interjections," are excluded by definition from language proper. Thus, the dichotomy between language and nonverbal communication, as vehicles respectively of proposition and affect, is not — as Myers claims — revealed in the material of human vocalization but is inscribed in the very conceptual framework by which we come to apprehend it. The dichotomy is of course grounded in an ontological dualism at the heart of Western thought, between reason and emotion (Ingold 1992a).

5. Lieberman's interpretation of the fossil evidence (originally proposed in Lieberman and Crelin 1971) is highly controversial and has been strongly criticized by many paleoanthropologists (notably Carlisle and Siegel 1974, Falk 1975, LeMay 1975, DuBrul 1977, Laitman 1983). For an excellent review, see Richards (1987, pp. 264–269).

PART IV

The Rise and Spread of Sapient Culture

Introduction

What is the feature of human culture that distinguishes it most sharply from, for instance, chimpanzee or bonobo culture? Many — probably most — anthropologists would agree that it is language. It is true that since the 1960s apes have made great strides in understanding the signal systems devised by human beings in order to communicate with them — greater, certainly, than we have made toward understanding the natural signal systems of those apes, if progress is measured by success in communicating across species and cooperating toward the achievement of shared goals! But we are a long way from sitting down outside the lab, say, on the corner of 4th and Elm or under a palm tree and sharing views on what it is like, how it feels, to be a member of one species and not the other.

One can object, of course, that human beings of different sexes, ethnic groupings, economic classes, villages, and schools don't do such a great job of sharing (as opposed to expounding) different perspectives. Our point, however, is that a formidable obstacle lies in the way of attempts at conversation with chimpanzees, at least for those of us who have not been trained in reading gestural languages, because chimpanzees do not talk and, like other nonhuman primates, apparently cannot be trained to produce vocal signals that come close to approximating human speech. Students of language and communication are still not in agreement as to whether chimpanzees and bonobos comprehend grammatical structure, whether the arrangement of a properly ordered sentence means anything to them, but there is little doubt that they lack the anatomical prerequisites for effective vocal production of sentences of any kind or, for that matter and with few exceptions, simple vocal signs (words) as pronounced by normal human beings.

So apparently we cannot learn much about the character of vocal language or, presumably, its origins (glottogenesis) by studying

317

monkeys and apes. There are limits to what investigations of primate behavior can tell us about human evolution, and one of the more productive applications of primate behavior studies may lie in defining those limits with ever-increasing rigor and clarity.

Did glottogenesis occur at the outset of hominid evolution or later on? There is no community of opinion on this matter. Until recently, the great majority of scientists concerned with the nature of language and willing to speculate about its origins probably would have answered at the outset. The main evidence for that view, a current strong proponent of which is Falk (1984, 1987), is drawn from cranial anatomy and depends primarily on interpretations of the significance of features in endocranial casts of early Pleistocene hominids associated with what would be the speech areas in modern brains. A problem with this evidence is that interpretations vary considerably and that nothing is revealed about internal organization that could inform us whether we are dealing in these extinct species with the neurocranial bases of spoken language per se or with neural substrates responsible for processes common to either spoken or gestural language.

Hewes (1973b, 1977) has argued that as there is no apparent link between nonhuman primate call systems and human vocal language, we would do well to investigate the possibility of an evolutionary connection between other action systems involved in communication. His proposal that glottogenesis must have followed an intervening period of syntactically governed gestural communication has led many to see the idea of a late origin as tenable, an idea that finds support both in anatomical and cultural evidence. These evidence bodies, which Milo and Quiatt treat in Chapter 10, are as problematic in many ways as are those that give rise to interpretations of a very early glottogenesis (few are prepared, it would seem, to argue that spoken language was "invented" a million years or so ago by *Homo erectus*), and we assume that readers will take all current theories of language origins *cum grano salis* — while attending, in their examination of each, to the quality of evidence produced and the heuristic properties of proposals set forth.

The final chapter of our book should be read not as a speculative scenario that its authors are convinced is the true story of events in late Pleistocene prehistory, nor as a passionate argument for the very late evolution of human vocal language. True, they think it is likely that vocal language did appear at a late date, and they think it likely also that whenever glottogenesis did occur it would have significantly restructured the contexts of human selection, changing in important ways the character of competition among individuals, groups, populations, and hominid species (if, at the outset, species differences obtained), and transforming radically the means by and the extent to which human beings process information about their social and "natural" environments.

10
Language in the Middle and Late Stone Ages: Glottogenesis in Anatomically Modern Homo Sapiens

RICHARD G. MILO AND DUANE QUIATT

ABSTRACT

Recent debate over a single geographical versus a multiregional origin for anatomically modern (a.m.) *Homo sapiens* and, consequently, over the manner of replacement by a.m. *H. sapiens* of prior forms of hominids, provides a context for asking rather more specific questions than hitherto have been raised about the role of language in those biocultural processes. In this chapter we review the evidence from paleontology and archeology that supports the argument for a late origin of vocal language. Finding that argument sound, we conclude, following Hewes (1973a, 1973b, 1976a), that a prolonged period of gradually evolving, relatively efficient gestural language must have preceded glottogenesis. A fully vocal language, phonemicized, syntactical, and infinitely open and productive, appears to have clear advantages over a language consisting primarily of gestural elements. We discuss these advantages and note implications for group selection at this particular juncture in human evolution.[1]

INTRODUCTION

In recent years anthropologists have evidenced increased interest in a cladogenetic interpretation of human evolution in its late as well as its earlier phases. We believe that a.m. *H. sapiens* first emerged in a single locale (in sub-Saharan Africa or, just possibly, in Asia) much earlier than had been supposed on the basis of the European fossil record — perhaps between 150 and 200 kya. This view has gained support from several lines of broadly interdisciplinary evidence, including human paleontology, archeology, molecular genetics, and new dating of key fossil materials in Africa, Europe, and the Levant. An important implication of this model is that contact between indigenous archaic hominid communities and in-migrating *H. sapiens* groups — for example, between Neanderthals and modern humans in Europe — may have involved direct competition for resources with no genetic intermixture. This possibility, in our opinion, entails some interesting implications for the nature and role of language in such a competition, particularly if phonemicization of language can be assumed to have occurred late in human evolution. We assume that it did.

We are interested, then, in the question whether and how language may be implicated in (a) hominid differentiation and the evolution of human culture, (b) the differentiation of a.m. *H. sapiens*, and (c) the replacement by a.m. *H. sapiens* of prior forms of hominid (read Neanderthals and their archaic *H. sapiens* cousins in other parts of the world). We acknowledge at the beginning that language behavior, like many other forms of social behavior, can rarely if ever be studied directly in the fossil and archeological records. Nevertheless, we take the various lines of evidence to which we referred just above as supportive of a model of glottogenesis following from and, we presume, contingent upon the appearance of modern human morphology. By glottogenesis, we mean specifically the origin of rapidly spoken human language based on semantically arbitrary phonemes.

LANGUAGE AND ITS IMPLICATIONS IN THE ORIGIN OF HUMAN CULTURE

Asking how, in hominid evolution, nature and culture are intertwined necessarily raises the question of language in its relation to natural and cultural selection. Recognizably human language, as contrasted, for example, with the communicative behaviors evidenced by the extant great apes, may not have been essential to the initial hominid subsistence economy; this we assume to have included "tolerated scrounging" of food (and, eventually, reciprocal exchange of foods and services), the making of simple tools, and cooperation in foraging, scavenging, and reproduction. Language may not have been essential to transmission of learning over generations, nor even perhaps to the stage transformation of behaviors such as these, manifested occasionally and incompletely by a forerunner species, into a complex of activities habitually practiced by hominids (cf. Isaac 1978a, 1978b). However, the role of language in the construction of culture beyond this initial base of integrated behaviors seems undeniable. We may note that of the two hominids that evidently coexisted during the terminal Pliocene and earliest Pleistocene — robust *Australopithecus* (which includes *A.* — alternatively, *Paranthropus* — *robustus* and *A. boisei*) and *H. habilis* — only the latter appears to have possessed the beginnings of the neurological structures, especially Broca's area, which we associate with language (Falk 1984), in conjunction with the first signs of significant cerebral expansion.

It seems reasonable to suggest that the socioecological environment of *H. habilis* included areally extensive foraging/scavenging, with an ecotonal diet in which meat protein played some as yet indeterminate role; a technological base that included stone tool use and manufacture; some degree of socially mediated sharing of information and resources but as yet no sign of division of labor by sex or otherwise; and perhaps differential responsibilities and activity schedules but no regular segregation of group members into geographically detached subunits reassembling at a home base or central location. Under these conditions, demands for efficient spatial analysis and toolmaking appear to have selected for increased cerebral lateralization of sequential motor function and visuospatial coordination (Blumenberg 1983, p. 598; Falk 1987, p. 26). These changes arguably laid the neurological basis for increasingly precise and specific — that is, elementary, rule-governed, syntactic — communicative behavior (Lieberman 1985).

Relatively advanced cultural behaviors appear to be strongly implicated in the dispersion of *H. erectus* (*sensu lato*) out of Africa after 1.5 mya. We conclude that these behaviors must have included language, which we regard as indispensable in maintaining a social environment capable of buffering the cooperating group against the exigencies of an increasingly large, heterogeneous, and unpredictable natural environment. It is difficult to think how a full-blown, dual subsistence system — a plant/animal economy with reciprocal exchange at its center — could have been maintained without language;

and it is reasonable to suppose that the transition to that economy was effected at least no later than the onset of the middle Pleistocene, when we find an efficient though broadly omnivorous hominid in place in a wide range of habitats in Africa and Eurasia.

In a closely reasoned argument drawing on evidence from a wide range of disciplines, Gordon Hewes (1973a, 1977) has suggested that language had its origins in gesture. It will be useful at this point to distinguish sememes, units of meaning carried by morphemes of unspecified form, from phememes, that is, vocal signs of distinct semantic and phonological character — as well as phememes from phonemes, that is, vocal units with no inherent semanticity. Gestural language may have accrued, and probably did accrue, phememes along with sememic vocal qualifiers overlying gesture; but in Hewes's view glottogenesis proper must have involved a second and probably much later phase in the evolution of language — through phonemi-cization. We are persuaded by Hewes's argument and will not speculate about the specific character of language in its presumed gestural phase, primarily because we cannot think how to improve on it. The two-phase aspect of his theory of language evolution seems to fit the facts of morphological change and, insofar as they can be inferred from the archeological record, general changes in subsistence, patterns of foraging, and, we presume, social organization. We associate the initiation of Hewes's first phase of language evolution with the appearance of *H. habilis*, and we note the possible implications of that hominid's suite of behavioral innovations for the subsequent disappearance of *A. (P.) robustus/boisei* from the fossil record.

Following the appearance of *H. erectus*, hominid evolution appears to be anagenetic in character and conservative — at least within regional populations (see Clarke 1990) — for almost a million years with regard to both physical morphology and artifacts evidencing culture (Howells 1959, Kennedy 1980, but see Clarke 1990 for a critique of this view). During that period the brain increased in size gradually, on the order of 50 cc per 100,000 years (Leigh 1988). Then, with apparent suddenness, cranial capacity jumped 200 cc or more to 1350–1600 cc for so-called archaic H. sapiens in the late middle and early upper Pleistocene. With this increase we find ourselves contemplating a situation roughly similar to that in the Plio-Pleistocene, a situation in which, brain size apart, qualitative differences in certain skeletal features and bimodal quantitative differences in others suggest cladogenetic divergence of (at least) two species; differential selection and adaptation to different circumstances; and (in western Europe, for example) eventual direct competition culminating in the disappearance of one from the fossil record. The main difference, one might argue, is that in this latter period, events are speeded up, as little as 100,000 years elapsing before the extinction of Neanderthals (see Zubrow 1989). It may well be that the resolution to this interspecific competition was, as seems likely in the prior case, at least partly contingent on differential achievement of language capabilities.

We attribute the success of middle Pleistocene hominids to two related factors: the ability to make use of a wide variety of vegetal foods and substantial amounts of meat, made possible by an efficient extractive technology based on regular tool use; and the ability to survive as internally cooperating social groups in environments in which no individual could survive alone. Language must have played a pivotal role in organizing the subsistence and reproductive behavior of these social groups and in mediating the social tensions arising therefrom. However, in view of the persistent homogeneity of material culture within if not across continental regions, and in the absence of archeological evidence for large social groups or strategic exploitation of resources, we speculate that this was a hominid with culture neither highly structured nor highly differentiated by group (or deme or, to begin with and in basic form, regional population). As Hewes has pointed out, the maintenance of such a culture would not have required a highly elaborated language; a relatively simple and largely iconic gestural language, probably supplemented with slowly articulated vocalizations, should have served adequately.

The situation clearly changes in the closing millennia of the middle Pleistocene. In contrast to the conventional view of regional transitions from archaic to modern human morphology midway through the late Pleistocene, recent discoveries in Africa and southwest Asia suggest (a) that anatomically modern morphology appeared in sub-Saharan Africa at least 120 kya, and (b) that populations of archaic hominids (Neanderthals and their cousins) and a.m. *H. sapiens* existed sympatrically for thousands to tens of thousands of years. Thermoluminescence dating of cultural materials from Qafzeh in the Levant (Valladas et al. 1988) places a.m. *H. sapiens* in that area at about 92 kya, concurrent with or before the appearance of Neanderthals. In western Europe, the discovery of a classic Neanderthal at St. Césaire dated to 30–35 kya, well after the appearance in Europe of fully anatomically modern *H. sapiens* (Mellars 1989), strongly suggests the presence of sympatric populations in that region as well. The perception of distinct sympatric populations is further strengthened by quantitative and qualitative analyses of cranial and postcranial remains that argue for a greater morphological discontinuity between Neanderthal and a.m. *H. sapiens* than previously had been recognized (cf. Tattersall 1991).

Brauer and Rimbach (1990), in multivariate analyses of archaic and modern crania from Europe, the Near East, and Africa, demonstrated discrete clustering of the two hominid forms. They argued that their results were inconsistent with regional continuity in Europe and supported an African origin of modern humans and subsequent replacement of prior archaic forms. Trinkhaus (1983) remarked "a significant change in human postcranial morphology and proportions" when comparing fossil specimens from Zuttiyeh, Tabun, Shanidar, and other "generalized" Neanderthal sites with those from Qafzeh and Skhul. The latter he placed fully within the range of modern *H. sapiens*. Rak's description of the nearly complete Neanderthal pelvis from Kebara (Rak and Arensburg 1987) reinforces this perception of significant differences in postcranial morphology between coeval Neanderthal and anatomically modern hominid forms in the Near East. Rak suggested that Neanderthal's characteristic locomotion may have differed substantially from that of a.m. *H. sapiens*.

Without taking sides on the issue of species recognition in hominid systematics (see Kimball 1991, Tattersall 1991), we believe that the apparent persistence of characteristic Neanderthaloid morphology in Europe and southwest Asia raises serious doubts about any significant genetic intermixture between archaic and modern hominids during their period of sympatry. This position presents the challenge of characterizing niche differentiation, both behavioral and ecological, between the two forms. Klein (1989) has noted that the rugosity and intermembral proportions of Neanderthal skeletons, their large prognathus faces, and the distinctive wear patterns of their anterior dentition, when contrasted with the reversal or absence of these traits in modern human skeletons, suggest that the latter hominid accomplished by cultural means tasks that the former had accomplished physically. Future archeological research is needed to clarify the ways in which archaic and modern humans differed with respect to their exploitation of the subsistence environment (Clark 1989). We contend that the growing corpus of morphological, archeological, and behavioral evidence bearing on the appearance of a.m. *H. sapiens* and the disappearance of archaic forms bear witness to phase two of Hewes's theory of language evolution: glottogenesis.

EVIDENCE FOR GLOTTOGENESIS IN THE LATE PLEISTOCENE

Morphological Evidence

The earliest putative examples of modern human morphology discovered to date derive from the lowest archeological levels at Klasies River Mouth Cave in South Africa (Singer and Wymer 1982). These remains have been dated to between about 90 and at least 120 kya by correlation with the 6–8-meter raised beach and by oxygen isotope, uranium disequilibrium, aspartic acid, and sedimentological methods (Rightmire and Deacon 1991, p. 153). The hominid remains evidence a high degree of phenotypic variability (e.g., rather robust beside very gracile individuals, chins with prominent mental eminences beside chins with very weak eminences), within which the orthognathus face, high forehead, rounded vault, reduced brow ridges, absence of a retromolar space, and details of postcranial morphology diagnostic of a.m. *H. sapiens* are strongly represented. Perhaps significant is that one of the morphological constants within this highly variable fossil population is teeth that are reduced buccolingually (Singer and Wymer 1982, p. 148).

From this material, and from broadly similar fossil remains from Border Cave, Die Kelders Cave, and others (Rightmire 1979, Klein 1989), we draw two tentative conclusions relative to the topic at hand. The first is that selection was operating on a very early, genetically variable population in sub-Saharan Africa to reduce and rotate the mouth and jaw and reform the calvarium and basicranium. We speculate that the shortened jaw and narrowed teeth are related to the appearance of the relatively short, broad, and rounded human tongue. Our second conclusion is that reduction, rotation, and retraction of the face contributes to the production of that marked flexure of the basicranium that Lieberman (1983) and Laitman (1984) have associated with a modern human supralaryngeal conformation. Both of these features are argued to be morphological preconditions for the production of rapidly spoken language (Lieberman 1989).

Lieberman, in his well-known reconstruction of the vocal tract of Neanderthal (Lieberman et al. 1972), seriously questioned whether that hominid could articulate sounds with the speed and facility and of the variety characteristic of spoken language today. Laitman (1984, 1985) emphasized more strongly the connection between basicranial flexion in modern human skulls and the uniquely low position of the human larynx in the esophopharyngeal tract. Both Laitman and Lieberman have correlated this position of the larynx with the ability to produce certain nonnasalized phonemes, particularly those that appear to play a crucial role in the comprehension of rapidly spoken language; and both workers have remarked on the disadvantages of the low larynx for respiration and food intake (Laitman 1983, Lieberman 1989). Since the restructured mouth, tongue, and supralaryngeal tract of modern humans are functionally related only to the production of speech, positive selection for the communicative advantages conferred by these changes must significantly have outweighed the potential negative consequences for the individual. We address some of the evolutionary issues raised by this perception in a following section.

It is tempting to read into the increased doming of the cranial vault some evidence for neural reorganization. For one thing, it seems unlikely that a potentially deleterious morphological change could have been selected for had not the cognitive-behavioral potential for its use already existed. Kien (1991) and others have speculated that neural reorganization involved the ability to access memory via reiterative neural symbols, and her suggestion accords with Hewes's proposal that phonemic speech facilitates memory and lexical access via acoustic symbols. At this time, however, direct evidence for neurological reorganization is highly equivocal, and it is not in any case essential to our basic argument. It is likely, and enough, that from the aforementioned morphological (and motor) adaptations — to socioecological circumstances as yet ill defined — phonemicized, rapidly spoken language emerged as an exaptive cultural/behavioral innovation.

It would appear then, on the basis of present evidence (and we do not mean to suggest that the evidence is either unequivocal or unanimously accepted), that characters distinctive of anatomically modern humans are present in some African populations by at least the early late Pleistocene, and that among these are just those characters reputed to make possible rapidly spoken phonemicized language.

The nature of the morphological changes just described — rotation of the face, exaggeration of basicranial flexion, and hints concerning modification of the tongue and supralaryngeal tract — suggests motor-oriented rather than or in addition to, let us say, cognitively oriented adaptations. Because of the well-recognized connection between the lips and tongue and the hand, and the fingers in particular, we are tempted to include refinement of small motor acuity among whatever behavioral adjustments are reflected in these early but anatomically modern remains (but see Susman 1988).

Archeological Evidence

The archeological record from the early upper Pleistocene does not precisely mirror the fossil record. The earliest manifestations of modern morphology, both in southern Africa and in the Near East, appear in a middle Stone Age/middle Paleolithic technological context, and there is no corresponding evidence of a sudden increase in symbolic behavior and art; that efflorescence comes later. On this account, some workers (e.g., Lindley and Clark 1990) have argued that the archeological record does not support the hypothesized expansion of a.m. *H. sapiens* out of Africa, successively replacing indigenous archaic populations. It is our position, however, that one ought not to expect direct correspondence between morphological and technological change. Among other things, such a position trivializes the evolutionary issues involved in a major transformation of behavioral morphology. Neither should the appearance of the anatomical prerequisites for rapidly spoken language be taken as necessary and sufficient evidence to argue for the contemporaneous appearance of that or associated behaviors (cf. Ambrose and Lorenz 1990, p. 28). The same anatomical modifications and their presumed motor associations might have contributed initially to refinement of an existing, gesturally based linguistic modality, for example, by the accretion of a growing set of arbitrary vocal and gestural elements (cf. Petitto and Marentette 1991). We should not expect to discover, if such were the case, paleontological evidence of an "explosion" of this more "oral" hominid into the territory of its neighbors. The point is that the morphological changes that made possible rapid spoken phonemic communication did not in themselves produce that behavior. Once a phonemic language had evolved, assuring (in potential) significant advantages for its speakers in cultural competition with gesturing neighbors (if such an oversimplification of evolutionary possibilities is appropriate), it seems unlikely still that immediate, explosive territorial expansion could ensue. It would take time, for example, to conventionalize phonemes, time as well to build a lexicon in the new modality. Furthermore, we note that rapidly spoken language and associated cultural and behavioral innovations likely provided the potential but not the impetus for geographical expansion; the latter more likely arose from ecological or demographic factors (cf. Clark 1989, Deacon 1989, Ambrose and Lorenz 1990).

Nevertheless, within a few thousand years, or a few tens of thousands, the autocatalytic nature of spoken, conventional language on the one hand and cultural elaboration on the other might well have produced a population — a culture — that was, for the first time, conscious of its own distinction from other hominid groups in ways that previous populations were not. Culture, as the term is understood by most cultural anthropologists today, resides in language; it may be that the human recognition of cultural distinctions specific to relatively circumscribed populations and based in large part on language differences had its origins among the first anatomically modern humans.

By the time the upper Paleolithic *sensu stricto* is recognized in the archeological record, about 40 kya, there is a well-established connection between anatomically modern human fossils, elaborate and differentiated blade tools, and complex symbolic behavior including art and, possibly, calendrical notation (cf. Marshack 1979b). In Europe, at least, the contemporaneous appearance of a.m. *H. sapiens*, upper Paleolithic blade tool technology, and art appear to support arguments for an intrusive population. This conclusion is strengthened by the identification of the late, classic Neanderthal at St. Césaire in a Châtelperronian tool context. The Châtelperronian has been interpreted as a basically Mousterian industry to which certain blade tools appear to have been added, as by borrowing (Harrold 1989, Mellars 1989, Farizy 1990; see also Allsworth-Jones 1990). There are, however, no unequivocal indications of advanced symbolic behavior (read art) in the Châtelperronian. We take that circumstance as suggesting that archaic *sapiens* populations, while they were able to adopt elements of a new technology, were unable to adopt the "symbolic consciousness" that was irreducibly rooted in a new way of making language (cf. Lindley and Clark 1990 both for a counterargument and for commentary in rebuttal.)

Behavioral Evidence

The European record documents substantial changes in behavior between middle Paleolithic (archaic) and upper Paleolithic (modern) populations in such areas as site organization, faunal exploitation, settlement patterns and land use, and territorial expansion (Gamble 1986, Farizy 1990, Stiner 1990, Soffer 1991), as well as in the areas of technological and symbolic behavior. All of these changes evidence new levels of social organization and differentiation and new ways of conceptualizing the natural and social environments that are arguably contingent upon, or at least are intimately related to, a fundamental change in linguistic behavior. The same changes are, however, congruent with the appearance of humans of modern morphology. In southern Africa, where both middle Stone Age (MSA) and later Stone Age (LSA) industries are associated with a.m. *H. sapiens*, the situation is more complex.

Klein (1989) found that in MSA faunal assemblages the predominant large mammal remains are those of animals such as eland, which are relatively safe to hunt. In contrast, LSA assemblages contain faunal remains in rough proportion to their relative abundance in the environment, with a much higher frequency of dangerous game like buffalo and bush pig, suggesting that LSA people were more proficient hunters than their MSA forebears. Furthermore, fur seal remains in MSA levels represent a wide range of ages, while those in LSA levels are heavily biased toward animals of weaning age, suggesting a greater degree of scheduling by LSA people. Finally, the average diameter of limpet shells in LSA shell middens is smaller than the average size of those in MSA middens, suggesting more intensive harvesting — or higher population densities — by LSA people. From these findings, Klein provisionally concluded that MSA people were anatomically modern but not yet fully behaviorally modern (Klein 1989, p. 543). Klein acknowledged, however, that his interpretation was based on a patchwork database and that future research and larger sample sizes will allow a more informed investigation of faunal exploitation patterns. With respect to the large mammalian fauna, it may be that the increased proportion of dangerous game in LSA contexts is more a reflection of a new hunting technology (i.e., the bow) than of a change in behavioral potential (Ambrose and Lorenz 1990, p. 28).

Deacon (1989), in contrast, has emphasized evidence that MSA people strategically exploited carbohydrate-rich vegetal foods (geophytes), implying scheduling, resource management, and the ability to make fire at will. He also argued that the Howieson's Poort stone tool industry (dated to ca. 70 kya), characterized by small elements made of exotic raw materials obtained at some distance and interstratified between typical MSA assemblages, reflects the ability of MSA people to employ stylistic markers for local groups and group boundaries under stressful ecological conditions. "The conclusion is that Middle Stone Age people did not differ from the Later Stone Age people in their basic subsistence ecology, and that subsistence behavior was essentially modern" (Deacon 1989, p. 557).

In their socioecological analysis of southern African MSA people, Ambrose and Lorenz (1990) noted with Klein an apparent change in hunting behavior across the MSA-LSA boundary; but they found in site distributions and in the Howieson's Poort industry a marked advancement in the ability of anatomically modern MSA hominids to respond flexibly and strategically to changing ecological circumstances compared to previous archaic hominids. In their opinion, "the Howieson's Poort marks the first time in human history when there was a significant change in human territorial organization," implying more information sharing and a home range mode of socioterritorial organization (Ambrose and Lorenz 1990, p. 26).

From these analyses, we draw two tentative conclusions related to human use of language. First, anatomically modern people possessed almost from the outset an ability to organize their social and territorial behavior and to respond flexibly to changing environments, an ability lacking in their archaic predecessors. We note that the possession of articulate language would greatly enhance the ability of modern humans to synthesize individual experiences and insights into a meaningful (and abstracted?) whole and to employ such constructions as guides for individual and group behavior, for example in organizing and scheduling their entire seasonal round of subsistence activities. Second, the apparent pace of behavioral and technological innovation following the appearance of a.m. *H. sapiens* is unprecedented in previous human evolution. This suggests to us that the ability to produce rapidly spoken articulate speech led to profoundly different and powerful ways of conceptualizing the self, the social group, and the potentialities of the natural and subsistence environments, but that the cognitive potential to capitalize on this new linguistic modality was already in place. In sum, the apparent changes in subsistence behavior between archaic and modern populations may be explained by positing the possession by the latter of rapidly spoken language and a concomitant ability to organize both social activity and the environment into more coherent and meaningful patterns, rather than — or at least in addition to — any substantial advance in individual cognitive capacity. We contend that the morphological and motor changes contingent on the appearance of anatomically modern humans made possible the innovation of an advanced form of social behavior — fully phonemic speech — that enhanced but did not create the ability to manipulate symbols and to think abstractly.

Neanderthals and perhaps other similarly large-brained "archaic" populations, for example, *H. sapiens daliensis* (Wu and Olsen 1985) in China, may have performed intentional burials, and they appear to have engaged in other practices indicative of cultural complexity and abstract thought. To deny to Neanderthal a functional language backed by well-developed neural and behavioral structures strikes us as folly, and just plain mean. We speculate, therefore, that the linguistic potential of archaic hominids (meaning by that all of the requisite structures and circuitry for abstract, symbol-based, metaphorical language) had outstripped the expressive capacity of a communication system in which primary informational content was still conveyed via gesture. We might expect that vocalization had already become an increasingly rich supplementary modality and that sensory and cross-sensory neurological capacities, at least, differed in no significant way from our own (although the apparent expansion of frontal lobes in the brain of a.m. *H. sapiens* demands consideration). Yet for the archaic descendants of *H. erectus*, oral and esophopharyngeal limitations could have made it impossible for speech to supplant gesture as the primary linguistic modality.

Glottogenesis and Population Replacement

Assuming that phonemicization occurred in the late Pleistocene, referent groups of particular concern are archaic and anatomically modern populations residing in proximity, and questions critical to acceptance of a multiregional versus a local-origins theory of *H. sapiens* ancestry are: did they interbreed, and, if they did interbreed, did they produce viable, fertile offspring? The first question has an obvious logical priority. We suggest that they must have interbred rarely if at all; we argue a minimum flow of genes consequent on language barriers between, for example, Neanderthal and *H. sapiens* groups in a context of competition differing markedly from that between contemporary or recent groups of *H. sapiens* (which, as we should hardly need to point out, is itself greatly exacerbated by linguistic differences. See, in this connection, Barbujani 1991 and Barbujani and Sokal, 1990, 1991, cited by Lieberman 1992). We offer a parallel suggestion that close cooperation within sapient groups (and inclusive small communities) was facilitated by a (phonemic) language that promoted a common perspective and a habit of within-group unity in consciousness and in action more efficiently than did Neanderthal language (whether phonemic or not). Surely the relevant question with regard to whether a.m. *H. sapiens* interbred with archaic hominids is not whether they could but whether they would. We cannot say that they could not; we do argue that they did not.

EVOLUTIONARY ISSUES IN GLOTTOGENESIS

It seems reasonable to assume that the capacity to cooperate socially, especially with regard to subsistence and reproductive behavior, was under strong selective pressure from the start in hominid evolution; it seems reasonable that the ability to exchange and maintain, within groups, information concerning subsistence and reproduction would have been a part of the suite of social behaviors under selection. Language is a quintessentially social behavior; it is of inestimable value where group cooperation and cohesion are concerned, and it is of value to the individual primarily to the degree that it enhances his or her fitness within the context of that of the group.

Is it important to distinguish between cultural selection and natural selection in human evolution? The main thing that culture does, it seems to us, is increase the difficulty of distinguishing between individual and group selection. Culture undoubtedly establishes new contexts for individual selection. At the same time, culture as an activity (as opposed to a catalog of technological items, relationships, and ideas) is primarily *group* activity; and, except in a narrow, contrived sense, it is an activity that reinforces group selection. If we were to separate out something called "cultural" selection it would be in order to emphasize the influence, in this sense, of collective memory, shared traditions, and sanctioned behavioral prescriptions and proscriptions as effective constraints on behavior. We especially would want to emphasize the corporate influence of cultural entities such as household, clan, lineage, and community on population structure. Although geneticists interested in population structure have tended to emphasize the importance of random events, following Sewall Wright (Maynard Smith 1988), anthropologists are or should be interested as well in the determinative influence of culture, perhaps especially in the way that culturally constrained individual and corporate decisions influence whether and how genes migrate across group boundaries (Quiatt and Kelso 1988).

Culture appears to exert a kind of extrabiological force, selective in character, that suppresses individuality in favor of behaviors that enhance group cohesion. Culture also, whether selecting for group fitness or not, may work in ways sometimes contrary to "natural" selection for individual fitness; cultural sanctions against proscribed behavior may be as immediate and as severe in effect — they include, after all, ostracism and homicide — as any others in nature. This factor seems often to be overlooked in discussions of group selection in hominid evolution (e.g., Irons 1979). Thus, whether or not cultural selection is to be considered as requiring a separate level of analysis, a distinct tier analogous to those recognized by Stephen Gould (1985), group selection is, if not exactly characteristic of culture in certain stages of *its* evolution, at least more likely to occur within cultural contexts of selection in the prehistoric evolution of hominids (contra Wynne-Edwards 1962).

Alleles carried by a few individuals, alleles that might in other circumstances enhance individual fitness (and that might be predominant, in the present instance, in neighboring communities or demes) could if phenotypically expressed encounter negative selection through corporate action of the group (as we still tend to view departure from group norms, physical or mental, with jaundiced eye). Such selection could be expected to be especially strong in the case of behavior clearly important to group cohesion, and it seems not unreasonable to think of linguistic behavior in this light. In emergent conditions, for example, with language undergoing phonemicization, the ability to conform closely to speech norms of the group could be critical to an individual's fitness — and we note that the overall genic consequences for, say, hopeful immigrants, would be similar whether the disability were a matter of genetic or cultural inheritance.

Punctuationalists, incidentally, might well see evidence for macroevolutionary change under such circumstances. Gould (1985) has stressed the randomicity of evolutionary change at one level with respect to evolutionary trends at other levels. Morphological changes to the tongue and esophopharyngeal tract, when seen in the context of microevolutionary pressure for manual small-motor elaboration, may meet the criterion of second-tier evolution.

DISCUSSION

We do not wish to make too much of what many undoubtedly will regard as baseless speculation to begin with. Our aim is to call attention to increasing varieties of evidence bearing on glossogenetics from the fossils themselves and to point out that linguistic distinctions such as we have suggested, if they obtained, must have conditioned in important ways relations among groups, populations, or species to which they applied.

The advantages of a fully vocal language, phonemicized, syntactical, and infinitely open and productive, over a language consisting simply or primarily of a vocabulary of gestural/vocal sememes seem obvious. Among other things, the vocal-auditory channel in itself affords broadcast communication and multidirectional reception by day or night and largely independent of habitat and climatic variables (Hockett and Altmann 1968).

But while we argue for the late appearance of rapidly spoken phonemic speech, we do not wish to seem to be arguing that spoken human language is emergent (*sensu* Chomsky 1968, p. 60). We stress that except for the ability to produce regularly distinguishable, arbitrary phonemes at a rapid enough rate to accommodate the inherent limits of short-term memory (Lieberman 1989), such linguistic features as syntax and the cognitive capacity to employ language were evolved over more than 1.5 million years.

We assume that the importance of behaviors acquired through social learning, including language, was significant in prolonging the period of hominid postnatal development. We note that the predisposition to and timing of language acquisition — whether gestural or vocal — appears to be a biological constant in *H. sapiens* (Petitto and Marentette 1991). Yet most of the neurological structures for language are acquired after birth (Walker 1981), and the same neural structures associated with the production and processing of speech (e.g., Broca's and Wernicke's areas) are implicated in the production and processing of gestural language (Hewes 1978, Meier 1991). Recent research into the acquisition of linguistic skills by deaf children suggests that rudimentary articulation of phonemes occurs slightly later than the production of rudimentary gestures: the former make greater demands on small motor ability and aural symbol processing (Meier 1991, p. 65; Petitto and Marentette 1991, p. 1495). From these insights we draw three inferences relative to the evolution of human linguistic capacity. First, the postnatal acquisition of learned social behavior, including linguistic behavior — was under strong selection beginning early in human evolution. Second, the basic neurological structures for language were in place prior to the appearance of a.m. *H. sapiens*. Finally, phonemicized speech arose as an exaptive behavioral innovation following the evolutionary appearance of a new morphotype under selection for more efficient neural processing and advanced small motor behavior.

Acheulean hunters/foragers may not have required vocabularies larger than a few hundred "words." If not, achievement of language performance acceptable by community standards should have been simple enough. Middle Pleistocene culture may not have varied greatly across wide regions judging from relative homogeneity of the central tool-manufacturing tradition, and, assuming that language at that time was primarily gestural in form, component units may well have been tied rather closely to species-characteristic body form and movement patterns (cf. Hewes 1973b, 1977), for the most part perhaps to a highly limited subset associated with subsistence and technological processes immediately recognizable throughout the species. Thus, for these hominids, translation and learning of language across communities, however widely separated geographically, may have been akin to picking up a dialect.

By the late Pleistocene, artifacts and bone assemblages evidence greatly increased proficiency in hunting and, to varying degrees according to habitat, concentration on a limited number of preferred species of prey (Gamble 1986). In some regions (e.g., western Europe around 40,000 years ago) these changes are correlated with increased population density, territorial restrictions on group migration, and more intensive exploitation of local resources. Ramification of tool types and local diversity of form and stylistic elaboration in the upper Paleolithic evidence increasing heterogeneity of information arrays relating to basic subsistence, not just across but within regions. They suggest as well increase by an order of magnitude in the store of information appropriate to the kind of distributed diet and scheduled collection of foods necessary for groups who must, in the face of increasing population density, engage in more intensive exploitation of an expanded list of food species.

If the language of middle Pleistocene hominids was, as has been suggested, gestural at base and sememic in character, it would have been hard pressed to serve communities in these new circumstances. Gradual increase in brain size may reflect selection for greater memory capacity and greater efficiency of information processing in species in which cultural information was organized in a sememic system, the "natural" boundaries of which (see above) are highly limited. The jump in brain size in the late Pleistocene suggests sudden, marked change in the intensity of that selection, possibly in conjunction with changing climate. Strong selection, in conjunction with temporary isolation of some populations and reduced migration overall, after the first dissemination of *H. erectus* throughout the Old World, could well have resulted in one or several avenues of neural development.

The characteristic form and size of (classic) Neanderthal crania, frequently accounted for in terms of adaptation to cold or other external environmental conditions, may represent one such avenue: an increase in the number of neurons and physical expansion of the brain. An increasingly large brain, however, was doubtless energetically very expensive to maintain and may conceivably have had deleterious obstetrical consequences. Any innovation that provided a faster, more conventional and less limited system of organizing information for storage and processing would have been advantageous. In this context, we recall Kien's suggestion of a new way of organizing and accessing memory via reiterative neural symbols — a method of data reduction (Kien 1990). Such a cytoarchitectonic reorganization may well have relaxed selection for increasing brain size, and it would also have laid the neurological basis for phonemic processing of speech.

The advantage of phonemicized speech over straight sememic vocalizations arguably has largely to do with economy in the storage and retrieval of information (Hewes 1983). Hewes argued that phonemicization would facilitate filing words and phrases by initial sound, in a manner analogous to an alphabetical system, even before the invention of alphabetic writing. This strikes us as reasonable. So does the likelihood that phonemicization, in conjunction with syntactic structure and a rhythmic line, would provide a superior mnemonic system of end-rhymes, front-rhymes, and internal sound correspondences of many sorts, useful in the memorization of rituals, recipes, formulas, and longer recitatives central to the cultural life of a group. Such a system would be useful not only for individual storage and retrieval of organized sets of essential information ("Thirty days hath September . . .") but for interindividual transmission across generations and, above all perhaps, for synchronized retrieval, iteration, and restorage (perhaps with modifications subject to common approval) in formal contexts by several or all members of a group as a way of affirming group identity and maintaining the common cultural store of information. We suggest that the extensive use of extrasomatic symbols (viz. art) as markers for individuals, groups, and group boundaries, and as crystallizations of collective understandings, is a predictable — if not an inevitable — outcome of the use of phonetic symbols in language.

The organizational capacity of rapidly spoken, phonemic language in the realms of subsistence and society, to say nothing of the consequences for technological innovation, would have given speaking hominids a clear advantage over hominids still communicating in a more "antique" linguistic system. Not only would hominids of archaic morphology (and perhaps neural architecture) have been precluded from acquiring the new language, but there would have existed formidable linguistic barriers between competing communities. The fossil record bears ample testimony to the outcome of that competition.

NOTE

A version of this chapter was published with commentary and reply by the authors in *Current Anthropology,* vol. 34, No. 5, December 1993.

References

Acredolo, L. P., and S. W. Goodwyn. 1985. Symbolic gesturing in language development. *Human Development* 29:40–49.

Alexander, C. 1964. *Notes on the Synthesis of Form*. Cambridge: Harvard University Press.

Alexander, R. D. 1987. *The Biology of Moral Systems*. New York: Aldine de Gruyter.

Allen, C. A., and G. D. Stephens (eds.). 1962. *Satire: Theory and Practice*. Belmont, CA: Wadsworth.

Allsworth-Jones, P. 1990. The Szeletian and the stratigraphic succession in central Europe and adjacent areas: Main trends, recent results, and problems for resolution. In: P. Mellars (ed.), *The Emergence of Modern Humans: An Archaeological Perspective*, pp. 160–242. Edinburgh: Edinburgh University Press.

Altman, J. 1978. Three levels of mentation and the hierarchic organization of the human brain. In: G. A. Miller and E. Lenneberg (eds.), *Psychology and Biology of Language and Thought*, pp. 87–109. New York: Academic Press.

Ambrose, S. H., and K. G. Lorenz. 1990. Social and ecological models for the Middle Stone Age in southern Africa. In: P. Mellars (ed.), *The Emergence of Modern Humans: An Archaeological Perspective*, pp. 3–33. Edinburgh: Edinburgh University Press.

American Anthropological Association. 1989. Symposium, "The Tasaday controversy: An assessment of the evidence." 15–19 November 1989. Washington, D. C.

Anderson, I. 1986. Tasaday: Stone age survivors or a modern hoax? *New Scientist*, 23 October, p. 18.

Anderson, J. R. 1986. Mirror-mediated finding of hidden food by monkeys (*Macaca tonkeana* and *Macaca fascicularis*). *Journal of Comparative Psychology* 100:237–242.

Anderson, J. R., and E. A. Williamson. 1983. Chimpanzees of Sapo Forest, Liberia: Density, nests, tools, and meat eating. *Primates* 24(4):594–601.

André, J. L. 1882–1883. On puns and rebuses in history and archaeology. *Reliquary* 23:169–173.

Armstrong, D. F. 1985. Comment on Parker, "A social-technological model for the evolution of language." *Current Anthropology* 26:626–627.

Atran, S. 1982. Constraints on a theory of hominid tool-making behavior. *L'Homme* 22:35–68.

Australian Encyclopaedia. 1983. s.v. Australian aborigines, Vol. 1. Sydney: Grolier Society of Australia.

Badrian, A., and N. Badrian. 1984. Social organization of *Pan paniscus* in the Lomako Forest, Zaire. In: R. Susman (ed.), *The Pygmy Chimpanzee: Evolutionary Biology and Behavior*, pp. 325–346. New York: Plenum.

Badrian, A., N. Badrian, and R. M. Susman. 1981. Preliminary observations on the feeding behavior of *Pan paniscus* in the Lomako forest of central Zaire. *Primates* 22:173–181.

Baker, C. 1977. Regulators and turn-taking in ASL discourse. In: L. Friedman (ed.), *On the Other Hand*, pp. 215–236. New York: Academic Press.

References

Barbujani, G. 1991. What do languages tell us about human microevolution? *Tree* 6:151–156.

Barbujani, G., and R. R. Sokal. 1990. Zones of sharp genetic change in Europe are also linguistic boundaries. *Proceedings of the National Academy of Sciences, U.S.A.* 187:1816–1819.

Barbujani, G., and R. R. Sokal. 1991. Genetic population structure of Italy: Physical and cultural barriers to gene flow. *American Journal of Human Genetics* 48:398–411.

Baron–Cohen, S. 1987. Autism and symbolic play. *British Journal of Developmental Psychology* 5:139–148.

Baron–Cohen, S. 1989. Perceptual role taking and protodeclarative pointing in autism. *British Journal of Developmental Psychology* 7:113–127.

Bates, E. 1976. *Language and Context: The Acquisition of Pragmatics.* New York: Academic Press.

Bates, E. 1979a. The biology of symbols: Some concluding thoughts. In: E. Bates, L. Benigni, I. Bretherton, L. Camaioni, and V. Volterra (eds.), *The Emergence of Symbols: Cognition and Communication in Infancy*, pp. 315–370. New York: Academic Press.

Bates, E. 1979b. Intentions, conventions, and symbols. In: E. Bates, L. Benigni, I. Bretherton, L. Camaioni, and V. Volterra (eds.), *The Emergence of Symbols: Cognition and Communication in Infancy*, pp. 33–68. New York: Academic Press.

Bates, E., L. Benigni, I. Bretherton, L. Camaioni, and V. Volterra (eds.). 1979. *The Emergence of Symbols: Cognition and Communication in Infancy.* New York: Academic Press.

Bates, E., I. Bretherton, C. Shore, and S. McNew. 1983. Names, gestures, and objects: Symbolization in infancy and aphasia. In: K. E. Nelson (ed.), *Children's Language*, Vol. 4, pp. 59–123. Hillsdale, NJ: Erlbaum.

Bates, E., D. Thal, and V. Marchman. 1991. Symbols and syntax: A Darwinian approach to language development. In: N. Krasnegor, D. Rumbaugh, R. Schiefelbusch, and M. Studdert-Kennedy (eds.), *Biological and Behavioral Determinants of Language Development*, pp. 29–65. Hillsdale, NJ: Erlbaum.

Bates, M. 1950. *The Nature of Natural History.* New York: Charles Scribner's Sons.

Bates, M. 1960. Ecology and evolution. In: S. Tax (ed.), *Evolution After Darwin*, Vol. 1, *The Evolution of Life: Its Origin, History and Future*, pp. 547–568. Chicago: University of Chicago Press.

Bateson, G. 1955/1972. A theory of play and fantasy. In: G. Bateson, *Steps to an Ecology of Mind*, pp. 177–193. New York: Ballantine Books.

Bateson, G. 1956. The message "This is play." In: B. Schaffner (ed.), *Group Processes: Transactions of the Second Conference*, pp. 145–242. Madison, NJ: Josian Macy, Jr., Foundation.

Bateson, G. 1971/1972. Style, grace, and information in primitive art. In: G. Bateson, *Steps to an Ecology of the Mind*, pp. 128–152. New York: Ballantine Books.

Bavelas, J. B., A. Black, C. R. Lemery, and J. Mullett. 1987. Motor mimicry as primitive empathy. In: N. Eisenberg and J. Strayer (eds.), *Empathy and Its Development*, pp. 317–338. Cambridge: Cambridge University Press.

Beck, B. B. 1974. Baboons, chimpanzees, and tools. *Journal of Human Evolution* 3:509–516.
Beck, B. B. 1975. Primate tool behavior. In: R. H. Tuttle (ed.), *Primate Socioecology and Psychology of Primates*, pp. 413–447. The Hague: Mouton.

Beck, B. B. 1980. *Animal Tool Behavior: The Use and Manufacture of Tools and Animals.* New York: Garland STPM Press.

Bekoff, M. 1975. The communication of play intention: Are play signals functional? *Semiotica* 15:231–240.

Bellugi, U., A. Bihrle, and D. Corina. 1991. Linguistic and spatial development: Dissociations between cognitive domains. In: N. Krasnegor, D. Rumbaugh, R. Schiefelbusch, and M.

References

Studdert-Kennedy (eds.), *Biological and Behavioral Determinants of Language Development*, pp. 363–393.

Bennett, M. 1989. Children's self-attribution of embarrassment. *British Journal of Developmental Psychology* 7:207–217.

Bernatzik, H. A., with E. Bernatzik. 1951. *The Spirits of the Yellow Leaves*. E. W. Dockes, trans. London: Robert Hale. lst German ed.: *Die Geister der Gelben Blätter*, Munich: Verlag F. Bruckmann, 1938.

Bernatzik, H. A. (ed.). 1954. *Die Neue Grosse Völkerkunde*. Frankfurt: Herkul Verlagsanstalt.

Bicchieri, M. G. 1990. From Eurocentric to geocentric: The perspective from small-scale societies. *Anthropos* 85:1–9.

Bickerton, D. 1981. *Roots of Language*. Ann Arbor, MI: Karoma.

Bickerton, D. 1990. *Language and Species*. Chicago: University of Chicago Press.

Biesele, M. 1983. Interpretation in rock art and folklore: Communication systems in evolutionary perspective. *South African Archaeological Society Goodwin Series* 4:54–60.

Binford, L. R. 1981. *Bones: Ancient Men and Modern Myths*. New York: Academic Press.

Binford, L.R. 1982. Commentary on "Rethinking the Middle/Upper Paleolithic transition." *Current Anthropology* 23:177–181.

Blades, M., and C. Spencer. 1987. The use of maps by 4–6-year-old children in a large-scale maze. *British Journal of Developmental Psychology* 5:19–24.

Blaschke, M., and G. Ettlinger. 1987. Pointing as an act of social communication by monkeys. *Animal Behaviour* 35:1520–1523.

Bloch, J. 1952. L'Andaman. In: A. Meillet and M. Cohen (eds.), *Les langues du monde*, pp. 513–521. Paris: Centre National de la Recherche Scientifique.

Bloch, M. 1991. Language, anthropology, and cognitive science. *Man* (N.S.) 26:183–198.

Bloom, L., and M. Lahey. 1978. *Language Development and Language Disorders*. New York: John Wiley and Sons.

Blumenberg, B. 1983. The evolution of the advanced hominid brain. *Current Anthropology* 24:589–623.

Blumenschine, R. J. 1987. Characteristics of an early hominid scavenging niche. *Current Anthropology* 28:383–407.

Bock, W. J. 1980. The definition and recognition of biological adaptation. *American Zoology* 20:217–227.

Bock, W. J. 1989. Organisms as functional machines: A connectivity explanation. *American Zoology* 29:1119–1132.

Bock, W. J., and G. von Wahlert. 1965. Adaptation and the form-function complex. *Evolution* 19:269–299.

Boehm, C. 1989. Vocal communication of *Pan troglodytes*: Possibilities and prospects for explaining human language. Cortona: NATO Advanced Studies Institute.

Boehm, C. 1992. Vocal communication of *Pan troglodytes*: "Triangulating" to the origin of spoken language. In: J. Wind et al. (eds.)., *Language Origin: A Multidisciplinary Approach*. Dordrecht: Kluwer.

Boeles, J. J. 1963. Second expedition to the Mrabri ("Khon-pa") of north Thailand. *Journal of the Siam Society* 51(2).

Boelkins, R. C., and A. P. Wilson. 1972. Intergroup dynamics of the Cayo Santiago rhesus (*Macaca mulatta*) with special reference to changes in group membership by males. *Primates* 14:359–68.

Boesch, C. 1991a. Handedness in wild chimpanzees. *International Journal of Primatology* 12:541–558.

Boesch, C. 1991b. Teaching among wild chimpanzees. *Animal Behavior* 41:530–532.

Boesch, C., and H. Boesch. 1981. Sex differences in the use of natural hammers by wild chimpanzees: A preliminary report. *Journal of Human Evolution* 10:585–593.

Boesch, C., and H. Boesch. 1983. Optimization of nut-cracking with natural hammers by wild chimpanzees. *Behaviour* 83:265–286.

Boesch, C., and H. Boesch. 1984. Mental map in chimpanzee: an analysis of hammer transports for nut cracking. *Primates* 25(2):160–170.

Boesch, C., and H. Boesch. 1989. Hunting behavior of wild chimpanzees in the Tai National Park. *American Journal of Physical Anthropology* 78:547–573.

Boesch, C., and H. Boesch. 1990. Tool use and tool making in wild chimpanzees. *Folia Primatologia* 54:86–99.

Boesch, C., and H. Boesch. 1993. Aspects of transmission of tool use in wild chimpanzees. In: K. R. Gibson and T. Ingold (eds.), *Tools, Language and Cognition in Human Evolution.* Cambridge: Cambridge University Press.

Boesch, C., and H. Boesch-Achermann. 1991. Dim forest, bright chimps. *Natural History*, September, pp. 50–56.

Boon, C. S. c. 1962–1963. *The Hill Tribes of Siam.* Bangkok: Khun Aroon.

Borchert, C. M., and A. L. Zihlman. 1990. The ontogeny and phylogeny of symbolizing. In: M. L. Foster and L. J. Botscharow (eds.), *The Life of Symbols*, pp. 15–44. Boulder, CO: Westview Press.

Bouton, L. 1913. Le pseudo-langage. Observations effectuées sur un anthropoïde: le gibbon (*Hylobates leucogenys*-Ogilby). *Actes de la Société Linnaeus de Bordeaux* 67:5–80.

Bouton, L. 1914. Les deux methodes de l'enfant. *Actes de la Société Linnaeus de Bordeaux* 68:3–146.

Boyce, M. S. (ed.). 1988. *Evolution of Life Histories of Mammals: Theory and Pattern.* New Haven, CT: Yale University Press.

Bradshaw, J. L., and N. C. Nettleton. 1982. Language lateralization to the dominant hemisphere: Tool use, gesture and language in hominid evolution. *Current Psychological Reviews* 2:171–192.

Brain, C. K. 1981. *The Hunters or the Hunted? An Introduction to African Cave Taphonomy.* Chicago: Aldine.

Brauer, G., and K. W. Rimbach. 1990. Late archaic and modern *Homo sapiens* from Europe, Africa, and Southwest Asia: Craniometric comparisons and phylogenetic implications. *Journal of Human Evolution* 19:789–807.

Braunwald, S. R. 1978. Context, word and meaning: Towards a communicational analysis of lexical acquisition. In: A. Lock (ed.), *Action, Gesture and Symbol: The Emergence of Language*, pp. 485–527. London: Academic Press.

Bretherton, I. 1984. Representing the social world in symbolic play: Reality and fantasy. In: I. Bretherton (ed.), *Symbolic Play: The Development of Social Understanding*, pp. 3–41. Orlando: Academic Press.

Breuggeman, J. A. 1973. Parental care in a group of free-ranging rhesus monkeys (*Macaca mulatta*). *Folia Primatologica* 20:178–210.

Breuggeman, J. A. 1978. The function of adult play in free-ranging *Macaca mulatta*. In: E. O. Smith (ed.), *Social Play in Primates*, pp. 169–191. New York: Academic Press.

Brewer, S., and W. C. McGrew. 1990. Chimpanzee use of a tool-set to get honey. *Folia Primatologica* 54:100–104.

Brockelman, W. Y. 1984. Social behavior of gibbons: Introduction. In: H. Preuschoft et al. (eds.), *The Lesser Apes: Evolutionary and Behavioral Biology*, pp. 285–290. Edinburgh: Edinburgh University Press.

References

Brockelman, W. Y., and U. Treesucon. 1986. Observations on social changes in the gibbon *Hylobates lar* and *H. lar–H. pileatus* hybrids in Thailand. *Primate Report* 14:72.

Brown, R. 1973. *A First Language: The Early Stages*. Cambridge: MIT Press.

Brown, R. W., A. H. Black, and A. E. Horowitz. 1955. Phonetic symbolism in natural languages. *Journal of Abnormal and Social Psychology* 50:388–393.

Bruner, J. 1975. The ontogenesis of speech acts. *Journal of Child Language*, 2:1–19.

Bruton, M. N. (ed.). 1989. *Alternative Life-History Styles of Animals*. Boston: Kluwer Academic Publishers.

Bunge, M. 1973. *Method, Model, and Matter*. Boston: Reidel.

Bunney, S. 1990. Ancient artists painted with human blood. *New Scientist* 31:31.

Byers, J. A. 1987. Why the deer and the antelope play. *Natural History*, May, pp. 54–60.

Byrne, R. W., and A. Whiten. 1985. Tactical deception of familiar individuals in baboons (*Papio ursinus*). *Animal Behaviour* 33:669–673.

Byrne, R. W., and A. Whiten (eds.). 1988a. *Machiavellian Intelligence: Social Expertise and the Evolution of Intellect in Monkeys, Apes and Humans*. Oxford: Clarendon Press.

Byrne, R. W., and A. Whiten (eds.). 1988b. Towards the next generation in data quality: A new survey of primate tactical deception. *Behavioral and Brain Sciences* 11:233–273.

Byrne, R. W., and A. Whiten. 1992. Cognitive evolution in primates: Evidence from tactical deception. *Man* 27:609–627.

Calvin, B. 1983. A stone's throw and its launch window: Timing precision and its implications for language and hominid brains. *Journal of Theoretical Biology* 104:121–135.

Cambefort, J. P. 1981. A comparative study of culturally transmitted patterns of feeding habits in the chacma baboons *Papio ursinus* and the vervet monkeys *Cercopithecus aethiops*. *Folia Primatologica* 36:243–263.

Capelli, C. A., N. Nakagawa, and C. M. Madden. 1990. How children understand sarcasm: The role of context and intonation. *Child Development* 61:1824–1841.

Carlisle, R., and M. Siegel. 1974. Some problems in the interpretation of Neanderthal speech capabilities: A reply to Lieberman. *American Anthropologist* 76:319–322.

Carpenter, C. R. 1940. A field study in Siam of the behavior and social relations of the gibbon (*Hylobates lar*). *Comparative Psychology Monographs* 16:1–212.

Carrithers, M. 1985. *The Category of the Person*. Cambridge: Cambridge University Press.

Carrithers, M. 1990. Why humans have cultures. *Man* (N.S.) 25:189–206.

Cartmill, M. 1974. Rethinking primate origins. *Science* 184:436–443.

Case, R. 1985. *Intellectual Development from Birth to Adulthood*. London: Academic Press.

Caselli, M. 1983. Communication to language: Deaf children's and hearing children's development compared. *Sign Language Studies* 39:113–144.

Cash, T. F., and T. Pruzinsky (eds.). 1990. *Body Images: Development, Deviance, and Change*. New York: Guilford Press.

Cazden, C. B., and R. Brown. 1975. The early development of the mother tongue. In: E. H. Lenneberg and E. Lenneberg, *Foundations of Language Development*, Vol. 1, pp. 299–309. New York: Academic Press.

Chagnon, N. 1988. Male Yanomami manipulations of kinship classifications of female kin for reproductive advantage. In: L. Betzig et al. (eds.), *Human Reproductive Behavior*, pp. 23–48. Cambridge: Cambridge University Press.

Chapais, B. 1983. Dominance relatedness and the structure of female relationships in rhesus monkeys. In: R. A. Hinde (ed.), *Primate Social Relationships: An Integrated Approach*, pp. 209–217. Oxford: Blackwell.

References

Chapman, C. A., and L. J. Chapman. 1990. Dietary variability in primate populations. *Primates* 31(1):121–128.

Cheney, D. L. 1984. Category formation in vervet monkeys. In: R. Harre and V. Reynolds (eds.), *The Meaning of Primate Signals*, pp. 58–72. Cambridge: Cambridge University Press.

Cheney, D. L., and R.M. Seyfarth. 1988. Social and non–social knowledge in vervet monkeys. In: R. W. Byrne and A. Whiten (eds.), *Machiavellian Intelligence: Social Expertise and the Evolution of Intellect in Monkeys, Apes and Humans*. Oxford: Clarendon Press.

Cheney, D. L., and R. M. Seyfarth. 1990. *How Monkeys See the World*. Chicago: University of Chicago Press.

Cheney, D. L., and R. M. Seyfarth. 1991. Truth and deception in animal communication. In: C. A. Ristau (ed.), *Cognitive Ethology: The Minds of Other Animals*, pp. 127–151. Hillsdale, NJ: Erlbaum.

Chevalier–Skolnikoff, S. 1986. An exploration of the ontogeny of deception in human beings and nonhuman primates. In: R. W. Mitchell and N. S. Thompson (eds.), *Deception: Perspectives on Human and Nonhuman Deceit*, pp. 205–220. Albany: State University of New York Press.

Chivers, D. J., and J. J. Raemaekers. 1980. Long-term changes in behaviour. In: D. J. Chivers (ed.), *Malayan Forest Primates*, pp. 209–260. New York: Plenum.

Chomsky, N. 1957. *Syntactic Structures*. The Hague: Mouton.

Chomsky, N. 1965. *Aspects of a Theory of Syntax*. Cambridge: MIT Press.

Chomsky, N. 1968. *Language and Mind*. New York: Harcourt, Brace and World.

Chomsky, N. 1976. On the nature of language. In: S. Harnard, H. Steklis, and J. Lancaster (eds.), *Annals of the New York Academy of Sciences: Origins of Evolution and Language and Speech* 280:46–57.

Chomsky, N. 1980. *Rules and Representations*. Oxford: Blackwell.

Clark, H., and E. Clark. 1977. *Psychology and Language: An Introduction to Psycholinguistics*. San Diego, CA: Harcourt Brace Jovanovich.

Clark, H., and C. Marshall. 1978. Reference diaries. In: D. L. Waltz (ed.), *Theoretical Issues in Natural Language Processing 2*. New York: Association for Computing Machinery.

Clark, H. H., and R. J. Gerrig. 1990. Quotations as demonstrations. *Language* 66:765–805.

Clark, J. D. 1989. The origins and spread of modern humans: A broad perspective on the African evidence. In: P. Mellars and C. Stringer (eds.), *The Human Revolution: Behavioral and Biological Perspectives on the Origins of Modern Humans*, pp. 565–588. Edinburgh: Edinburgh University Press.

Clarke, R. J. 1990. The Ndutu cranium and the origin of *Homo sapiens*. *Journal of Human Evolution* 19:699–736.

Clementson-Mohr, D. 1982. Towards a social-cognitive explanation of imitation development. In: G. Butterworth and P. Light (eds.), *Social Cognition: Studies of the Development of Understanding*, pp. 53–74. Chicago: University of Chicago Press.

Clutton-Brock, T. H. (ed.). 1988. *Reproductive Success: Studies of Individual Variation in Contrasting Breeding Systems*. Chicago: University of Chicago Press.

Cocking, R. R., and C. E. Copple. 1987. Social influences on representational awareness: Plans for representing and plans as representations. In: S. L. Friedman, E. K. Skolnick, and R. R. Cocking (eds.), *Blueprints for Thinking: The Role of Planning in Cognitive Development*, pp. 428–465. Cambridge: Cambridge University Press.

Colvin, J. 1986. Proximate causes of male emigration at puberty in rhesus monkeys. In: R. G. Rawlins and M. J. Kessler (eds.), *The Cayo Santiago Macaques*, pp. 131–157. Albany: State University of New York Press.

References

Conkey, M. W. 1983. On the origins of Paleolithic art: A review of some critical thoughts. In: E. Trinkaus (ed.), *The Mousterian Legacy: Human Biocultural Change in the Upper Pleistocene*, pp. 201–227. International Series 164. Oxford: British Archaeological Reports.

Conkey, M. W. 1984. To find ourselves: Art and social geography of prehistoric hunter gatherers. In: C. Schrire (ed.), *Past and Present in Hunter Gatherer Studies*, pp. 253–276. Orlando: Academic Press.

Conkey, M. W. 1985. Ritual communication, social elaboration, and the variable trajectories of Paleolithic material culture. In: T. D. Price and J. A. Brown (eds.), *Prehistoric Hunter–Gatherers: The Emergence of Cultural Complexity*, pp. 299–323. Orlando: Academic Press.

Cooley, C. H. 1902/1983. *Human Nature and the Social Order*. New Brunswick, NJ: Transaction Books.

Coolidge, H. J. 1984. Foreword: Historical remarks bearing on the discovery of *Pan paniscus*. In: R. L. Susman (ed.), *The Pygmy Chimpanzee: Evolutionary Biology and Behavior*, pp. ix–xiii. New York: Plenum.

Count, E. W. 1976. Language of organism: Requisite fabric for an evolution of the speech function: some theoretical considerations. In: S. R. Harnard, H. Steklis, and J. Lancaster (eds.), *Annals of the New York Academy of Sciences: Origins and Evolution of Language and Speech* 280:456–466.

Crawford, M. P. 1937. The cooperative solving of problems by young chimpanzees. *Comparative Psychology Monographs* 14:1–88.

Crider, C., and L. Cirillo. 1991. Systems of interpretation and the function of metaphor. *Journal for the Theory of Social Behaviour* 21:171–195.

Cunningham, C. E. 1966. Urgent research in northern Thailand. *Bulletin of the International Committee on Urgent Anthropological and Ethnological Research* 8:57–74.

Dahl, J. F., R. D. Nadler, and D. C. Collins. 1991. Monitoring the ovarian cycles of *Pan troglodytes* and *P. paniscus*: A comparative approach. *American Journal of Primatology* 24(3):195–209.

Dale, P. S. 1976. *Language Development*. 2nd ed. New York: Holt, Rinehart and Winston.

Dandan, V. B. (ed.). 1989. *Readings on the Tasaday*. Manila: Tasaday Community Care Foundation.

Darwin, C. 1874. *The Descent of Man and Selection in Relation to Sex*. 2nd ed. London: John Murray.

Datta, S. 1986. The role of alliances in the acquisition of rank. In: J. G. Else and P. Lee (eds.), *Primate Ontogeny, Cognition, and Social Behaviour*. Cambridge: Cambridge University Press.

Daugherty, J.W.D., and C. M. Keller. 1982. Taskonomy: A practical approach to knowledge structures. *American Ethnologist* 5:763–764.

Davidson, I., and W. Noble. 1989. The archaeology of perception: Traces of depiction and language. *Current Anthropology* 30:125–155.

Davidson, I., and W. Noble. 1993. Tools, language and evolution: The relevance of the Upper Palaeolithic. In: K. R. Gibson and T. Ingold (eds.), *Tools, Language and Cognition in Human Evolution*. Cambridge: Cambridge University Press.

Davies, D. 1973. *The Last of the Tasmanians*. London: Frederick Muller.

Davis, W. 1986. The origins of image making. *Current Anthropology* 27:193–215.

Davis, W. 1987. Replication and depiction in Paleolithic art. *Representations* 19:111–147.

Davis, W. 1988. Comment on "The signs of all times." *Current Anthropology* 29:201–245.

Davis, W. 1989a. Comment on "The archaeology of perception." *Current Anthropology* 30:140–141.

Davis, W. 1989b. Finding symbols in history. In: H. Morphy (ed.), *Animals into Art*, pp. 179–189. London: Unwyn Hyman.

Dawkins, R., and J. R. Krebs. 1978. Animal signals: Information or manipulation? In: J. R. Krebs and N. B. Davies (eds.), *Behavioural Ecology: An Evolutionary Approach*, pp. 282–309. Oxford: Blackwell.

Deacon, H. J. 1989. Late Pleistocene palaeoecology and archaeology in the Southern Cape, South Africa. In: P. Mellars and C. Stringer (eds.), *The Human Revolution: Behavioral and Biological Perspectives on the Origins of Modern Humans*, pp. 547–564. Edinburgh: Edinburgh University Press.

Deacon, T. W. 1989. The neural circuitry underlying primate calls and human language. *Human Evolution* 4:367–401.

de Laguna, G. A. 1927. *Speech: Its Function and Development*. New Haven, CT: Yale University Press.

DeLoache, J. S. 1990. Young children's understanding of models. In: R. Fivush and J. A. Hudson (eds.), *Knowing and Remembering in Young Children*, pp. 94–126. Cambridge: Cambridge University Press.

Dennett, D. C. 1978. *Brainstorms: Philosophical Essays on Mind and Psychology*. Cambridge, MA: Bradford Books.

Dennett, D. C. 1983. Intentional systems in cognitive ethology: The "Panglossian paradigm" defended. *Behavioral and Brain Sciences* 6:343–390.

Dettwyler, K. A. 1991. Can paleopathology provide evidence for "compassion"? *American Journal of Physical Anthropology* 84:375–384.

Deuchar, M. 1990. Are the signs of language arbitrary? In: H. Barlow, C. Blakemore, and M. Weston–Smith (eds.), *Images and Understanding: Thoughts About Images, Ideas About Understanding*, pp. 168–179. Cambridge: Cambridge University Press.

de Villiers, J., and P. de Villiers. 1978. *Language Acquisition*. Cambridge: Harvard University Press.

de Waal, F.B.M. 1982. *Chimpanzee Politics: Power and Sex Among Apes*. New York: Harper and Row.

de Waal, F.B.M. 1986. Deception in the natural communication of chimpanzees. In: R. W. Mitchell and N. S. Thompson (eds.), *Deception: Perspectives on Human and Nonhuman Deceit*, pp. 221–244. Albany: State University of New York Press.

de Waal, F.B.M. 1988. The communicative repertoire of captive bonobos (*Pan paniscus*), compared to that of chimpanzees. *Behaviour* 106:183–251.

de Waal, F.B.M. 1989a. Behavioral contrasts between bonobo and chimpanzee. In: P. G. Heltne and L. A. Marquardt (eds.), *Understanding Chimpanzees*, pp. 154–175. Cambridge: Harvard University Press.

de Waal, F.B.M. 1989b. Food sharing and reciprocal obligations among chimpanzees. *Journal of Human Evolution* 18:433–459.

de Waal, F.B.M. 1991. The chimpanzee's sense of social regularity and its relation to the human sense of justice. *American Behavioral Scientist* 34:335–349.

de Waal, F.B.M., and A. van Roosmalen. 1979. Reconciliation and consolation among chimpanzees. *Behavioral Ecology and Sociobiology* 5:55–66.

Diamond, A. S. 1959. *The History and Origin of Language*. London: Methuen.

Dibble, H. 1989. The implications of stone tool types for the presence of language during the Lower and Middle Pleistocene. In P. Mellars and C. Stringer (eds.), *The Human Revolution: Behavioral and Biological Perspectives on the Origins of Modern Humans*, pp. 415–432. Edinburgh: Edinburgh University Press.

Donald, M. 1991. *Origins of the Modern Mind: Three Stages in the Evolution of Culture and Cognition.* Cambridge: Harvard University Press.

Dore, J. 1974. A pragmatic description of early language development. *Journal of Psycholinguistic Research* 3:343–350.

Drickamer, L. C., and S. H. Vessey. 1973. Group changing in free-ranging rhesus monkeys. *Primates* 14:359–368.

DuBrul, E. L. 1977. Origin of the speech apparatus and its reconstruction in fossils. *Brain and Language* 4:365–381.

Dumont, J.–P. 1987. Quels Tasaday? de la découverte et de l'invention d'autrui. *L'Homme* 103(273):27–42.

Dunbar, R.I.M. 1988. *Primate Social Systems.* London: Croom Helm.

Dutton, B. P. 1983. *American Indians of the Southwest.* Albuquerque: University of New Mexico Press.

Duval, S., and R. A. Wicklund. 1972. *A Theory of Objective Self Awareness.* New York: Academic Press.

Eco, U. 1979. *A Theory of Semiotics.* Bloomington: Indiana University Press.

Efron, D. 1972/1978. Gesture, race and culture: A tentative study of some of the spatio-temporal and "linguistic" aspects of the gestural behavior of Eastern Jews and southern Italians in New York City, living under similar as well as different environmental conditions. In: T. Polhemus (ed.), *The Body Reader: Social Aspects of the Human Body*, pp. 225–247. New York: Pantheon Books.

Ekman, P. 1977/1980. Biological and cultural contributions to body and facial movement in the expression of emotions. In: A. O. Rorty (ed.), *Explaining Emotions*, pp. 103–126. Berkeley: University of California Press.

Ellefson, J. O. 1968. Territorial behavior in the common white-handed gibbon, *Hylobates lar*. In: P. C. Jay (ed.), *Primates: Studies in Adaptation and Variability*, pp. 180–199. New York: Holt, Rinehart and Winston.

Elliot, R. C. 1977. Cross-modal recognition in three primates. *Neuropsychologia* 15:183–186.

Engels, F. 1934. *Dialectics of Nature.* Moscow: Progress.

Epstein, R., R. Lanza, and B. Skinner. 1980. Symbolic communication between two pigeons (*Columba livia domestica*). *Science* 207:543–545.

Ervin–Tripp, S. 1970. Discourse agreement: How children answer questions. In: J. Hayes (ed.), *Cognition and the Development of Language*, pp. 79–107. New York: John Wiley and Sons.

Falk, D. 1975. Comparative anatomy of the larynx in man and the chimpanzee: Implications for language in Neanderthal. *American Journal of Physical Anthropology* 43:123–132.

Falk, D. 1980. Language, handedness and primate brains: Did the Australopithecines sign? *American Anthropologist* 82:72–78.

Falk, D. 1984. The petrified brain. *Natural History* 93(9):36, 38–39.

Falk, D. 1987. Hominid paleoneurology. In: B. J. Siegel, A. R. Beals, and S. A. Taylor (eds.), *Annual Review of Anthropology* 16:13–20. Palo Alto, CA: Annual Review.

Falk, D. 1990. Brain evolution in *Homo*: The "radiator" theory. *Behavioral and Brain Sciences* 13:333–381.

Falk, D. 1992. *Braindance.* New York: Henry Holt.

Fallon, A. 1990. Culture in the mirror: Sociocultural determinants of body image. In: T. F. Cash and T. Pruzinsky (eds.), *Body Images: Development, Deviance, and Change*, pp. 80–109. New York: Guilford Press.

Fant, L. 1972. *Ameslan: An Introduction to American Sign Language*. Northridge, CA: Joyce Motion Picture Company.

Farizy, C. 1990. The transition from Middle to Upper Paleolithic at Arcy-sur-Cure (Yonne, France): Technological, economic and social aspects. In: P. Mellars (ed.), *The Emergence of Modern Humans: An Archaeological Perspective*, pp. 303–326. Edinburgh: Edinburgh University Press.

Fein, G. G., and N. Apfel. 1979. Some preliminary observations on knowing and pretending. In: M. Smith and M. B. Franklin (eds.), *Symbolic Functioning in Childhood*, pp. 87–100. Hillsdale, NJ: Erlbaum.

Feldman, H., S. Goldin–Meadow, and L. Gleitman. 1978. Beyond Herodotus: The creation of language by linguistically deprived deaf children. In: A. Lock (ed.), *Action, Gesture, and Symbol: The Emergence of Language*, pp. 351–414. London: Academic Press.

Ferguson, C., D. Peizer, and T. Weeks. 1973. Model-and-replica phonological grammar of a child's first words. *Lingua* 31:35–65.

Ferguson, T. J., H. Stegge, and I. Damhuis. 1991. Children's understanding of guilt and shame. *Child Development* 62:827–839.

Fergusson, F. 1949/1972. *The Idea of Theater: A Study of Ten Plays; the Art of Drama in Changing Perspective*. Princeton: Princeton University Press.

Fobes, J. L., and J. E. King, (eds.). 1982. *Primate Behavior*. New York: Academic Press.

Fodor, J. A. 1981. Imagistic representation. In: N. Block (ed.), *Imagery*, pp. 63–86. Cambridge: MIT Press.

Fodor, J. A. 1983. *The Modularity of Mind: An Essay on Faculty Psychology*. Cambridge: MIT Press.

Foley, R. 1987. *Another Unique Species: Patterns in Human Evolutionary Ecology*. New York: John Wiley and Sons.

Fortes, M. 1983. *Rules and the Emergence of Society*. Royal Anthropology Institute Occasional Paper 39.

Fossey, D. 1983. *Gorillas in the Mist*. Boston: Houghton-Mifflin.

Foster, M.L.C. 1980. The growth of symbolism in culture. In: M.L.C. Foster and S. H. Brandes (eds.), *Symbol as Sense: New Approaches to the Analysis of Meaning*, pp. 371–397. New York: Academic Press.

Foster, M.L.C. 1990. Analogy, language, and the symbolic process. In: M.L.C. Foster and L. J. Botscharow (eds.), *The Life of Symbols*, pp. 81–100. Boulder, CO: Westview Press.

Foster, M.L.C., and S. H. Brandes. 1980. Introduction. In: M.L.C. Foster and S. H. Brandes (eds.), *Symbol as Sense: New Approaches to the Analysis of Meaning*, pp. 71–75. New York: Academic Press.

Fouts, R. 1973. Acquisition and testing of gestural signs in four young chimpanzees. *Science* 180:978–980.

Fouts, R. 1974. Language: Origins, definitions, and chimpanzees. *Journal of Human Evolution* 3:475–482.

Fouts, R. 1978. Sign language in chimpanzees: Implications for the visual mode and the comparative approach. In: F. Peng (ed.), *Sign Language and Language Acquisition in Man and Ape: New Dimensions in Comparative Pedolinguistics*, pp. 121–136. Boulder, CO: Westview Press.

Fouts, R., B. Chown, and L. Goodin. 1976. Transfer of signed responses in American Sign Language from vocal English stimuli to physical object stimuli by a chimpanzee (*Pan*). *Learning and Motivation* 7:458–475.

Fouts, R., W. Chown, G. Kimball, and J. Couch. 1976. Comprehension and production of American Sign Language by a chimpanzee (*Pan*). Paper presented at the Twenty-first International Congress of Psychology, Paris.

References

Fouts, R., and D. Fouts. 1989. Loulis in conversation with the cross–fostered chimpanzees. In: R. Gardner, B. Gardner, and T. Van Cantfort (eds.), *Teaching Sign Language to Chimpanzees*, pp. 293–307. Albany: State University of New York Press.

Fouts, R., D. Fouts, and D. Schoenfeld. 1984. Sign language conversational interactions between chimpanzees. *Sign Language Studies* 34:1–12.

Fouts, R., D. Fouts, and T. Van Cantfort. 1989. The infant Loulis learns signs from cross–fostered chimpanzees. In: R. Gardner, B. Gardner, and T. Van Cantfort (eds.), *Teaching Sign Language to Chimpanzees*, pp. 280–292. Albany: State University of New York Press.

Fouts, R., A. Hirsch, and D. Fouts. 1982. Cultural transmission of a human language in a chimpanzee mother-infant relationship. In: H. Fitzgerald, J. Mullins, and P. Gage (eds.), *Child Nurturance*, Vol. 3, pp. 159–193. New York: Plenum.

Fox, M. W. 1962. Observations on paw raising and sympathy lameness in the dog. *Veterinary Record* 74:895–896.

Fox, R. 1980. *The Red Lamp of Incest*. London: Hutchinson.

Freire–Marreco, B. 1915. A note on kinship terms compounded with the postfix 'e in the Hano dialect of Tewa. *American Anthropologist* 17:198–202.

Freud, S. 1900/1963. *The Interpretation of Dreams*. New York: Wiley.

Freud, S. 1930/1962. *Civilization and Its Discontents*. New York: W. W. Norton.

Fridlund, A. J., J. P. Sabini, L. E. Hedlund, J. A. Schaut, J. I. Shenker, and M. J. Knauer. 1990. Audience effects on solitary faces during imagery: Displaying to the people in your head. *Journal of Nonverbal Behavior* 14:113–137.

Frisch, J. E. 1973. The Hylobatid dentition. In: D. Rumbaugh (ed.), *Gibbon and Siamang: A Series of Volumes on the Lessor Apes*, Vol. 2, pp. 56–95. Basel: Karger.

Frishberg, N. 1975. Arbitrariness and iconicity: Historical change in American Sign Language. *Language* 51:696–719.

Fromkin, V., and R. Rodman. 1983. *An Introduction to Language*. 3rd ed. New York: Holt, Rinehart and Winston.

Furness, W. 1916. Observations on the mentality of chimpanzees and orang-utans. *Proceedings of the American Philosophical Society* 55:281–290.

Gallup, G. G., Jr. 1977. Self–recognition in primates. *American Psychologist* 32:329–338.

Gallup, G. G., Jr. 1985. Do minds exist in species other than our own? *Neurosciences and Biobehavioral Review* 9:631–641.

Gallup, G. G., Jr. 1987. Self-awareness. In: G. Mitchell and J. Erwin, (eds.), *Primate Biology*, Vol. 2B, *Behavior, Cognition, and Motivation*, pp. 3–16. New York: Alan R. Liss.

Gamble, C. 1986. *The Palaeolithic Settlement of Europe*. Cambridge: Cambridge University Press.

Gardner, B., and R. Gardner. 1974. Comparing the early utterances of child and chimpanzee. In: A. Pick (ed.), *Minnesota Symposium on Child Psychology*, Vol. 8, pp. 3–23. Minneapolis: University of Minnesota Press.

Gardner, B., and R. Gardner. 1978. Comparative psychology and language acquisition. *Annals of the New York Academy of Sciences* 309:37–76.

Gardner, B., and R. Gardner. 1989. A test of communication. In: R. Gardner, B. Gardner, and T. Van Cantfort (eds.), *Teaching Sign Language To Chimpanzees*, pp. 181–197. Albany: State University of New York Press.

Gardner, B., R. Gardner, and S. Nichols. 1989. The shapes and uses of signs in a cross–fostering laboratory. In: R. Gardner, B. Gardner, and T. Van Cantfort (eds.), *Teaching Sign Language to Chimpanzees*, pp. 55–180. Albany: State University of New York Press.

Gardner, R., and B. Gardner. 1969. Teaching sign language to a chimpanzee. *Science* 165:664–672.

351

Gardner, R., and B. Gardner. 1980. Two comparative psychologists look at language acquisition. In K. Nelson (ed.), *Children's Language*, Vol. 2. New York: Halsted Press.

Gardner, R., and B. Gardner. 1989. A cross-fostering laboratory. In: R. Gardner, B. Gardner, and T. Van Cantfort (eds.), *Teaching Sign Language to Chimpanzees*, pp. 1–28. Albany: State University of New York Press.

Gardner, R., and B. Gardner. 1991. Absence of evidence and evidence of absence. *Behavioral and Brain Sciences* 14(4):558–560.

Gardner, R., B. Gardner, and T. Van Cantfort (eds.). 1989. *Teaching Sign Language to Chimpanzees*. Albany: State University of New York Press.

Gargett, R. H. 1989. Grave shortcomings: The evidence for Neanderthal burial. *Current Anthropology* 30:157–190.

Garvey, C. 1977. *Play*. Cambridge: Harvard University Press.

Gentner, D. 1989. The mechanisms of analogical reasoning. In: S. Vosniadou and A. Ortony (eds.), *Similarity and Analogical Reasoning*, pp. 199–241. Cambridge: Cambridge University Press.

Ghiglieri, M. 1984. *The Chimpanzees of Kibale Forest: A Field Study of Ecology and Social Structure*. New York: Columbia University Press.

Gibbons, A. 1992. Chimps: More diverse than a barrel of monkeys. *Science* 255:287–288.

Gibson, K. R. 1983. Comparative neurobehavioral ontogeny and the constructivist approach to the evolution of the brain, object manipulation, and language. In: E. deGrolier (ed.), *Glossogenetics*. New York: Harwood Academic.

Gibson, K. R. 1990. New perspectives on instincts and intelligence: Brain size and the emergence of hierarchical mental constructional skills. In: S. T. Parker and K. R. Gibson (eds.), *"Language" and Intelligence in Monkeys and Apes*. Cambridge: Cambridge University Press.

Gibson, K. R. Forthcoming. The ontogeny and evolution of the brain, cognition and language. In: A. Lock and C. R. Peters (eds.), *Handbook of Human Symbolic Evolution*. Oxford: Oxford University Press.

Gladwin, T. 1970. *East Is a Big Bird: Navigation and Logic on Puluwat Atoll*. Cambridge: Harvard University Press.

Gleitman, H. 1990. Some reflections on drama and the dramatic experience. In: I. Rock (ed.), *The Legacy of Solomon Asch: Essays in Cognition and Social Psychology*, pp. 127–141. Hillsdale, NJ: Erlbaum.

Goffman, E. 1959. *The Presentation of Self in Everyday Life*. Garden City, NY: Doubleday/Anchor Books.

Gomez, J. C. 1991. Visual behavior as a window for reading the minds of others in primates. In: A. Whiten (ed.), *Natural Theories of Mind*, pp. 195–207. Oxford: Blackwell.

Goodall, J. 1965. Chimpanzees of the Gombe Stream Reserve. In: I. DeVore (ed.), *Primate Behavior*, pp. 425–473. New York: Holt, Rinehart and Winston.

Goodall, J. 1968. The behavior of free-living chimpanzees in the Gombe Stream Reserve. *Animal Behavior Monographs* 1:161–311.

Goodall, J. 1986. *The Chimpanzees of Gombe: Patterns of Behavior*. Cambridge, Mass.: Belknap Press of Harvard University.

Goodenough, W. H. 1990. Evolution of the human capacity for beliefs. *American Anthropologist* 92:597–612.

Goodman, N. 1968. *Languages of Art: An Approach to a Theory of Symbols*. Indianapolis: Bobbs–Merrill.

Goody, E. 1978. Towards a theory of questions. In: E. Goody (ed.) *Questions and Politeness: Strategies in Social Interaction*, pp. 17–43. Cambridge: Cambridge University Press.

References

Gopnik, A. 1982. Words and plans: Early language and the development of intelligent action. *Journal of Child Language* 9:303–318.

Gordon, R. M. 1986. Folk psychology as simulation. *Mind and Language* 1:158–171.

Gough, E. K. 1959. The Nayars and the definition of marriage. *Journal of the Royal Anthropology Institute*, 50:23–24.

Gould, J. L., and P. M. Marler. 1987. Learning by instinct. *Scientific American*, January, pp. 74–85.

Gould, S. J. 1977. *Ontogeny and Phylogeny*. Cambridge: Harvard University Press.

Gould, S. J. 1979. Comment on S. T. Parker and K. R. Gibson, "A developmental model for the evolution of language and intelligence in early hominids." *Behavioral and Brain Sciences* 2:385–386.

Gould, S. J. 1985. The paradox of the first tier: An agenda for paleobiology. *Paleobiology* 11:2–12.

Gowlett, J.A.J. 1984. *Ascent to Civilization*. London: Collins.

Graves, P. 1991. The persistence of memory: Dynamics of sociocultural evolution. *Cultural Dynamics* 4:290–320.

Greenberg, G., and E. Tobach (eds.). 1988. *Evolution of Social Behavior and Integrative Levels*. Hillsdale, N.J.: Erlbaum.

Greenfield, P. M. 1978. Structural parallels between language and action in development. In: A. Lock (ed.), *Action, Gesture and Symbol: The Emergence of Language*, pp. 415–445. London: Academic Press.

Greenfield, P. M. 1991. Language, tools, and the brain: The ontogeny and phylogeny of hierarchically organized sequential behavior. *Behavioral and Brain Sciences* 14:531–595.

Greenfield, P. M., and E. S. Savage-Rumbaugh. 1990. Grammatical combination in *Pan paniscus*: Processes of learning and invention in the evolution and development of language. In: S. Parker and K. Gibson (eds.), *"Language" and Intelligence in Monkeys and Apes: Comparative Developmental Perspectives*, pp. 540–578. New York: Cambridge University Press.

Greenfield, P. M., and E. S. Savage-Rumbaugh. 1991. Imitation, grammatical development, and the invention of protogrammar by an ape. In: N. Krasnegor, D. Rumbaugh, R. Schiefelbusch, and M. Studdert-Kennedy (eds.), *Biological and Behavioral Determinants of Language Development*, pp. 235–262. Hillsdale, NJ: Erlbaum.

Greenfield, P. M., and J. Smith. 1976. *The Structure of Communication in Early Language Development*. New York: Academic Press.

Grene, M. 1987. Hierarchies in biology. *American Scientist* 75:504–510.

Grice, H. P. 1957. Meaning. *Philosophical Review* 66:377–388.

Grice, H. P. 1968. Utterer's meaning, sentence-meaning, and word-meaning. *Foundations of Language* 4:225–242.

Grice, H. P. 1982. Meaning revisited. In: N. V. Smith (ed.), *Mutual Knowledge*, pp. 223–243. London: Academic Press.

Guilmet, G. 1977. The evolution of tool-using and tool-making behavior. *Man* (N.S.) 12:33–47.

Hadamard, J. 1945/1954. *An Essay on the Psychology of Invention in the Mathematical Field*. New York: Dover.

Hall, R. L. 1978. Variability and speciation in canids and hominids. In: R. L. Hall and H. S. Sharp (eds.), *Wolf and Man: Evolution in Parallel*, pp. 153–177. New York: Academic Press.

Hallowell, A. I. 1956. The structural and functional dimensions of human existence. *Quarterly Review of Biology* 31:88–101.

References

Hallowell, A. I. 1960. Self, society, and culture in phylogenetic perspective. In: S. Tax (ed.), *Evolution After Darwin*, Vol. 2, *The Evolution of Man*, pp. 309–372. Chicago: University of Chicago Press.

Hallowell, A.I. 1962. The protocultural foundations of human adaption. In: S. L. Washburn (ed.), *Social Life of Early Man*, pp. 236–255. London: Methuen.

Hamilton, W. D. 1963. The evolution of altruistic behavior. *American Naturalist* 97:354–356.

Hamilton, W. D. 1964. The genetical theory of social behavior. *Journal of Theoretical Biology* 7:1–52.

Hamilton, W. D. 1975. Innate social aptitudes of man: An approach from evolutionary genetics. In: R. Fox (ed.), *Biosocial Anthropology*, pp. 133–155. London: Malaby Press.

Hammond, P., and P. Hughes. 1978. *Upon the Pun: Dual Meaning in Words and Pictures*. London: W. H. Allen.

Haraway, D. J. 1989. *Primate Visions: Gender, Race, and Nature in the World of Modern Science*. New York: Routledge, Chapman and Hall.

Harris, P. L. 1989. *Children and Emotion: The Development of Psychological Understanding*. Oxford: Blackwell.

Harrold, F. B. 1989. Mousterian, Chatelperronian and early Aurignacian in Western Europe: Continuity or discontinuity. In: P. Mellars and C. Stringer (eds.), *The Human Revolution: Behavioral and Biological Perspectives on the Origins of Modern Humans*, pp. 677–713. Edinburgh: Edinburgh University Press.

Hausfater, G., and S. B. Hrdy (eds.). 1984. *Infanticide: Comparative and Evolutionary Perspectives*. New York: Aldine.

Hayes, C. 1951. *The Ape in Our House*. New York: Harper & Brothers.

Headland, T. N. (ed.). 1992. *The Tasaday Controversy: Assessing the Evidence*. Washington, DC: American Anthropological Association.

Heine–Geldern, R. (ed.). 1958. Introduction. Urgent anthropological research. *Bulletin of the International Committee on Urgent Anthropological and Ethnological Research* 1:31–32.

Herdt, G. 1990. Sambia nosebleeding rites and male proximity to women. In: J. W. Stigler, R. A. Shweder, and G. Herdt (eds.), *Cultural Psychology: Essays on Comparative Human Development*, pp. 366–400. Cambridge: Cambridge University Press.

Hewes, G. W. 1961. Food transport and the origin of hominid bipedalism. *American Anthropologist* 63:687–710.

Hewes, G. W. 1973a. An explicit formulation of the relationship between tool-using, tool-making and the emergence of language. *Visible Language* 7:101–127.

Hewes, G. W. 1973b. Primate communication and the gestural origin of language. *Current Anthropology* 14:5–24.

Hewes, G. W. 1976a. The current status of the gestural theory of language origin. In: S. F. Harnard, H. D. Steklis, and J. Lancaster (eds.), *Annals of the New York Academy of Sciences: Origins and Evolution of Language and Speech* 280:482–504.

Hewes, G. W. 1976b. Visual learning, thinking, and communication in human biosocial learning. In: B. S. Randhawa and W. W. Coffman (eds.), *Visual Learning, Thinking, and Communication*. New York: Academic Press.

Hewes, G. W. 1977. A model for language evolution. *Sign Language Studies* 15:97–168.

Hewes, G. W. 1978. The phylogeny of sign language. In: I. M. Schlesinger and L. Namir (eds.), *Sign Language of the Deaf: Psychological, Linguistic, and Sociological Perspectives*, pp. 11–56. New York: Academic Press.

Hewes, G. W. 1989. Comment on "The archaeology of perception." *Current Anthropology* 30:145–146.

References

Hiatt, L. 1968. Gidjingali marriage arrangements. In: R. B. Lee and I. DeVore (eds.), *Man the Hunter*, pp. 165–175. New York: Aldine.

Hinde, R. A. 1976. Interactions, relationships and social structure. *Man* (N.S.) 11:1–17.

Hinde, R. A. 1991. A biologist looks at anthropology. *Man* (N.S.) 26:583–608.

Hiraiwa, M. 1975. Pebble-collecting behavior by juvenile Japanese monkeys. *Monkey* 19(5-6):24–25. (in Japanese)

Hitch, G. J., M. S. Halliday, A. M. Schaafstal, and T. M. Heffernan. 1991. Speech, "inner speech," and the development of short-term memory: Effects of picture-labeling on recall. *Journal of Experimental Child Psychology* 51:220–234.

Ho, M.-W. 1988. Genetic fitness and natural selection: Myth and metaphor. In: G. Greenberg and E. Tobach (eds.), *Evolution of Social Behavior and Integrative Levels*, pp. 85–111. Hillsdale, NJ: Erlbaum.

Hockett, C. F. 1959. Animal "languages" and human language. In: J. N. Spuhler (ed.), *The Evolution of Man's Capacity for Culture*. Detroit: Wayne State University Press.

Hockett, C. F. 1960a. Logical considerations in the study of animal communication. In: W. Lanyon and W. Tavolga (eds.), *Animal Sounds and Communication*. Washington, DC: American Institute of Biological Sciences.

Hockett, C. F. 1960b. The origin of speech. *Scientific American* 103:88–111.

Hockett, C. F. 1963. The problem of universals in language. In: J. H. Greenberg (ed.), *Universals of Language*, pp. 1–29. Cambridge: MIT Press.

Hockett, C. F., and S. A. Altmann. 1968. A note on design features. In: T. A. Sebeok (ed.), *Animal Communication*, pp. 61–72. Bloomington: Indiana University Press.

Hoffman, M. L. 1977. Empathy, its development and prosocial implications. *Nebraska Symposium on Motivation* 25:169–218.

Hoffmeister, R., and D. Moores. 1973. *The Acquisition of Specific Reference in the Linguistic System of a Deaf Child of Deaf Parents*. Washington, DC: U.S. Office of Education.

Hoffmeister, R., D. Moores, and R. Ellenberger. 1975. Some procedural guidelines for the study of the acquisition of sign language. *Sign Language Studies*, 7:121.

Holland, P. C. 1990. Event representation in Pavlovian conditioning: Image and action. *Cognition* 37:105–131.

Holloway, R. L. 1969. Culture, a *human* domain. *Current Anthropology* 10:395–412.

Holloway, R. L. (ed.). 1974. *Primate Aggression, Territoriality and Xenophobia*. New York: Academic Press.

Holloway, R. L. 1981. Culture, symbols, and human brain evolution: A synthesis. *Dialectical Anthropology* 5:287–303.

Horn, A. D. 1980. Some observations on the ecology of the bonobo chimpanzee (*Pan paniscus* Schwarz, 1929) near Lake Tumba, Zaire. *Folia Primatologica* 34:145–169.

Horr, D. A. 1977. Orang-utan maturation: Growing up in a female world. In: S. Chevalier–Skolnikoff and F. E. Poirier (eds.), *Primate Bio-social Development: Biological, Social, and Ecological Determinants*, pp. 289–322. New York: Garland.

Howells, W. 1959. *Mankind in the Making: The Story of Human Evolution*. Garden City, NY: Doubleday.

Hrdy, S. B. 1977. *The Langurs of Abu: Female and Male Strategies of Reproduction*. Cambridge: Harvard University Press.

Huffman, M. A. 1984. Stone-play of *Macaca fuscata* in Arashiyama B troop: transmission of a non-adaptive behavior. *Journal of Human Evolution* 13:725–735.

Huffman, M.A. and D. Quiatt. 1986. Stone handling by Japanese macaques (*Macaca fuscata*): Implications for tool use of stone. *Primates* 27(4):413–423.

References

Huffman, T. N. 1983. The trance hypothesis and the rock art of Zimbabwe. *South African Archaeological Society Goodwin Series* 4:49–53.

Humphrey, N. 1976. The social function of intellect. In: P. Bateson and R. Hinde (eds.), *Growing Points in Ethology*, pp. 303–317. Cambridge: Cambridge University Press.

Hunt, K. D. 1991. Mechanical implications of chimpanzee positional behavior. *American Journal of Physical Anthropology* 86:521–536.

Hunt, K. D. 1992. Positional behavior of *Pan troglodytes* in the Mahale Mountains and Gombe Stream National Parks, Tanzania. *American Journal of Physical Anthropology* 87:83–105.

Hutchinson, G. E. 1959. A speculative consideration of certain possible forms of sexual selection in man. *American Naturalist* 92:81–91.

Imanishi, K. 1952. Evolution of humanity. In: K. Imanishi (Ed), *Man*, pp. 36–44. Tokyo: Mainichi-Shinbunsha.

Ingold, T. 1986. *The Appropriation of Nature: Essays on Human Ecology and Social Relations.* Manchester: Manchester University Press.

Ingold, T. 1988. Tools, minds and machines: An excursion in the philosophy of technology. *Techniques et culture* 12:151–176.

Ingold, T. 1990a. An anthropologist looks at biology. *Man* (N.S.) 25:208–229.

Ingold, T. 1990b. Society, nature and the concept of technology. *Archaeological Review from Cambridge* 9:5–17.

Ingold, T. 1993a. Technology, language, intelligence: A reconsideration of basic concepts. In: K. R. Gibson and T. Ingold (eds.), *Tools, Language and Cognition in Human Evolution.* Cambridge: Cambridge University Press.

Ingold, T. 1993b. Tool-use, sociality and intelligence. In: K. R. Gibson and T. Ingold, *Tools, Language and Cognition in Human Evolution.* Cambridge: Cambridge University Press.

Inobe, H. 1990. Interspecific interaction between wild pygmy chimpanzees (*Pan paniscus*) and red colobus (*Colobus badius*). *Primates* 31(1):109–112.

Irons, W. 1979. Natural selection, adaptation, and human social behavior. In: N. A. Chagnon and W. Irons (eds.), *Evolutionary Biology and Human Social Behavior: An Anthropological Perspective*, pp. 4–39. North Scituate, MA: Duxbury Press.

Isaac, G. L. 1976. Stages of cultural elaboration in the Pleistocene: Possible archaeological indicators of the development of language capabilities. In: S. Harnad, H. Steklis, and J. Lancaster (eds.), *Annals of the New York Academy of Sciences: Origins and Evolution of Language and Speech* 280:275–288.

Isaac, G. L. 1978a. Food sharing and human evolution: Archaeological evidence from the plio-Pleistocene of East Africa. *Journal of Anthropological Research* 34:311–325.

Isaac, G. L. 1978b. The food-sharing behavior of protohuman hominids. *Scientific American* 238(4):90–108.

Ishiguro, H. 1980. The primitiveness of the concept of a person. In: Z. van Straaten (ed.), *Philosophical Subjects: Essays presented to P. F. Strawson*, pp. 62–75. Oxford: Clarendon Press.

Itani, J. 1958. On the acquisition and propagation of a new food habit in the troop of Japanese monkeys at Takasakiyama. *Primates* 1–2:84–98.

Itani, J. 1974. Communication systems among primates. *Annual Review of Social Psychology.* Japan 15:31–54.

Itani, J. 1985. The evolution of primate social structures. *Man* (N.S.) 20:593–611.

Itani, J., and N. Hazawa. 1953. Development and mother-baby relationship of Japanese monkeys. *Physiology and Ecology* 5(3–4):42–51.

Itani, J., and A. Nishimura. 1973. The study of infrahuman culture in Japan. In: E. Menzel (ed.), *Symposia of the Fourth Congress of the International Primatological Society*, Vol. 1, *Precultural Primate Behavior*. Basel: Karger.

Itani, J., and A. Nishimura. 1975. The study of infrahuman culture in Japan. A review. In Menzel, E. W., and S. Brook (eds.), *Symposia of the Fourth International Congress of Primatology*, Vol. 1, pp. 26–50. Tokyo:

Itani, J., and A. Suzuki. 1967. The social unit of wild chimpanzees. *Primates* 8:355–381.

Izawa, K. 1979. Foods and feeding behavior of wild black-capped capuchin (*Cebus apella*). *Primates* 21:57–76.

Jakobson, R. 1960/1990. Why "mama" and "papa"? In: L. R. Waugh and M. Monville-Burston (eds.), *On Language*, pp. 305–311. Cambridge: Harvard University Press.

Jakobson, R. 1966/1990. Quest for the essence of language. In: L. R. Waugh and M. Monville-Burston (eds.), *On Language*, pp. 407–421. Cambridge: Harvard University Press.

Jakobson, R., and L. Waugh. 1979. *The Sound Shape of Language*. Bloomington: Indiana University Press.

Jarvis, M. J., and G. Ettlinger. 1977. Cross-modal recognition in chimpanzees and monkeys. *Neuropsychologia* 15:499–506.

Jay, P. 1965. The common langur of north India. In: I. DeVore (ed.), *Primate Behavior: Field Studies of Monkeys and Apes*, pp. 197–249. New York: Holt, Rinehart and Winston.

Jerison, H. J. 1976. Discussion paper: the paleoneurology of language. In: S. Harnad, H. Steklis, and J. Lancaster (eds.), *Annals of the New York Academy of Sciences: Origins and Evolution of Language and Speech* 280.

Johanson, D., and T. White. 1979. A systematic assessment of early African hominids. *Science* 203:321–330.

Jolly, A. 1966. Lemur social behavior and primate intelligence. *Science* 153:501–506.

Jolly, A. 1985. *The Evolution of Primate Behavior*. 2nd ed. New York: Macmillan.

Jolly, A. 1988. The evolution of purpose. In: R. Byrne and A. Whiten (eds.), *Machiavellian Intelligence: Social Expertise and the Evolution of Intellect in Monkeys, Apes and Humans*, pp. 363–378. Oxford: Clarendon Press.

Jurmain, R. 1989. Trauma, degenerative disease, and other pathologies among the Gombe chimpanzees. *American Journal of Physical Anthropology* 80:229–237.

Kano, T. 1971. The chimpanzee of Filabanga, western Tanzania. *Primates* 12:229–246.

Kano, T. 1979. A pilot study on the ecology of pygmy chimpanzees. In: D. A. Hamburg and E. R. McCown (eds.), *The Great Apes*, pp. 123–135. Menlo Park, CA: Benjamin/Cummings.

Kano, T. 1980. Social behavior of wild pygmy chimpanzees (*Pan paniscus*) of Wamba: A preliminary report. *Journal of Human Evolution* 9:243–260.

Kano, T. 1982a. The social group of pygmy chimpanzees (*Pan paniscus*) of Wamba. *Primates* 23:171–189.

Kano, T. 1982b. The use of leafy twigs for rain cover by the pygmy chimpanzees of Wamba. *Primates* 23:453–457.

Kano, T. 1983. An ecological study of the pygmy chimpanzees (*Pan paniscus*) of Yalosidi, Republic of Zaire. *International Journal of Primatology* 4:1–31.

Kano, T. 1984. Feeding ecology of the pygmy chimpanzee (*Pan paniscus*) of Wamba. In: R. Susman (ed.), *The Pygmy Chimpanzee: Evolutionary Biology and Behavior*, pp. 233–274. New York: Plenum.

Kano, T. 1987. Social organization of the pygmy chimpanzee and the common chimpanzee: Similarities and differences. In: S. Kawano, J. H. Connell, and T. Hidaka (eds.), *Evolution and Coadaptation in Biotic Communities*. Tokyo: University of Tokyo Press.

Kano, T. 1989. The sexual behavior of pygmy chimpanzees. In: P. G. Heltne and L. A. Marquardt (eds.), *Understanding Chimpanzees*, pp. 176–183. Cambridge: Harvard University Press.

Kantor, R. 1982. Communication interaction: Mother modification and child acquisition of American Sign Language. *Sign Language Studies* 36:233–282.

Kaplan, B. 1979. Symbolism: From the body to the soul. In: N. R. Smith and M. B. Franklin (eds.), *Symbolic Functioning in Childhood*, pp. 219–228. Hillsdale, NJ: Erlbaum.

Kaplan, J. R. 1977. Patterns of fight interference in free-ranging rhesus monkeys. *American Journal of Physical Anthropology* 47:279–288.

Kaplan, J. R. 1978. Fight interference and altruism in rhesus monkeys. *American Journal of Physical Anthropology* 49:449–456.

Katz, D. 1937/1953. *Animals and Men*. Melbourne: Penguin Books.

Kawai, M. 1958 (English trans. 1965). On the system of social ranks in a natural troop of Japanese monkeys. In: S. Altmann (ed.), *Japanese Monkeys*. Alberta: University of Alberta Press.

Kawai, M. 1965. Newly acquired pre-cultural behavior of the natural troop of Japanese monkeys on Koshima Islet. *Primates* 1:1–30.

Kawamura, S. 1959. The process of subcultural propagation among Japanese macaques. *Primates* 2:43–60.

Keeley, L. H., and N. Toth. 1981. Microwear polishes on early stone tools from Koobi Fora, Kenya. *Nature* 293:464–465.

Keller, H. 1902/1965. *The Story of My Life*. Clinton, MA: Airmont Books.

Kendon, A. 1975. Gesticulation, speech and the gesture theory of language origins. *Sign Language Studies* 9:349–373.

Kendon, A. 1980. Gesticulation and speech: Two aspects of the process of utterance. In: M. R. Key (ed.), *The Relationship of Verbal and Nonverbal Communication*, pp. 207–227. The Hague: Mouton.

Kendon, A. 1991. Some considerations for a theory of language origins. *Man* 26:199–221.

Kennedy, G. E. 1980. *Paleoanthropology*. New York: McGraw-Hill.

Kien, J. 1991. The need for data reduction may have paved the way for the evolution of language ability in hominids. *Journal of Human Evolution* 20:157–165.

Killen, M., and I. C. Uzgiris. 1981. Imitation of actions with objects: The role of social meaning. *Journal of Genetic Psychology* 138:219–229.

Kimball, W. H. 1991. Species, species concepts and hominid evolution (review of *A Theory of Human and Primate Evolution* by Colin Groves). *Journal of Human Evolution* 20:355–371.

Kitahara-Frisch, J. 1978. Stone tools as indicators of linguistic abilities in early man. *Annals of the Japan Association for Philosophy of Science* 5:101–109.

Kitamura, K. 1983. Pygmy chimpanzee association patterns in ranging. *Primates* 24:1–12.

Klein, N. 1978. Review of Nance, John, *The Gentle Tasaday: A Stone Age People in the Philippine Rain Forest. Journal of American Folklore* 91:976–977.

Klein, R. G. 1989. Biological and behavioral perspectives on modern human origins in southern Africa. In: P. Mellars and C. Stringer (eds.), *The Human Revolution: Behavioral and Biological Perspectives on the Origins of Modern Humans*, pp. 529–546. Edinburgh: Edinburgh University Press.

References

Klima, E., and U. Bellugi. 1979. *The Signs of Language*. Cambridge: Harvard University Press.

Kline, M. 1980. *Mathematics: The Loss of Certainty*. Oxford: Oxford University Press.

Knowlton, J. Q. 1966. On the definition of "picture." *Audio-visual Communication Review* 14:157–182.

Köhler, W. 1925/1959. *The Mentality of Apes*. 2nd ed. New York: Viking.

Kohn, A. J. 1989. Natural history and the necessity of the organism. *American Zoologist* 29:1095–1103.

Kohts, N. 1921. *Report of the Zoopsychological Laboratory of the Darwinian Museum in Moscow* (in Russian). Moscow: Darwinian Museum.

Kohts, N. 1923. *Untersuchungen über die Erkenntnisfähigkeiten des Schimpansen. Aus dem Zoopsychologischen Laboratorium des Museum Darwinianum in Moskau* (in Russian with a German translation of the summary). Moscow: Darwinian Museum.

Kortlandt, A. 1986. The use of stone tools by wild chimpanzees and earliest hominids. *Journal of Human Evolution* 15:77–132.

Kortlandt, A., and E. Holzhaus. 1987. New data on the use of stone tools by chimpanzees in Guinea and Liberia. *Primates* 28(94):473–496.

Kortlandt, A., and M. Kooij. 1963. Protohominid behavior in primates (preliminary communication). *Symposium of the Zoological Society, London* 10:61–88.

Krafft-Ebing, R. von. 1893/1978. *Psychopathia Sexualis*. New York: Stein and Day.

Kraisri, N. 1963. The Mrabri language. *Journal of the Siam Society* 51(2):179–183.

Krebs, J. R., and R. Dawkins. 1984. Animal signals and manipulation. In: J. R. Krebs and N. B. Davies (eds.), *Behavioural Ecology: An Evolutionary Approach*, pp. 380–402. Oxford: Blackwell.

Kroeber, A. L. 1928. Sub-human culture beginnings. *Quarterly Review of Biology* 3:325–342.

Kroeber, A. L., and C. Kluckhohn. 1952. *Culture: A Critical Review of Concepts and Definitions*. Papers of the Peabody Museum of American Archaeology and Ethnology, Harvard University, 47, 1.

Kurland, J. A. 1977. *Kin Selection in the Japanese Monkey*. Contributions to Primatology 12. Basel: Karger.

Kuroda, S. 1979. Grouping of the pygmy chimpanzees. *Primates* 20:161–183.

Kuroda, S. 1980. Social behavior of the pygmy chimpanzees. *Primates* 21:181–197.

Kuroda, S. 1984. Interaction over food among pygmy chimpanzees. In: R. Susman (ed.), *The Pygmy Chimpanzee: Evolutionary Biology and Behavior*, pp. 233–274. New York: Plenum.

Laidler, K. 1978. Language in the orang-utan. In: A. Lock (ed.), *Action, Gesture and Symbol*, pp. 133–155. New York: Academic Press.

Laitman, J. 1983. The evolution of the hominid upper respiratory system and implications for the origins of speech. In: E. de Grolier, E. (ed.), *Glossogenetics: The Origin and Evolution of Language*, pp. 63–90. New York: Harwood Academic.

Laitman, J. 1984. The anatomy of human speech. *Natural History* 93:20–27.

Laitman, J. 1985. Evolution of the hominid upper respiratory tract: The fossil evidence. In: P. V. Tobias (ed.), *Hominid Evolution: Past, Present and Future*, pp. 281–286. New York: Alan R. Liss.

Laitman, J., R. Heimbuch, and E. Crelin. 1978. Developmental change in a basicranial line and its relationship to the upper respiratory system in living primates. *American Journal of Anatomy* 152:467–483.

Lamendella, J. T. 1980. Neurofunctional foundations of symbolic communication. In: M.L.C. Foster and S. H. Brandes (eds.), *Symbol as Sense: New Approaches to the Analysis of Meaning*, pp. 147–174. New York: Academic Press.

Lancaster, J. 1968. Primate communication systems and the emergence of human language. In: P. Jay (ed.), *Primates: Studies in Adaptation and Variability*. New York: Holt, Rinehart and Winston.

Lancaster, J., and C. Lancaster. 1983. Parental investment: The hominid adaptation. In: D. J. Ortner (ed.), *How Humans Adapt: A Biocultural Odyssey*. Washington, DC: Smithsonian Institution Press.

Landry, M. O., and K. Lyons–Ruth. 1980. Recursive structure in cognitive perspective taking. *Child Development* 51:386–394.

Langer, J. 1986. *The Origins of Logic: One to Two Years*. Orlando: Academic Press.

Langer, S. K. 1942. *Philosophy in a New Key*. New York: Mentor.

Langer, S. K. 1967. *Mind: An Essay on Human Feeling*. Baltimore: Johns Hopkins University Press.

Lave, J. 1990. The culture of acquisition and the practice of understanding. In: J. Stigler, R. A. Shweder, and G. Herdt (eds.), *Cultural Psychology: Essays on Comparative Human Development*, pp. 309–327. Cambridge: Cambridge University Press.

Leach, E. R. 1990. The social anthropology of marriage and mating. In: V. Reynolds (ed.), *Mating and Marriage*, pp. 91–110. Oxford: Oxford University Press.

Leakey, M. D. 1971. *Olduvai Gorge: Excavations in Beds I and II, 1960–1963*. Cambridge: Cambridge University Press.

Leakey, R., and M. Leakey. 1986. A new Miocene hominoid from Kenya. *Nature* 324:143–146.

Leakey, R., and R. Lewin. 1977. *Origins*. New York: E. P. Dutton.

Leask, J., R. N. Haber, and R. B. Haber. 1969. Eidetic imagery in children: II. Longitudinal and experimental results. *Psychonomic Monograph Supplements* 3:25–48.

Lebar, F. M., G. C. Hickey, and J. Musgrave. 1964. *Ethnic Groups of Mainland Southeast Asia*. New Haven, CT: Human Relations Area Files.

Lee, P. C. 1991. Adaptations to environmental change: An evolutionary perspective. In: H. O. Box (ed.), *Primate Response to Environmental Change*, pp. 39–56. New York: Chapman and Hall.

LeGros Clark, W. E. 1959. *The Antecedents of Man*. Edinburgh: Edinburgh University Press.

Leigh, S. R. 1988. Comparisons of rates of evolutionary change in cranial capacity in *Homo erectus* and early *Homo sapiens*. *American Journal of Physical Anthropology* 75:237–238.

LeMay, J. 1975. The language capability of Neanderthal man. *American Journal of Physical Anthropology* 42:9–14.

Lenneberg, E. H. 1967. *Biological Foundations of Language*. New York: John Wiley and Sons.

Leslie, A. M. 1988. The necessity of illusion: Perception and thought in infancy. In: L. Weiskrantz (ed.), *Thought Without Language*, pp. 185–210. Oxford: Clarendon Press.

Leung, E.H.L., and H. L. Rheingold. 1981. Development of pointing as a social gesture. *Developmental Psychology* 17:215–220.

Lévi–Strauss, C. 1949. *Les Structures élémentaires de la parente*. Paris: Presses Universitaires de France.

Levinson, J. 1990. *Music, Art, and Metaphysics*. Ithaca, NY: Cornell University Press.

Lewis, D. K. 1969. *Convention: A Philosophical Study*. Cambridge: Harvard University Press.

Lewis, M., and J. Brooks-Gunn. 1979. *Social Cognition and the Acquisition of Self*. New York: Plenum.

References

Lewis, M., M. Sullivan, C. Stanger, and M. Weiss. 1989. Self development and self–conscious emotions. *Child Development* 60:146–156.

Lewis, M. M. 1936/1951. *Infant Speech: A Study of the Beginnings of Language*. London: Routledge and Kegan Paul.

Lewis, O. J. 1989. *Functional Morphology of the Evolving Hand and Foot*. Oxford: Clarendon Press.

Lewis-Williams, J. D. 1981. *Believing and Seeing: Symbolic Meanings in Southern San Rock Paintings*. London: Academic Press.

Lewis-Williams, J. D. 1983. Introductory essay: Science and rock art. *South African Archaeological Society Goodwin Series* 4:3–13.

Lewis-Williams, J. D. 1986. Cognitive and optical illusions in San rock art research. *Current Anthropology* 27:171–178.

Liben, L. S., and R. M. Downs. 1989. Understanding maps as symbols: The development of map concepts in children. *Advances in Child Development and Behavior* 22:145–201.

Liberman, A. M. 1979. An ethological approach to language through the study of speech perception. In: M. von Cranach, K. Foppa, W. Lepenies, and D. Ploog (eds.), *Human Ethology*, pp. 682–704. Cambridge: Cambridge University Press.

Lieberman, D., D. Pilbeam, and B. Wood. 1988. A probabilistic approach to the problem of sexual dimorphism in *Homo habilis*: A comparison of KNM-ER-1740 and KNM-ER-1813. *Journal of Human Evolution* 17:503–511.

Lieberman, P. 1975. *On the Origins of Language*. New York: Macmillan.

Lieberman, P. 1983. On the nature and evolution of the biological bases of language. In: E. de Grolier (ed.), *Glossogenetics: The Origin and Evolution of Language*, pp. 91–114. New York: Harwood Academics.

Lieberman, P. 1984. *The Biology and Evolution of Language*. Cambridge: Harvard University Press.

Lieberman, P. 1985a. Comment on S. T. Parker, "A social-technological model for the evolution of language." *Current Anthropology* 26:628.

Lieberman, P. 1985b. On the evolution of human syntactic ability: Its pre-adaptive bases — motor control and speech. *Journal of Human Evolution* 14:657–668.

Lieberman, P. 1989. The origins of some aspects of human language and cognition. In: P. Mellars and C. Stringer (eds.), *The Human Revolution: Behavioral and Biological Perspectives on the Origins of Modern Humans*, pp. 391–414. Edinburgh: Edinburgh University Press.

Lieberman, P. 1992. On Neanderthal speech and Neanderthal extinction. *Current Anthropology* 33:409–410.

Lieberman, P., and E. S. Crelin. 1971. On the speech of Neanderthal man. *Linguistic Inquiry* 2:203–222.

Lieberman, P., E. S. Crelin, and D. H. Klatt. 1972. Phonetic ability and related anatomy of the newborn and adult human, Neandertal man, and the chimpanzee. *American Anthropologist* 74:287–307.

Liem, K. F. 1990. Key evolutionary innovations, differential diversity, and symecomorphosis. In: M. H. Nitecki (ed.), *Evolutionary Innovations*, pp. 147–170. Chicago: University of Chicago Press.

Limber, J. 1977. Language in child and chimp? *American Psychologist* 32:280–295.

Linden, E. 1981. *Apes, Men, and Language*. New York: Penguin Books.

Lindley, J. M., and G. A. Clark. 1990. Symbolism and modern human origins. *Current Anthropology* 31:233–261.

References

Liska, J. 1986. Symbols: The missing link? In: J. Else and P. Lee (eds.), *Primate Ontogeny, Cognition, and Social Behaviour*, pp. 169–178. Cambridge: Cambridge University Press.

Liska, J. 1987. Variations in the arbitrariness of ASL: An assessment of the symbolicity of simial signs. *Human Evolution* 2:205–212.

Liska, J. 1988. Dominance-seeking strategies in primates: An evolutionary perspective. Paper presented at the Twelfth Congress of the International Primatological Society, Brasília.

Lock, A. 1978. *Action, Gesture and Symbol: The Emergence of Language*. London: Academic Press.

Lock, A. 1980. *The Guided Re-invention of Language*. London: Academic Press.

Lorenz, K. 1957. The past twelve years in the comparative study of behavior. In: C. H. Schiller (ed.), *Instinctive Behavior*, pp. 288–310. New York: International Universities Press.

Lovejoy, C. O. 1981. The origin of man. *Science* 211:341–350.

Lovejoy, C. O. 1988. Evolution of human walking. *Scientific American* November, 118–125.

Low, B. S. 1979. Sexual selection and human ornamentation. In: N. A. Chagnon and W. Irons (eds.), *Evolutionary Biology and Human Social Behavior: An Anthropological Perspective*, pp. 462–487. North Scituate, MA: Duxbury Press.

Lumsden, C. J., and E. D. Wilson. 1981. *Genes, Mind, and Culture*. Cambridge: Harvard University Press.

Lynch, F. 1984. The Tasaday: Cave-dwelling food gatherers of South Cotabato, Mindanao. In A. A. Yengoyan and P. Makil (eds.), *Philippine Society and the Individual: Selected Essays of Frank Lynch, 1949–1976*. Ann Arbor: University of Michigan Center for South and Southeast Asian Studies.

Lyons, J. 1981. *Language and Linguistics*. Cambridge: Cambridge University Press.

Macdonald, D. 1960. *Parodies: An Anthology from Chaucer to Beerbohm — and After*. New York: Da Capo.

MacLeish, K. 1972. Stone Age cavemen of Mindanao. *National Geographic Magazine*, August, pp. 219–246.

MacMahon, J. A., D. L. Phillips, J. V. Robinson, and D. J. Schimpf. 1978. Levels of biological organization: An organism-centered approach. *BioScience* 28:700–704.

MacNamara, J. 1972. The cognitive basis of language learning in infants. *Psychological Review* 79:1–13.

Maestas y Moores, J. 1980. Early linguistic environment: Interactions of deaf parents with their infants. *Sign Language Studies* 26:1–13.

Magnusson, M., and H. Palsson. 1960. *Njal's Saga*. Harmondsworth: Penguin.

Mair, L. 1972. *An Introduction to Social Anthropology*. 2nd ed. Oxford: Oxford University Press.

Malenky, R. K., and E. W. Stiles. 1991. Distribution of terrestrial herbaceous vegetation and its consumption by *Pan paniscus* in the Lomako Forest, Zaire. *American Journal of Primatology* 23:153–169.

Malmi, W. 1976. Chimpanzees and language evolution. In: S. Harnard, H. Steklis, and J. Lancaster (eds.), *Annals of the New York Academy of Sciences: Origins and Evolution of Language and Speech* 280:598–603.

Mandel, B., and R. Fouts. 1976. Human-chimpanzee conversation in a social setting. Paper presented at the annual meeting of the American Sociological Association in New York, N.Y.

Mandler, J. M. 1983. Representation. In: P. H. Mussen (ed.), *Handbook of Child Psychology*, Vol. 3, *Cognitive Development*, pp. 420–494. New York: Wiley.

Maple, T. 1980. *Orang-utan Behavior*. New York: Van Nostrand Reinhold.

Marchant, L., and W. C. McGrew. 1991. Laterality of function in apes: A meta-analysis of methods. *Journal of Human Evolution* 21:425–438.

Margolis, J. 1978. Culture and technology. *Research in Philosophy and Technology* 1:23–37.

Marks, L. E., R. J. Hammeal, and M. H. Bornstein. 1987. *Perceiving Similarity and Compre-hending Metaphor*. Monographs of the Society for Research in Child Development 52(1, serial no. 215).

Marschark, M. 1985. Metaphoric competence in cognitive and language development. *Advances in Child Development and Behavior* 19:49–81.

Marshack, A. 1976. Some implications of the Paleolithic symbolic evidence for the origin of language. *Current Anthropology* 17:274–282.

Marshack, A. 1991. *The Roots of Civilization*. Rev. ed. Mount Kisco, NY: Moyer Bell.

Marshack, A. 1979a. Comment on S. T. Parker and K. R. Gibson, "A developmental model for the evolution of language and intelligence in early hominids." *Behavioral and Brain Sciences* 2:394–396.

Marshack, A. 1979b. European Upper Paleolithic–Mesolithic symbolic continuity: A cognitive, comparative study of ritual marking. In: M. A. Beltran et al. (eds.), *International Symposium on the Intellectual Expressions of Prehistoric Man: Art and Religion; Valcamonica 7/28–8/3/79*. Milan: Centro di Studi Preistorici.

Marten, K. 1991. Self-recognition in the bottlenosed dolphin. Paper presented at symposium, "Cognitive correlates of self-awareness and self-recognition in monkeys, apes, and humans," Sonoma State University, Rohnert Park, California.

Martin, R. D. 1990. *Primate Origins and Evolution: A Phylogenetic Reconstruction*. Lawrenceville, NJ: Princeton University Press.

Masataka, N., and K. Fujita. 1989. Vocal learning of Japanese and rhesus monkeys. *Behaviour* 109:191–199.

Matsuzawa, T. 1990. Spontaneous sorting in human and chimpanzee. In: S. Parker and K. Gibson (eds.), *"Language" and Intelligence in Monkeys and Apes: Comparative Developmental Perspectives*, pp. 451–468. New York: Cambridge University Press.

Mawby, R., and R. W. Mitchell. 1986. Feints and ruses: An analysis of deception in sports. In: R. W. Mitchell and N. S. Thompson (eds.), *Deception: Perspectives on Human and Nonhuman Deceit*, pp. 313–322. Albany: State University of New York Press.

Maxwell, M. M. 1983. Language acquisition in a deaf child of deaf parents: Speech, sign variations, and print variations. In: E. K. Nelson (ed.), *Children's Language*, Vol. 4, pp. 283–313. Hillsdale, NJ: Erlbaum.

Maynard Smith, J. 1988. Sewall Wright (1889–1988). *Nature* 332–492.

Mayr, E. 1982. *The Growth of Biological Thought: Diversity, Evolution and Inheritance*. Cambridge: Harvard University Press.

Mayr, E. 1988. *Toward a New Philosophy of Biology: Observations of an Evolutionist*. Cambridge: Harvard University Press.

McArthur, D. 1991. Language as technology and skill. *Deafness and Development* 2:15–20.

McCabe, M. A., and I. C. Uzgiris. 1983. Effects of model and action on imitation in infancy. *Merrill-Palmer Quarterly* 29:69–82.

McCune–Nicolich, L. 1981. Toward symbolic functioning: Structure of early pretend games and potential parallels with language. *Child Development* 52:785–797.

McDowell, J. 1980. Meaning, communication, and knowledge. In: Z. van Straaten (ed.), *Philosophical Subjects: Essays Presented to P. F. Strawson*, pp. 117–139. Oxford: Clarendon Press.

References

McFarland, D. (ed.). 1987. *The Oxford Companion to Animal Behavior*. Oxford: Oxford University Press.

McGrew, W. C. 1974. Tool use by wild chimpanzees in feeding upon driver ants. *Journal of Human Evolution* 3:501–508.

McGrew, W. C. 1987. Tools to get food: The subsistence of Tasmanian aborigines and Tanzanian chimpanzees compared. *Journal of Anthropological Research* 43(3):247–258.

McGrew, W. C. 1992. *Chimpanzee Material Culture: Implications for Human Evolution*. Cambridge: Cambridge University Press.

McGrew, W. C., and L. Marchant. 1992. Chimpanzees, tools, and termites: Hand preference or handedness? *Current Anthropology* 33:114–119.

McGrew, W. C., C.E.G. Tutin, and P. J. Baldwin. 1978. Evidence for a social custom in wild chimpanzees. *Man* 13:234–251.

McGrew, W. C., C.E.G. Tutin, and P. J. Baldwin. 1979. Chimpanzees, tools, and termites: Cross–cultural comparisons of Senegal, Tanzania, and Rio Muni. *Man* (N.S.) 14:185–214.

McMahan, P. 1978. Natural history of the coyote. In: R. L. Hall and H. S. Sharp (eds.), *Wolf and Man: Evolution in Parallel*, pp. 41–54. New York: Academic Press.

McMillan, C., and C. Duggleby. 1981. Interlineage genetic differentiation among rhesus macaques on Cayo Santiago. *American Journal of Physical Anthropology* 56:305–312.

McNeill, D. 1979. Language origins. In: M. von Cranach, K. Foppa, W. Lepenies, and D. Ploog (eds.), *Human Ethology*, pp. 715–728. Cambridge: Cambridge University Press.

McNeill, D. 1985. So you think gestures are nonverbal? *Psychological Review* 92:350–371.

Mead, G. H. 1934/1974. *Mind, Self, and Society from the Standpoint of a Social Behaviorist*. Chicago: University of Chicago Press.

Meddin, J. 1979. Chimpanzees, symbols, and the reflective self. *Social Psychology Quarterly* 42:99–109

Medin, D., and A. Ortony. 1989. Comment on Part I: Psychological essentialism. In: S. Vosniadou and A. Ortony (eds.), *Similarity and Analogical Reasoning*, pp. 183–195. Cambridge: Cambridge University Press.

Meier, R. P. 1991. Language acquisition by deaf children. *American Scientist* 79:60–70.

Meikle, D. B., and S. H. Vessey. 1981. Nepotism among rhesus monkey brothers. *Nature* 94:160–161.

Mellars, P. A. 1973. The character of the middle-upper Palaeolithic transition in southwest France. In: C. Renfrew (ed.), *The Explanation of Culture Change: Models in Prehistory*, pp. 255–276. Pittsburgh: University of Pittsburgh Press.

Mellars, P. A. 1989. Major issues in the emergence of modern humans. *Current Anthropology* 30(3):349–385.

Meltzoff, A. N. 1988. The human infant as *Homo imitans*. In: T. R. Zentall and B. G. Galef, Jr. (eds.), *Social Learning: Psychological and Biological Perspectives*, pp. 319–341. Hillsdale, NJ: Erlbaum.

Meltzoff, A. N. 1990. Foundations for developing a concept of self: The role of imitation in relating self to other and the value of social mirroring, social modeling, and self practice in infancy. In: D. Cicchetti and M. Beeghly (eds.), *The Self in Transition: Infancy to Childhood*, pp. 139–164. Chicago: University of Chicago Press.

Meltzoff, A. N., and K. Moore. 1989. Imitation in newborn infants: Exploring the range of gestures imitated and the underlying mechanisms. *Developmental Psychology* 25:954–962.

Menzel, E. W., Jr. 1974. A group of young chimpanzees in a one-acre field. In: A. M. Schrier and F. Stollnitz (eds.), *Behavior of Nonhuman Primates*, Vol. 5, pp. 83–153. New York: Academic Press.

Menzel, E. W., Jr., D. Premack, and G. Woodruff. 1978. Map reading by chimpanzees. *Folia Primatologica* 29:241–249.

Menzel, E. W., Jr., E. S. Savage-Rumbaugh, and J. Lawson. 1985. Chimpanzee (*Pan troglodytes*) problem solving with the use of mirrors and televised equivalents of mirrors. *Journal of Comparative Psychology* 99:211–217.

Merleau–Ponty, M. 1960/1982. The child's relations with others. In: J. M. Edie (ed.), *The Primacy of Perception*, pp. 96–155. Illinois: Northwestern University Press.

Meyersohn, R., and E. Katz. 1957. Notes on a natural history of fads. *American Journal of Sociology* 62:588–593.

Mignault, C. 1985. Transition between sensorimotor and symbolic activities in nursery-reared chimpanzees (*Pan troglodytes*). *Journal of Human Evolution* 14:747–758.

Miles, H. L. 1975. Tool use and language in early hominids. Paper presented at the annual meeting of the Northeastern Anthropological Association, Potsdam, New York.

Miles, H. L. 1976. The communicative competence of child and chimpanzee. In: S. Harnard, H. Steklis, and J. Lancaster (eds.), *Annals of the New York Academy of Sciences: Origins of Evolution of Language and Speech* 280:592–597.

Miles, H. L. 1978. Conversations with apes: The use of sign language with two chimpanzees. *Dissertation Abstracts International* 39:11A.

Miles, H. L. 1980. Acquisition of gestural signs by an infant orang-utan (*Pongo pygmaeus*). *American Journal of Physical Anthropology* 52:256–257.

Miles, H. L. 1983. Apes and language: The search for communicative competence. In: J. de Luce and H. Wilder (eds.), *Language in Primates: Implications for Linguistics, Anthropology, Psychology and Philosophy*, pp. 43–61. New York: Springer-Verlag.

Miles, H. L. 1986. How can I tell a lie? Apes, language, and the problem of deception. In: R. W. Mitchell and S. Thompson (eds.), *Deception: Perspectives on Human and Nonhuman Deceit*, pp. 245–266. Albany: State University of New York Press.

Miles, H. L. 1990. The cognitive foundations for reference in a signing orangutan. In: S. T. Parker and K. Gibson (eds.), *"Language" and Intelligence in Monkeys and Apes: Comparative Developmental Perspectives*, pp. 511–539. New York: Cambridge University Press.

Miles, H. L. 1991a. The development of symbolic communication in apes and early hominids. In: W. von Raffler-Engel, J. Wind, and A. Jonker (eds.), *Studies in Language Origins*, Vol. 2, pp. 9–20. Amsterdam: John Benjamins.

Miles, H. L. 1991b. ME CHANTEK: The development of self-awareness in a sign-using orangutan. Paper presented at symposium, "Cognitive correlates of self-awareness and self-recognition in monkeys, apes, and humans," Sonoma State University, Rohnert Park, California.

Miles, H. L. 1993. Language and the orangutan: The old "person" of the forest. In: P. Singer and P. Cavalieri (eds.), *A New Equality: The Great Ape Project*, pp. 42–57. London: Fourth Estate.

Miles, H. L., R. Mitchell, and S. Harper. 1992. Imitation and self-awareness in a signing orangutan. Paper presented at the Fourteenth Congress of the International Primatological Society, Strasbourg.

Millikan, R. 1984. *Language, Thought, and Other Biological Categories*. Cambridge: MIT Press.

Millikan, R. 1986. Thoughts without laws; cognitive science with content. *Philosophical Review* 95:47–80.

References

Millikan, R. 1989a. Biosemantics. *Journal of Philosophy* 86:281–297.

Millikan, R. 1989b. In defense of proper functions. *Philosophy of Science* 56:288–302.

Milton, K. 1988. Foraging behavior and evolution of primate intelligence. In: R. Byrne and A. Whiten (eds.), *Machiavellian Intelligence: Social Expertise and the Evolution of Intellect in Monkeys, Apes, and Humans*, pp. 285–305. Oxford: Clarendon Press.

Mitchell, R. W. 1986. A framework for discussing deception. In: R. W. Mitchell and N. S. Thompson (eds.), *Deception: Perspectives on Human and Nonhuman Deceit*, pp. 3–40. Albany: State University of New York Press.

Mitchell, R. W. 1987. A comparative-developmental approach to understanding imitation. In: P.P.G. Bateson and P. H. Klopfer (eds.), *Perspectives in Ethology*, Vol. 7, pp. 183–215. New York: Plenum.

Mitchell, R. W. 1988. Ontogeny, biography, and evidence for tactical deception. *Behavioral and Brain Sciences* 11:259–260.

Mitchell, R. W. 1990. A theory of play. In: M. Bekoff and D. Jamieson (eds.), *Interpretation and Explanation in the Study of Animal Behavior*, Vol. 1, *Interpretation, Intentionality, and Communication*, pp. 197–227. Boulder, CO: Westview Press.

Mitchell, R. W. 1991a. Bateson's concept of "metacommunication" in play. *New Ideas in Psychology* 9:73–87.

Mitchell, R. W. 1991b. Deception and hiding in captive lowland gorillas (*Gorilla gorilla gorilla*). *Primates* 32:523–527.

Mitchell, R. W. 1992. Developing concepts in infancy: Animals, self–perception, and two theories of mirror-self-recognition. *Psychological Inquiry* 3:127–130.

Mitchell, R. W. 1993a. Animals as liars: The human face of nonhuman deception. In: M. Lewis and C. Saarni (eds.), *Deception and Lying in Everyday Life*, pp. 59–89. New York: Guilford Press.

Mitchell, R. W. 1993b. Humans, nonhumans, and personhood. In: P. Singer and P. Cavalieri (eds.), *A New Equality: The Great Ape Project*, pp. 237–247. London: Fourth Estate.

Mitchell, R. W. 1993c. Mental models of mirror self-recognition: Two theories. *New Ideas in Psychology* 11.

Mitchell, R. W., and N. S. Thompson (eds.). 1986. *Deception: Perspectives on Human and Nonhuman Deceit*. Albany: State University of New York Press.

Mitchell, R. W., and N. S. Thompson. 1991. Projects, routines, and enticements in dog-human play. In: P.P.G. Bateson and P. H. Klopfer (eds.), *Perspectives in Ethology*, Vol. 9, pp. 189–216. New York: Plenum.

Mohnot, S. M. 1971. Some aspects of social changes and infant-killing in the Hanuman langur, *Presbytis entellus* in western India. *Mammalia* 35:175–198.

Montagu, A. 1976. Toolmaking, hunting, and the origin of language. In: S. Harnad, H. Steklis, and J. Lancaster (eds.), *Annals of the New York Academy of Sciences: Origins and Evolution of Language and Speech* 280:266–274.

Moore, J. 1992. Comment on J. M. Sept's "Was there no place like home?" *Current Anthropology* 33:197–198.

Morbeck, M. E. 1975. *Dryopithecus africanus* forelimb. *Journal of Human Evolution* 4:39–46.

Morbeck, M. E. 1979. Forelimb use and positional adaptation in *Colobus guereza*: Integration of behavioral, ecological, and anatomical data. In: M. E. Morbeck, H. Preuschoft, and N. Gomberg (eds.), *Environment, Behavior, and Morphology: Dynamic Interactions in Primates*, pp. 95–117. New York: Gustav Fischer.

Morbeck, M. E. 1983. Miocene hominoid discoveries from Radabánya: Implications from the postcranial skeleton. In: R. L. Ciochon and R. S. Corruccini (eds.), *New Interpretations of Ape and Human Ancestry*, pp. 369–404. New York: Plenum.

References

Morbeck, M. E. 1991a. Biology, behavior, and evolution. *Reviews in Anthropology* 20:113–123.

Morbeck, M. E. 1991b. Bones, gender, and life history. In: D. Walde and N. Willows (eds.), *The Archaeology of Gender*, pp. 39–45. Calgary: University of Calgary.

Morbeck, M. E. Forthcoming a. Life history, the individual and evolution. In: M. E. Morbeck and A. L. Zihlman (eds.), *Life History, Females, and Evolution*.

Morbeck, M.E. Forthcoming b. Reading life history in teeth, bones and fossils. In: M. E. Morbeck and A. L. Zihlman (eds.), *Life History, Females, and Evolution*.

Morbeck, M. E., and A. L. Zihlman. 1988. Body composition and limb proportions. In: J. Schwartz (ed.), *Orang–utan Biology*, pp. 285–297. Oxford: Oxford University Press.

Morbeck, M. E., and A. L. Zihlman. 1989. Body size and proportions in chimpanzees, with special reference to *Pan troglodytes schweinfurthii* from Gombe National Park, Tanzania. *Primates* 30:369–382.

Morbeck, M. E., A. L. Zihlman, D. R. Sumner, Jr., and A. Galloway. 1991. Poliomyelitis and skeletal asymmetry in Gombe Chimpanzees. *Primates* 32:77–91.

Morgan, C. L. 1900/1970. *Animal Behaviour*. New York: Johnson Reprint Company.

Mori, A. 1983. Comparisons of the communicative vocalizations and behaviors of group ranging in mountain gorillas, chimpanzees, and pygmy chimpanzees. *Primates* 24:486–500.

Mori, A. 1984. An ethological study of pygmy chimpanzees in Wamba, Zaire. *Primates* 25:255–278.

Morin, A., and J. Everett. 1990. Inner speech as a mediator of self-awareness, self-consciousness, and self-knowledge: An hypothesis. *New Ideas in Psychology* 8:337–356.

Moses, J. 1990. The Tasaday (letter). *Anthropology Today* 6(2):20.

Mounin, G. 1976. Language, communication, chimpanzees. *Current Anthropology* 17:1–7.

Munn, C. 1986. The deceptive use of alarm calls by sentinel species in mixed-species flocks of neotropical birds. In: R. W. Mitchell and N. S. Thompson (eds.), *Deception: Perspectives on Human and Nonhuman Deceit*, pp. 169–175. Albany: State University of New York Press.

Murdock, G. P. 1935. *Our Primitive Contemporaries*. New York: Macmillan.

Murdock, G. P. 1949. *Social Structure*. New York: Macmillan.

Murdock, G. P. 1959. Cross-language parallels in parental kin terms. *Anthropological Linguistics* 1(9):1–6.

Murdock, G. P., C. S. Ford, A. E. Hudson, R. Kennedy, L. W. Simmons, and J.W.M. Whiting. 1971. *Outline of World Cultures*. 4th ed. New Haven, CT: Human Relations Area Files.

Mussen, P. 1967. Early socialization: Learning and identification. *New Directions in Psychology* 3:51–110.

Mydans, S. 1986. The Tasaday revisited: A hoax or social change at work? *New York Times*, 3 May, p. C3.

Mydans, S. 1987. In Mindanao, tribe or 70's hoax? *New York Times*, 7 December, p. A13.

Mydans, S. 1988. Twentieth-century lawsuit asserts Stone–age identity. *New York Times*, 29 October, p. 1.

Myers, R. E. 1978. Comparative neurology of vocalization and speech: Proof of a dichotomy. In: S. L. Washburn and E. R. McCown (eds.), *Human Evolution: Biosocial Perspectives*, Vol. 4. Menlo Park, CA: Benjamin/Cummings.

Myers Thompson, J. Forthcoming. A cross-cultural comparison of captive populations of *Pan paniscus*. M. A. thesis. University of Colorado at Denver.

Nagel, T. 1969/1979. Sexual perversion. In: T. Nagel (ed.), *Mortal Questions*, pp. 39–52. Cambridge: Cambridge University Press.

Nance, J. 1975. *The Gentle Tasaday: A Stone Age People in the Philippine Rain Forest*. New York: Harcourt Brace Jovanovich.

Napier, J. R., and P. H. Napier. 1967. *A Handbook of Living Primates*. New York: Academic Press.

Nelson, K. 1973. Structure and strategy in learning to talk. *Monograph of the Society for Research in Child Development* 38:1–2.

Nishida, T. 1968. The social group of wild chimpanzees in the Mahale Mountains. *Primates* 9:167–224.

Nishida, T. 1973. The ant-gathering behavior by the use of tools among wild chimpanzees of the Mahale Mountains. *Journal of Human Evolution* 2:357–370.

Nishida, T. 1979. The social structure of chimpanzees of the Mahale Mountains. In: D. A. Hamburg and E. R. McCown (eds.), *The Great Apes*, pp. 72–121. Menlo Park, CA: Benjamin/Cummings.

Nishida, T. 1982. Preliminary information on the pygmy chimpanzees (*Pan paniscus*) of the Congo Basin. *Primates* 13:415–425.

Nishida, T. 1987. Local traditions and cultural transmission. In: B. B. Smuts, D. Cheney, R. Seyfarth, R. Wrangham, and T. Struhsaker (eds.), *Primate Societies*. Chicago: University of Chicago Press.

Nishida, T. (ed.). 1990. *The Chimpanzees of the Mahale Mountains: Sexual and Life History Strategies*. Tokyo: University of Tokyo Press.

Nishida, T., and M. Hiraiwa. 1982. Natural history of tool using behavior by wild chimpanzees in feeding upon wood-boring ants. *Journal of Human Evolution* 11:73–99.

Nishida, T., and M. Hiraiwa-Hasegawa. 1987. Chimpanzees and bonobos: Cooperative relationships among males. In: B. B. Smuts, D. Cheney, R. Seyfarth, R. Wrangham, and T. Struhsaker (eds.), *Primate Societies*. Chicago: University of Chicago Press.

Nishida, T., M. Hiraiwa–Hasegawa, T. Hasegawa, and Y. Takahata. 1985. Group extinction and female transfer in wild chimpanzees in the Mahale National Park, Tanzania. *Zeitschrift für Tierpsychologie* 67:284–301.

Nishida, T., and K. Kawanaka. 1972. Inter-unit-group relationships among wild chimpanzees of the Mahali Mountains. *Kyoto University African Studies* 7:131–169.

Noble, W., and I. Davidson. 1991a. The evolutionary emergence of modern human behavior: Language and its archaeology. *Man* (N.S.) 26:223–253.

Noble, W., and I. Davidson. 1991b. Evolving remembrance of times past and future. *Behavioral and Brain Sciences* 14:572.

Noll, R. 1985. Mental imagery cultivation as a cultural phenomenon: The role of visions in shamanism. *Current Anthropology* 26:443–461.

Nottebohm, F. 1977. Asymmetries in neural control of vocalization in the canary. In: S. Harnad, R. Doty, L. Goldstein, J. Jaynes, and G. Krauthamer (eds.), *Lateralisation in the Nervous System*, pp. 23–44. New York: Academic Press.

Oakley, K. P. 1954. Skill as a human possession. In: C. Singer, E. J. Holmyard, and A. R. Hall (eds.), *A History of Technology, Vol. 1: From Early Times to the Fall of Ancient Empires*, pp. 1–37. Oxford: Oxford University Press.

O'Keefe, J. O., and L. Nadel. 1978. *The Hippocampus as a Cognitive Map*. Oxford: Clarendon Press.

Olson, J. 1972. A case for the use of sign language to stimulate language development during the critical period for learning in a congenitally deaf child. *American Annals of the Deaf* 117:397–400.

Ong, W. J. 1982. *Orality and Literacy: The Technologizing of the Word*. London: Methuen.

Osgood, C. E. 1963. On understanding and creating sentences. *American Psychologist* 18:735–751.

Osolsobe, I. 1986. Two extremes of iconicity. In: P. Bouissac, M. Herzfeld, and R. Posner, *Iconicity: Essays on the Nature of Culture*, pp. 95–117. Tübingen: Stauffenburg–Verlag.

O'Sullivan, C., and C. Yeager. 1989. Communicative context and linguistic competence: The effects of social setting on a chimpanzee's conversational skill. In: R. Gardner, B. Gardner, and T. Van Cantfort (eds.), *Teaching Sign Language to Chimpanzees*, pp. 269–279. Albany: State University of New York Press.

Oswalt, W. H. 1976. *An Anthropological Analysis of Food-Getting Technology*. New York: John Wiley and Sons.

Parker, S. T. 1985. A social-technological model for the evolution of language. *Current Anthropology* 26:617–639.

Parker, S. T. 1987. The origins of symbolic communication: An evolutionary cost–benefit model. In: J. Montangero, A. Tryphon, and S. Dionnet (eds.), *Symbolism and Knowledge*, pp. 7–27. Geneva: Jean Piaget Archives Foundation.

Parker, S. T., and K. R. Gibson. 1977. Object manipulation, tool use, and sensorimotor intelligence as feeding adaptations in cebus monkeys and great apes. *Journal of Human Evolution* 6:623–641.

Parker, S. T., and K. R. Gibson. 1979. A developmental model for the evolution of language and intelligence in early hominids. *Behavioral and Brain Sciences* 2:367–408.

Parker, S. T., and K. R. Gibson (eds.). 1990. *"Language" and Intelligence in Monkeys and Apes: Comparative Developmental Perspectives*. New York: Cambridge University Press.

Parker, S. T., and C. Milbrath. 1993. Higher intelligence, propositional language, and culture as adaptations for planning. In: K. R. Gibson and T. Ingold (eds.), *Tools, Language and Cognition in Human Evolution*, pp. 314–32. Cambridge: Cambridge University Press.

Passingham, R. E. 1981. Broca's area and the origins of human vocal skill. *Philosophical Transactions of the Royal Society of London* B, 292:167–175.

Patterson, F. 1978. Linguistic capabilities of a lowland gorilla. In: F. Peng (ed.), *Sign Language and Language Acquisition in Man and Ape: New Dimensions in Comparative Pedolinguistics*, pp. 161–201. Boulder, CO: Westview Press.

Patterson, F. 1990. Self-recognition in gorillas. Paper presented at symposium, "Gorilla Cognition and Behavior," American Society of Primatologists, Davis, California.

Patterson, F., and E. Linden. 1981. *The Education of Koko*. New York: Holt, Rinehart and Winston.

Patterson, T. 1979. The behavior of a group of captive pygmy chimpanzees (*Pan paniscus*). *Primates* 20:341–354.

Peirce, C. S. 1895–1910/1955. Logic as semiotic: The theory of signs. In: J. Buchler (ed.), *Philosophical Writings of Peirce*, pp. 98–119. New York: Dover.

Peirce, C. S. 1935–1966. C. Hartshorne, P. Weiss, and A. W. Burks (eds.), *Collected Papers of Charles Sanders Peirce*. Cambridge: Harvard University Press.

Perelman, S. J. 1960/1985. Who stole my golden metaphor? In: D. Macdonald (ed.), *Parodies: An Anthology from Chaucer to Beerbohm — and After*, pp. 452–457. New York: Da Capo.

Perner, J. 1988. Developing semantics for theories of mind: From propositional attitudes to mental representation. In: J. W. Astington, P. L. Harris, and D. R. Olson (eds.), *Developing Theories of Mind*, pp. 141–172. Cambridge: Cambridge University Press.

Peters, C. R. 1974. On the possible contribution of ambiguity of expression to the development of proto–linguistic performance. In: R. W. Wescott (ed.), *Language Origins*, pp. 83–102. Silver Spring, MD: Linstok Press.

Peters, R. 1978. Communication, cognitive mapping, and strategy in wolves and hominids. In: R. L. Hall and H. S. Sharp (eds.), *Wolf and Man: Evolution in Parallel*, pp. 95–107. New York: Academic Press.

Petitto, L. A., and P. F. Marentette. 1991. Babbling in the manual mode: Evidence for the ontogeny of language. *Science* 251:1493–1496.

Piaget, J. 1936/1979. *The Origins of Intelligence in Children*. New York: W. W. Norton.

Piaget, J. 1937/1954. *The Construction of Reality in the Child*. New York: Basic Books.

Piaget, J. 1945/1962. *Play, Dreams, and Imitation in Childhood*. New York: W. W. Norton.

Piaget, J. 1947/1972. *The Psychology of Intelligence*. Totowa, NJ: Littlefield, Adams & Co.

Piaget, J., and B. Inhelder. 1963. Mental images. In: P. Oléron, J. Piaget, B. Inhelder, and P. Gréco (eds.), *Experimental Psychology: Its Scope and Method*, pp. 85–143. New York: Basic Books.

Pilbeam, D. 1972. *The Ascent of Man: An Introduction to Human Evolution*. New York: Macmillan.

Pilbeam, D. 1988. Primate evolution. In: G. Harrison, et al. (eds.), *Human Biology*, pp. 76–103. New York: Oxford University Press.

Plooij, F. X. 1978. Some basic traits of language in wild chimpanzees? In: A. Lock (ed.), *Action, Gesture, and Symbol: The Emergence of Language*, pp. 111–131. London: Academic Press.

Plotkin, H. C., and F. J. Odling-Smee. 1981. A multiple-level model of evolution and its implications for sociobiology. *Behavioral and Brain Sciences* 4:225–268.

Poizner, H., E. S. Klima, and U. Bellugi. 1988. *What the Hands Reveal About the Brain*. Cambridge: MIT Press.

Polhemus, T. (ed.). 1978. *The Body Reader: Social Aspects of the Human Body*. New York: Pantheon Books.

Pookajorn, S. 1985. Ethnoarchaeology with the Phi Tong Luang (Mlabri): Forest hunters of northern Thailand. *World Archaeology* 17(2):206–221.

Pookajorn, S. 1988. Archaeological research of the Hoabinhian culture or technocomplex and its comparison with ethnoarchaeology of the Phi Tong Luang, a hunter-gatherer group of Thailand. In: H. Müller-Beck (ed.), *Archaeologica Venatoria*, Vol. 9. Tübingen: Verlag Institut für Urgeschichte der Universität Tübingen.

Potts, R. B. 1983. Foraging for faunal resources by early hominids at Olduvai Gorge, Tanzania. In: J. Clutton-Brock and C. Grigson (eds.), *Animals and Archaeology*, Vol. 1, *Hunters and Their Prey*, pp. 51–62. International Series 163. Oxford: British Archaeological Reports.

Potts, R. B. 1984. Home bases and early hominids. *American Scientist* 72:338–347.

Povinelli, D. and G. Cant. 1992. Orangutan clambering and the evolutionary origins of self-cognition. Paper presented at the Fourteenth Congress of the International Primatological Society, Strasbourg.

Povinelli, D. J., K. E. Nelson, and S. T. Boysen. 1992. Comprehension of role reversal in chimpanzees: Evidence of empathy? *Animal Behaviour* 43:633–640.

Povinelli, D. J., K. A. Parks, and M. A. Novak. 1992. Role reversal by rhesus monkeys, but no evidence of empathy. *Animal Behaviour* 44:269–281.

Premack, D. 1972. Language in chimpanzees. *Science* 172:808–822.

Premack, D. 1984. Pedagogy and aesthetics as sources of culture. In: M. S. Gazzaniga (ed.), *Handbook of Cognitive Neuroscience*. New York: Plenum.

Premack, D., and A. J. Premack. 1983. *The Mind of an Ape*. New York: W. W. Norton.

Prost, J. H. 1965. A definitional system for the classification of primate locomotion. *American Anthropologist* 67:1198–1214.

Prudhommeau, G. 1990. Representation of movement in upper Paleolithic figurative art. In: M. L. Foster and L. J. Botscharow (eds.), *The Life of Symbols*, pp. 195–220. Boulder, CO: Westview Press.

Pusey, A. E. 1979. Intercommunity transfer of chimpanzees in Gombe National Park. In: D. A. Hamburg and E. R. McCown (eds.), *The Great Apes*, pp. 465–479. Menlo Park, CA: Benjamin/Cummings.

Quiatt, D. 1984. Devious intentions of monkeys and apes? In: R. Harré and V. Reynolds (eds.), *The Meaning of Primate Signals*, pp. 9–40. Cambridge: Cambridge University Press.

Quiatt, D. 1987. Household structure in a group of gibbons (*Hylobates lar*). *American Journal of Primatology* 12:365.

Quiatt, D. 1988. Regulation of mating choice in nonhuman primates. In: C.G.N. Mascie-Taylor and A. J. Boyce (eds.), *Human Mating Patterns*. Society for the Study of Human Biology Symposium, Vol. 28, pp. 132–151. Cambridge: Cambridge University Press.

Quiatt, D., and M. Huffman. 1993. On home bases, nesting sites, activity centers, and new analytic perspectives. *Current Anthropology* 34:68–80.

Quiatt, D., and J. Kelso. 1985. Household economics and hominid origins. *Current Anthropology* 26:207–222.

Quiatt, D., and J. Kelso. 1987. The concept of the household: Linking behavior and genetic analyses. *Human Evolution* 2:429–435.

Quiatt, D., and V. Reynolds. 1992. *Primate Behaviour: Information, Social Knowledge, and the Evolution of Culture*. Cambridge: Cambridge University Press.

Quine, W. V. 1960. *Word and Object*. Cambridge: MIT Press.

Quine, W. V. 1969. *Ontological Relativity and Other Essays*. New York: Columbia University Press.

Radcliffe-Brown, A. R., and D. Forde. 1950. *African Systems of Kinship and Marriage*. Oxford: Oxford University Press.

Rak, Y., and B. Arensburg. 1987. Kebara 2 Neanderthal pelvis: First look at a complete inlet. *American Journal of Physical Anthropology* 73:227–231.

Rappaport, R. A. 1979. *Ecology, Meaning, and Religion*. Richmond, CA: North Atlantic Books.

Redfern, W. 1984. *Puns*. Oxford: Blackwell.

Reichard, G. A. 1938. Social life. In: F. Boas (ed.), *General Anthropology*, pp. 409–486. Boston: D. C. Heath.

Révész, G. 1956. *The Origins and Prehistory of Language*. London: Longman, Green.

Reynolds, P. C. 1981. *On the Evolution of Human Behavior: The Argument from Animals to Man*. Berkeley: University of California Press.

Reynolds, V. 1991. The biological basis of human patterns of mating and marriage. In: V. Reynolds and J. Kellett (eds.), *Mating and Marriage*, pp. 46–90. Oxford: Oxford University Press.

Reynolds, V., and R. Tanner. 1983. *The Biology of Religion*. Harlow: Longman.

Richards, G. 1987. *Human Evolution: An Introduction for the Behavioral Sciences*. London: Routledge and Kegan Paul.

Rightmire, G. P. 1979. Implications of Border Cave skeletal remains for Later Pleistocene human evolution. *Current Anthropology* 20:23–35.

References

Rightmire, G. P., and H. J. Deacon. 1991. Comparative studies of late Pleistocene human remains from Klasies River Mouth, South Africa. *Journal of Human Evolution* 20:131–156.

Rimpau, J., R. Gardner, and B. Gardner. 1991. Expression of person, place, and instrument in ASL utterances of children and chimpanzees. In: R. Gardner, B. Gardner, and T. Van Cantfort (eds.), *Teaching Sign Language to Chimpanzees*, pp. 240–268. Albany: State University of New York Press.

Rivers, A., U. Bartecku, J. V. Brown, and G. Ettlinger. 1983. An unexpected "epidemic" of a rare stereotype: Unidentified stress or imitation? *Laboratory Primate Newsletter* 22:5–7.

Rivers, W.H.R. 1912. The disappearance of useful arts. In *Festskrift tillägnad Edvard Westermarch, Helsingfors*, pp. 109–130.

Robinson, B. W. 1976. Limbic influences on human speech. In: S. Harnad, H. Steklis, and J. Lancaster (eds.), *Annals of the New York Academy of Sciences: Origins and Evolution of Language and Speech* 280:761–771.

Rollin, B. E. 1989. *The Unheeded Cry: Animal Consciousness, Animal Pain and Science*. Oxford: Oxford University Press.

Romanes, G. J. 1883/1906. *Animal Intelligence*. New York: D. Appleton & Co.

Roth, H. L. 1890. *The Aborigines of Tasmania*. London: Kegan, Paul, Tranch, Truebner and Company.

Rozin, P. 1976. The evolution of intelligence and access to the cognitive unconscious. In: J. M. Sprague and A. N. Epstein (eds.), *Progress in Psychobiology and Physiological Psychology*, Vol. 6, pp. 245–280. New York: Academic Press.

Rubin, S., and D. Wolf. 1980. The development of maybe: The evolution of roles into narrative. In: E. Winner (ed.), *New Directions in Child Development*, Vol. 6, pp. 15–28. San Francisco: Jossey-Bass.

Rumbaugh, D. 1965. Maternal care in relation to infant behavior in the squirrel monkey. *Psychological Report* 16:171–176.

Rumbaugh, D., T. Gill, and E. Von Glaserfeld. 1973. Reading and sentence completion by a chimpanzee *(Pan)*. *Science* 182:731–733.

Russon, A., and B. Galdikas. 1992. Imitation in rehabilitant orangutans: Model and action selectivity. Paper presented at the Fourteenth Congress of the International Primatological Society, Strasbourg.

Russon, A., and B. Galdikas. 1993. Imitation in free-ranging rehabilitant orangutans (*Pongo pygmaeus*). *Journal of Comparative Psychology* 107:47–161.

Ryan, L. 1981. *The Aboriginal Tasmanians*. Vancouver: University of British Columbia Press.

Saarni, C. 1989. Children's understanding of strategic control of emotional expression in social transactions. In: C. Saarni and P. L. Harris (eds.), *Children's Understanding of Emotion*, pp. 181–208. Cambridge: Cambridge University Press.

Sachner, G. 1975. Maturation and longevity in relation to cranial capacity in hominid evolution. In: T. Tuttle (ed.), *Primate Functional Morphology and Evolution*. The Hague: Mouton.

Sachs, J., R. Brown, and R. Salerno. 1976. Adults' speech to children. In: W. van Raffler Engel and Y. LeBrun (eds.), *Baby Talk and Infant Speech (Neurolingustics)*, pp. 240–245. Amsterdam: Swets and Zeitlinger.

Sacks, H. 1980. Button button who's got the button. *Sociological Inquiry* 50(3/4):318–327.

Samuels, A., J. B. Silk, and J. Altmann. 1987. Continuity and change in dominance relations among female baboons. *Animal Behavior* 35:785–793.

Sapir, E. 1924. Culture, genuine and spurious. *American Journal of Sociology* 29:401–429.

References

Saussure, F. de. 1916/1959. *Course in General Linguistics*. New York: McGraw-Hill.

Savage-Rumbaugh, E. S. 1984. *Pan paniscus* and *Pan troglodytes*: Contrasts in preverbal communicative competence. In: R. L. Susman (ed.), *The Pygmy Chimpanzee: Evolutionary Biology and Behavior*. New York: Plenum.

Savage-Rumbaugh, E. S. 1986. *Ape Language: From Conditioned Response to Symbol*. New York: Columbia University Press.

Savage-Rumbaugh, E. S. 1990. The invention of language. Wenner-Gren Foundation for Anthropological Research, Symposium No. 110.

Savage-Rumbaugh, E. S. 1991. Language learning in the bonobo: How truly they learn. In: N. Krasnegor, D. Rumbaugh, R. Schiefelbusch, and M. Studdert-Kennedy (eds.), *Biological and Behavioral Determinants of Language Development*, pp. 209–233. Hillsdale, NJ: Erlbaum.

Savage-Rumbaugh, E. S., and R. Bakeman. 1978. Sexual morphology and behavior in *Pan paniscus*. In: *Proceedings of the Sixth International Congress of Primatology, Cambridge, England*, pp. 613–616. New York: Academic Press.

Savage-Rumbaugh, E. S., and K. McDonald. 1988. Deception and social manipulation in symbol using apes. In: R. Byrne and A. Whiten (eds.), *Machiavellian Intelligence: Social Expertise and the Evolution of Intellect in Monkeys, Apes and Humans*, pp. 224–237. Oxford: Clarendon Press.

Savage-Rumbaugh, E. S., K. McDonald, R. A. Sevcik, W. Hopkins, and E. Rubert. 1986. Spontaneous symbol acquisition and communicative use by a pygmy chimpanzee (*Pan paniscus*). *Journal of Experimental Psychology: General* 115:211–235.

Savage-Rumbaugh, E. S., J. L. Pate, J. Lawson, S. T. Smith, and S. Rosenbaum. 1983. Can a chimpanzee make a statement? *Journal of Experimental Psychology: General* 112:457–492.

Savage-Rumbaugh, E. S., and E. Rubert. 1992. Language comprehension in ape and child: Evolutionary implications. In: Y. Christen and P. Churchland (eds.), *Neurophilosophy and Alzheimer's Disease*, pp. 30–48. Berlin: Springer-Verlag.

Savage-Rumbaugh, E. S., and B. J. Wilkerson. 1978. Socio-sexual behavior in *Pan paniscus* and *Pan troglodytes*: a comparative study. *Journal of Human Evolution* 1:327–334.

Schultz, A. H. 1969. The skeleton of the chimpanzee. In: G. H. Bourne (ed.), *The Chimpanzee*, Vol. 1, pp. 50–103. Basel: Karger.

Schwartz, R. 1974. Representation and resemblance. *Philosophical Forum* 5:499–512.

Sebeok, T. 1987. In what sense is language a "primary modeling system"? In: J. Deeley (ed.), *Semiotics*, pp. 15–27. Lanham, MD: University Press of America.

Sebeok, T., and J. Umiker-Sebeok (eds.). 1980. *Speaking of Apes: A Critical Anthology of Two-way Communication with Man*. New York: Plenum.

Seitz, J. A., and H. Beilin. 1987. The development of comprehension of physiognomic metaphor in photographs. *British Journal of Developmental Psychology* 5:321–331.

Seyfarth, R. M., D. L. Cheney, and P. Marler. 1980. Monkey responses to three different alarm calls: Evidence of predator classification and semantic communication. *Science* 210:801–803.

Shapiro, G. 1975. Teaching language to a juvenile orang–utan. *American Journal of Physical Anthropology* 42:329.

Sharp, H. S. 1978. Comparative ethnology of the wolf and the Chipewyan. In: R. L. Hall and H. S. Sharp (eds.), *Wolf and Man: Evolution in Parallel*, pp. 55–79. New York: Academic Press.

Shaw, H. L. 1989. Comprehension of the spoken word and ASL translation by chimpanzees. *Friends of Washoe* 9(1/2):8–19.

Shipley, J. 1972. *Word Play*. New York: Hawthorne.

Shipman, P. 1983. Early hominid lifestyle: Hunting and gathering or foraging and scavenging? In: J. Clutton-Brock and C. Grigson (eds.), *Animals and Archaeology*, Vol. 1, *Hunters and Their Prey*, pp. 31–49. International Series 163. Oxford: British Archaeological Reports.

Shore, B. 1991. Twice-born, once conceived: Meaning construction and cultural cognition. *American Anthropologist* 93:9–27.

Shotwell, J. M., D. Wolf, and H. Gardner. 1980. Styles of achievement in early symbol use. In: M.L.C. Foster and S. H. Brandes (eds.), *Symbol as Sense: New Approaches to the Analysis of Meaning*, pp. 175–199. New York: Academic Press.

Sibley, C. G., and J. E. Ahlquist. 1984. The phylogeny of the hominoid primates as indicated by DNA-DNA hybridization. *Journal of Molecular Evolution* 20:2–15.

Sibley, C. G., and J. E. Ahlquist. 1987. DNA hybridization as evidence of hominoid phylogeny: Results from an expanded data set. *Journal of Molecular Evolution* 26:99–121.

Sigaut, F. 1985. More (and enough) on technology! *History and Technology* 2:115–132.

Silk, J. B. 1982. Altruism among female *Macaca radiata*: Explanations and analysis of patterns of grooming and coalition formation. *Behaviour* 79:162–168.

Silverstein, S. 1964. *Uncle Shelby's Zoo: Don't Bump the Glump*. New York: Simon and Schuster.

Sinclair, A.R.E., M. D. Leakey, and M. Norton-Griffiths. 1986. Migration and hominid bipedalism. *Nature* 324:307–308.

Singer, D. G., and J. L. Singer. 1990. *The House of Make-Believe: Children's Play and the Developing Imagination*. Cambridge: Harvard University Press.

Singer, R., and J. Wymer. 1982. *The Middle Stone Age at Klasies River Mouth in South Africa*. Chicago: University of Chicago Press.

Sinha, C. 1985. Comment on S. T. Parker, "A social-technological model for the evolution of language." *Current Anthropology* 26:632.

Slobodchikoff, C. N., C. Fischer, and J. Shapiro. 1986. Predator-specific alarm calls of prairie dogs. *American Zoologist* 26:557.

Slobodchikoff, C. N., J. Kiriazin, C. Fischer, and E. Creef. 1991. Semantic information distinguishing individual predators in the alarm calls of Gunnison's prairie dogs. *Animal Behaviour* 42:713–719.

Sluckin, A. 1981. *Growing Up in the Playground: The Social Development of Children*. London: Routledge and Kegan Paul.

Smith, D. G., and M. F. Small. 1987. Mate choice by lineage in three captive groups of rhesus macaques (*Macaca mulatta*). *American Journal of Physical Anthropology* 73:185–191.

Smith, W. J., J. Chase, and A. K. Lieblich. 1981. In: A. Kendon (ed.), *Nonverbal Communication, Interaction, and Gesture*, pp. 503–548. The Hague: Mouton.

Smuts, B. B. 1985. *Sex and Friendship in Baboons*. New York: Aldine.

Smuts, B. B., D. Cheney, R. Seyfarth, R. Wrangham, and T. Struhsaker (eds.), *Primate Societies*. Chicago: University of Chicago Press.

Snow, C. 1972. Mothers' speech to children learning language. *Child Development* 43:549–565.

Snow, C., and C. Ferguson (eds.). 1977. *Talking to Children: Language Input and Acquisition*. Cambridge: Cambridge University Press.

Snowdon, C. T. 1990. Language capacities of nonhuman animals. *Yearbook of Physical Anthropology* 33:215–243.

Soffer, O. 1991. Ancestral lifeways in Eurasia — the middle and upper Paleolithic records. Paper presented at the Fourteenth Annual Spring Systematics Symposium, Chicago.

References

Southwick, C. H., M. A. Beg, and M. R. Siddiqi. 1965. Rhesus monkeys in north India. In: I. DeVore (ed.), *Primate Behavior: Field Studies of Monkeys and Apes*, pp. 111–159. New York: Holt, Rinehart and Winston.

Speth, J. D. 1989. Early hominid hunting and scavenging: The role of meat as an energy source. *Journal of Human Evolution* 18:329–343.

Spiker, D., and M. Ricks. 1984. Visual self-recognition in autistic children: Developmental relationships. *Child Development* 55:214–225.

Spuhler, J. N. 1977. Biology, speech, and language. *Annual Review of Anthropology* 6:509–561.

Staddon, J.E.R. 1989. *Limits to Action: The Allocation of Individual Behavior*. New York: Academic Press.

Stanford, C. B., and J. S. Allen. 1991. On strategic storytelling: Current models of human behavioral evolution. *Current Anthropology* 32:58–61.

Steele, J. 1989. Hominid evolution and primate social cognition. *Journal of Human Evolution* 18:421–434.

Stekel, W. 1923/1952. *Sexual Aberrations: The Phenomena of Fetishism in Relation to Sex*. New York: Liveright.

Steklis, H., and S. Harnad. 1976. From hand to mouth: Some critical stages in the evolution of language. In: S. Harnad, H. Steklis, and J. Lancaster (eds.), *Annals of the New York Academy of Sciences: Origins and Evolution of Language and Speech* 280:445–455.

Steklis, H., and M. Raleigh. 1979. Behavioral and neurobiological aspects of primate vocalization and facial expression. In: H. D. Steklis and M. J. Raleigh (eds.), *Neurobiology of Social Communication in Primates*, pp. 257–282. New York: Academic Press.

Stephens, M. E. 1990. Longevity factor in hominoid social organization. *American Journal of Physical Anthropology* 81:357–361.

Stiner, M. 1990. The use of mortality patterns in archaeological studies of hominid predatory adaptations. *Journal of Anthropology and Archaeology* 9:305–351.

Stokoe, W. 1972. *Semiotics and Human Sign Languages*. The Hague: Mouton.

Stoliar, A. D. 1977. On the genesis of depictive activity and its role in the formation of consciousness (toward a formulation of the problem). *Soviet Anthropology and Archaeology* 16:3–34.

Straus, L. G. 1989. On early hominid use of fire. *Current Anthropology* 30(4):488–491.

Straus, W. 1949. The riddle of man's ancestry. *Quarterly Review of Biology* 24:200. Reprinted in W. Howells (ed.), *Ideas on Human Evolution — Selected Essays*, pp. 69–105. New York: Atheneum, 1967.

Stringer, C. B. 1984a. Fate of the Neanderthal. *Natural History* 93:6–12.

Stringer, C. B. 1984b. Human evolution and biological adaptation in the Pleistocene. In: R. Foley (ed.), *Hominid Evolution and Community Ecology*, pp. 55–83. New York: Academic Press.

Stringer, C. B., and P. Andrews. 1988. Genetic and fossil evidence for the origin of modern humans. *Science* 239:1263–1268.

Struhsaker, T. T., and P. Hunkeler. 1971. Evidence of tool-using by chimpanzees in the Ivory Coast. *Folia Primatologica* 15:212–219.

Struhsaker, T. T., and L. Leland. 1987. Colobines: Infanticide by adult males. In: B. B. Smuts, D. Cheney, R. Seyfarth, R. Wrangham, and T. Struhsaker (eds.), *Primate Societies*. Chicago: University of Chicago Press.

Strum, S. 1987. *Almost Human*. New York: Random House.

Suarez, S. D., and G. G. Gallup, Jr. 1981. Self-recognition in chimpanzees and orangutans, but not gorillas. *Journal of Human Evolution* 10:175–188.

Sugiyama, Y. 1965. On the social change of Hanuman langurs (*Presbytis entellus*) in their natural condition. *Primates* 6:381–418.

Sugiyama, Y. 1985. The brush-sticks of chimpanzees found in southwest Cameroons and their cultural characteristics. *Primates* 26(4):361–374.

Sugiyama, Y., and J. Koman. 1979. Tool-using and -making behavior in wild chimpanzees at Boussou, Guinea. *Primates* 20:513–524.

Susman, R. L. (ed.). 1984. *The Pygmy Chimpanzee: Evolutionary Biology and Behavior*. New York: Plenum.

Susman, R. L. 1987. Pygmy chimpanzees and common chimpanzees: Models for the behavioral ecology of the earliest hominids. In: W. G. Kinzey (ed.), *The Evolution of Human Behavior: Primate Models*. Albany: State University of New York Press.

Susman, R. L. 1988. New postcranial fossils from Swartkrans Member 1: Implications for the behavior of *Paranthropus robustus* (abstract only). *American Journal of Physical Anthropology* 75(2):277–278.

Susman, R. L., N. L. Badrian, and A. J. Badrian. 1980. Locomotor behavior of *Pan paniscus* in Zaire. *American Journal of Physical Anthropology* 53:69–80.

Suzuki, A. 1969. An ecological study of chimpanzees in a savanna woodland. *Primates* 10:103–148.

Swartz, K. B., and S. Evans. 1991. Not all chimpanzees (*Pan troglodytes*) show self-recognition. *Primates* 32:483–496.

Symons, D. 1978. *Play and Aggression: A Study of Rhesus Monkeys*. New York: Columbia University Press.

Tanner, N. M. 1981. *On Becoming Human*. Cambridge: Cambridge University Press.

Tanner, N. M., and A. Zihlman. 1976. Discussion paper: The evolution of human communication: What can primates tell us? In: S. R. Harnad, H. D. Streklis, and J. Lancaster (eds.), *Origins and Evolution of Language and Speech*, pp. 467–480. New York: Academy of Sciences.

Tattersall, I. 1991. What was the human revolution? (Review of *The Human Revolution*, P. Mellars and C. Stringer, eds.). *Journal of Human Evolution* 20:77–83.

Tattersall, I., E. Delson, and J. van Couvering. 1988. *Encyclopedia of Human Evolution and Prehistory*. New York: Garland.

Tayler, C. K., and G. S. Saayman. 1973. Imitative behavior by Indian Ocean bottlenose dolphins (*Tursiops aduncus*) in captivity. *Behaviour* 44:286–298.

Taylor, E. B. 1893. On the Tasmanians as representatives of Palaeolithic man. *Journal of the Anthropological Institute* 23:141–153.

Temerlin, M. K. 1975. *Lucy: Growing Up Human*. London: Souvenir Press.

Tennant, N. 1984. Intentionality, syntactic structure and the evolution of language. In: C. Hookaway (ed.), *Mind, Machines and Evolution: Philosophical Studies*, pp. 73–104. Cambridge: Cambridge University Press.

Terrace, H. 1979. *Nim: A Chimpanzee Who Learned Sign Language*. New York: Knopf.

Terrace, H., L. Petitto, and T. Bever. 1976. Project Nim progress reports I and II. Manuscript.

Terrace, H., L. Petitto, R. Sanders, and T. Bever. 1979. Can an ape create a sentence? *Science*, 206:809–902.

Tervoort, B. 1961. Esoteric symbolism in the communicative behavior of young deaf children. *American Annals of the Deaf* 106:46–80.

Thackeray, J. F. 1983. Disguises, animal behavior and concepts of control in relation to rock art of southern Africa. *South African Archaeological Society Goodwin Series* 4:38–43.

References

Thompson-Handler, N., R. K. Malenky, and N. Badrian. 1984. Sexual behavior of *Pan paniscus* under natural conditions in the Lomako Forest, Equateur, Zaire. In: R. Susman (ed.), *The Pygmy Chimpanzee: Evolutionary Biology and Behavior*, pp. 347–367. New York: Plenum.

Thomson, K. S. 1988. *Morphogenesis and Evolution*. New York: Oxford University Press.

Thomson, K. S. 1992. Macroevolution: The morphological problem. *American Zoology* 32:106–112.

Tomasello, M. 1990. Cultural transmission in the tool use and communicatory signaling of chimpanzees. In: S. T. Parker and K. R. Gibson (eds.), *"Language" and Intelligence in Monkeys and Apes*. Cambridge: Cambridge University Press.

Tomasello, M., M. Davis-Dasilva, L. Camak, and K. Bard. 1987. Observational learning of tool-use by young chimpanzees. *Human Evolution* 2(2):175–183.

Tomasello, M., B. L. George, A. C. Kruger, M. J. Farrar, and A. Evans. 1985. The development of gestural communication in young chimpanzees. *Journal of Human Evolution* 14:175–186.

Tomasello, M., D. Gust, and G. T. Frost. 1989. A longitudinal investigation of gestural communication in young chimpanzees. *Primates* 30(1):35–50.

Tooby, J., and I. DeVore. 1987. The reconstruction of hominid behavioral evolution through strategic modeling. In: W. G. Kinzey (ed.), *The Evolution of Human Behavior: Primate Models*, pp. 183–237. Albany: State University of New York Press.

Toth, N. 1985. Archaeological evidence for preferential right-handedness in the Lower and Middle Pleistocene, and its possible implications. *Journal of Human Evolution* 14:607–614.

Toth, N. 1986. The Oldowan reassessed: A close look at early stone artifacts. *Journal of Archaeological Science* 12:101–120.

Trinkaus, E. 1983. *The Shanidar Neandertals*. New York: Academic Press.

Trinkaus, E., and F. H. Smith. 1985. The fate of the Neandertals. In: E. Delson (ed.), *Ancestors: The Hard Evidence*, pp. 325–333. New York: Alan R. Liss.

Tyack, P. 1991. If you need me, whistle. *Natural History* 8:60–61.

Uzgiris, I. C. 1981. Two functions of imitation during infancy. *International Journal of Behavioral Development* 4:1–12.

Valentine, C. W. 1930. The psychology of imitation with special reference to childhood. *British Journal of Psychology* 21:105–132.

Valladas, H., J. L. Reyss, J. L. Joron, G. Valladas, O. Bar-Yosef, and B. Vandermeersch. 1988. Thermoluminescence dating of Mousterian "proto-Cro-Magnon" remains from Israel and the origin of modern man. *Nature* 331:614–616.

Van Cantfort, T., and J. Rimpau. 1982. Sign language studies with children and chimpanzees. *Sign Language Studies* 34:15–72.

Van Meel, J. M. 1982. The nature and development of the kinetic representational system. In: B. de Gelder (ed.), *Knowledge and Representation*, pp. 210–216. London: Routledge and Kegan Paul.

Vasek, M. 1986. Lying as a skill: The development of deception in children. In: R. Mitchell and N. Thompson (eds.), *Deception: Perspectives on Human and Nonhuman Deceit*, pp. 271–292. Albany: State University of New York Press.

Velder, C. 1964. Die Geister der gelben Blätter — ein Urvolk Thailands? Überblick über fünfzig Jahre Phi-Tong–Lüang Forschung. *Zeitschrift für Ethnologie* 89:10–23.

Visalberghi, E., and D. M. Fragaszy. 1990. Do monkeys ape? In: S. T. Parker and K. R. Gibson (eds.), *"Language" and Intelligence in Monkeys and Apes*. Cambridge: Cambridge University Press.

Vogel, C., and H. Loch. 1984. Reproductive parameters, adult replacements, and infanticide among free-ranging langurs (*Presbytis entellus*) at Jodhpur (Rajasthan), India. In: G. Hausfater and S. B. Hrdy (eds.), *Infanticide: Comparative and Evolutionary Perspectives*, pp. 237–255. New York: Aldine.

Vygotsky, L. S. 1933/1976. Play and its role in the mental development of the child. In: J. S. Bruner, A. Jolly, and K. Sylva (eds.), *Play: Its Role in Development and Evolution*, pp. 537–554. New York: Basic Books.

Vygotsky, L. S. 1935/1978. The prehistory of written language. In: M. Cole, V. John-Steiner, S. Scribner, and E. Souberman (eds.), *Mind in Society: The Development of Higher Psychological Processes*, pp. 105–119. Cambridge: Harvard University Press.

Vygotsky, L. S. 1962. *Thought and Language*. Cambridge: MIT Press.

Wake, M. H. 1990. The evolution of integration of biological systems: An evolutionary perspective through studies on cells, tissues, and organs. *American Zoology* 30:897–906.

Walker, L. C. 1981. The ontogeny of the neural substrate for language. *Journal of Human Evolution* 10:429–441.

Wallace, A. R. 1870. *Contributions to the Theory of Natural Selection*. London: Macmillan.

Wallman, J. 1992. *Aping Language*. Cambridge: Cambridge University Press.

Wallon, H. 1954/1984. Kinesthesia and the visual body image in the child. In: G. Voyat (ed.), *The World of Henri Wallon*, pp. 115–131. New York: Jason Aronson.

Walton, K. L. 1990. *Mimesis as Make-Believe: On the Foundations of the Representational Arts*. Cambridge: Harvard University Press.

Washburn, S. L. 1960. Tools and human evolution. *Scientific American* 203(3):63–75.

Washburn, S. L. 1968. *The Study of Human Evolution*. Eugene: Oregon State System of Higher Education.

Washburn, S. L., and C. S. Lancaster. 1968. The evolution of hunting. In: R. B. Lee and I. DeVore (eds.), *Man the Hunter*, pp. 293–303. Chicago: Aldine.

Watkins, M. 1986. *Invisible Guests: The Development of Imaginal Dialogues*. Hillsdale, NJ: Analytic Press.

Wax, M. 1957. Themes in cosmetics and grooming. *American Journal of Sociology* 62:594–601.

Weaver, R. W. 1956. Through unknown Thailand. *Natural History* 65:289–295, 336.

Wellman, H. M., W. V. Fabricius, and C. Sophian. 1985. The early development of planning. In: H. M. Wellman (ed.), *Children's Searching: The Development of Search Skills and Spatial Representation*, pp. 123–149. Hillsdale, NJ: Erlbaum.

Werner, H., and B. Kaplan. 1963. *Symbol Formation*. New York: Wiley.

Wescott, R. W. 1971. Linguistic iconism. *Language* 47:416–428.

White, F. 1989. Social organization of pygmy chimpanzees. In: P. G. Heltne and L. A. Marquardt (eds.), *Understanding Chimpanzees*, pp. 194–207. Cambridge: Harvard University Press.

White, F. J., and R. W. Wrangham. 1988. Feeding competition and patch size in the chimpanzee species *Pan paniscus* and *Pan troglodytes*. *Behaviour* 105(1–2):148–164.

White, J. C. 1982. *Discovery of a lost Bronze Age, Ban Chiang*. Philadelphia: University Museum.

White, J. P., N. Modjeska, and I. Hipuya. 1977. Group definitions and mental templates: An ethnographic experiment. In: R.V.S. Wright (ed.), *Stone Tools as Cultural Markers: Change, Evolution and Complexity*. Canberra: Australian Institute of Aboriginal Studies.

White, L. A. 1960. Four stages in the evolution of minding. In: S. Tax (ed.), *Evolution After Darwin*, Vol. 2, *The Evolution of Man*, pp. 239–253. Chicago: University of Chicago Press.

White, R. 1982. Rethinking the Middle/Upper Paleolithic transition. *Current Anthropology* 23:169–192.

White, R. 1985. Thoughts on social relationships and language in hominid evolution. *Journal of Social and Personal Relationships* 2:95–115.

Whiten, A. 1976. Primate perception and aesthetics. In: D. Brothwell and C. H. Waddington (eds.), *Beyond Aesthetics: Investigations into the Nature of Visual Art*, pp. 18–40. London: Thames and Hudson.

Whiten, A. 1991. *Natural Theories of Mind: Evolution Development and Simulation of Everyday Mind Reading*. Oxford: Blackwell.

Whiten, A., and R. W. Byrne. 1988a. The Machiavellian intelligence hypothesis: Editorial. In: R. Byrne and A. Whiten (eds.), *Machiavellian Intelligence: Social Expertise and the Evolution of Intellect in Monkeys, Apes and Humans*. Oxford: Clarendon Press.

Whiten, A., and R. W. Byrne. 1988b. Tactical deception in primates. *Behavioral and Brain Sciences* 11:233–244.

Whiten, A., and R. Ham. 1992. On the nature and evolution of imitation in the animal kingdom: Reappraisal of a century of research. *Advances in the Study of Behavior* 21:239–283.

Whitesides, G. H. 1985. Nut-cracking by wild chimpanzees in Sierra Leone. *Primates* 26(1):91–94.

Wilson, E. O. 1972. Animal communication. *Scientific American* 227(3):52–60.

Wilson, P. J. 1983. *Man, the Promising Primate*. New Haven, CT: Yale University Press.

Wind, J. 1976. Phylogeny of the human vocal tract. In: S. R. Harnard, H. D. Steklis, and J. Lancaster (eds.), *Annals of the New York Academy of Sciences: Origins and Evolution of Language and Speech*, 280:612–630.

Winner, E., and G. Ettlinger. 1979. Do chimpanzees recognize photographs as representations of objects? *Neuropsychologia* 17:413–420.

Wittgenstein, L. 1921/1961. *Tractatus logico-philosophicus*. London: Routledge and Kegan Paul.

Wittgenstein, L. 1953. *Philosophical Investigations*. Oxford: Blackwell.

Wollheim, R. 1974. *On Art and the Mind*. Cambridge: Harvard University Press.

Wolpoff, M. H. 1988. Second cousins: The Eurasian perspective. *American Journal of Physical Anthropology* 75(2).

Wolpoff, M. H., X. Z. Wu, and A. G. Thorne. 1984. Modern *Homo sapiens* origins: A general theory of hominid evolution involving the fossil evidence from east Asia. In: F. H. Smith and F. Spencer (eds.), *Origins of Modern Humans: A World Survey of the Fossil Evidence*, pp. 411–484. New York: Alan R. Liss.

Woodburn, J. 1980. Hunters and gatherers today and reconstruction of the past. In: E. Gellner (ed.), *Soviet and Western Anthropology*, pp. 95–117. London: Duckworth.

Woodburn, J. 1982. Egalitarian societies. *Man* 17:431–451.

Woodruff, G., and D. Premack. 1979. Intentional communication in the chimpanzee: The development of deception. *Cognition* 7:333–362.

Woolfson, C. 1982. *The Labor Theory of Culture: A Re-examination of Engels's Theory of Human Origins*. London: Routledge and Kegan Paul.

Wrangham, R. 1979. Sex differences in chimpanzee dispersion. In: D. A. Hamburg and E. R. McCown (eds.), *The Great Apes*, pp. 481–489. Menlo Park, CA: Benjamin/Cummings.

Wrangham, R. 1987. The significance of African apes for reconstructing human evolution. In: W. G. Kinzey (ed.), *The Evolution of Human Behavior: Primate Models*, pp. 28–47. Albany: State University of New York Press.

Wright, R.V.A. 1972. Imitative learning of a flaked stone technology — the case of an orangutan. *Mankind* 8:296–306.

Wu, R., and J. W. Olsen. 1985. *Palaeoanthropology and Palaeolithic Archaeology in the People's Republic of China*. Orlando: Academic Press.

Wynn, T. 1988. Tools and the evolution of human intelligence. In: R. Byrne and A. Whiten (eds.), *Machiavellian Intelligence: Social Expertise and the Evolution of Intellect in Monkeys, Apes and Humans*, pp. 271–284. Oxford: Clarendon Press.

Wynn, T. 1993. Layers of thinking in tool behavior. In: K. R. Gibson and T. Ingold (eds.), *Tools, Language and Cognition in Human Evolution*. Cambridge: Cambridge University Press.

Wynn, T. Forthcoming. The evolution of tools and symbolic behavior. In: A. Lock and C. R. Peters (eds.), *Handbook of Human Symbolic Evolution*, pp. 389–406. Oxford: Oxford University Press.

Wynn, T., and W. C. McGrew. 1989. An ape's view of the Oldowan. *Man* (N.S.) 24:383–398.

Wynne-Edwards, V. C. 1962. *Animal Dispersion in Relation to Social Behavior*. Edinburgh: Oliver and Boyd.

Yamazaki, N. 1985. Primate bipedal walking: Computer simulation. In: S. Kondo (ed.), *Primate Morphophysiology, Locomotor Analyses and Human Bipedalism*, pp. 105–130. Tokyo: University of Tokyo Press.

Yen, D. E., and J. Nance. 1975. *Further studies on the Tasaday*. Makati, Rizal, Panamin Foundation, Panamin Foundation Research Series 2.

Yengoyan, A. A. 1977. Review of Nance, John, *The gentle Tasaday: A Stone Age People in the Philippine Rain Forest*. *American Scholar* 46:134–138.

Yerkes, R. M. 1916a. *The Mental Life of Monkeys and Apes: A Study of Ideational Behavior*. Behavioral Monographs 3.

Yerkes, R. M. 1916b. Provision for the study of monkeys and apes. *Science*. 43:231–234.

Yerkes, R. M. 1925. *Almost Human*. New York: Century.

Yerkes, R. M. 1927. *The Mind of a Gorilla*. Genetic Psychology Monographs 2 and 6.

Yerkes, R. M. 1943. *Chimpanzees: A Laboratory Colony*. New Haven, CT: Yale University Press.

Yerkes, R. M., and B. W. Learned. 1925. *Chimpanzee Intelligence and Its Vocal Expressions*. Baltimore: Williams and Wilkins.

Zihlman, A. L. 1978. Women and evolution, part 2: Subsistence and social organization among early hominids. *Signs* 4:4–20.

Zihlman, A. L., J. E. Cronin, D. L. Cramer, and V. M. Sarich. 1978. Pygmy chimpanzees as a possible prototype for the common ancestor of humans, chimpanzees and gorillas. *Nature* 275:744–746.

Zihlman, A. L., M. E. Morbeck, and J. Goodall. 1990. Skeletal biology and individual life history of Gombe chimpanzees. *Journal of Zoology* (London) 221:37–61.

Zubrow, E. 1989. The demographic modelling of Neanderthal extinction. In: P. Mellars and C. Stringer (eds.), *The Human Revolution: Behavioral and Biological Perspectives on the Origins of Modern Humans*, pp. 212–231. Edinburgh: Edinburgh University Press.

Index